INTRODUCTION TO THEORETICAL NEUROBIOLOGY
Volume 1

HENRY C. TUCKWELL
Monash University

Introduction to theoretical neurobiology: volume 1 linear cable theory and dendritic structure

CAMBRIDGE UNIVERSITY PRESS
Cambridge
New York New Rochelle
Melbourne Sydney

Published by the Press Syndicate of the University of Cambridge
The Pitt Building, Trumpington Street, Cambridge CB2 1RP
32 East 57th Street, New York, NY 10022, USA
10 Stamford Road, Oakleigh, Melbourne 3166, Australia

First published 1988

Printed in the United States of America

Library of Congress Cataloging-in-Publication Data

Tuckwell, Henry C. (Henry Clavering), 1943–
Introduction to theoretical neurobiology.
(Cambridge studies in theoretical biology; 8)
Includes bibliographies and index.
Contents: v. 1. Linear cable theory and dendritic structure—v. 2. Nonlinear and stochastic theories.
1. Neurons—Mathematical models. 2. Neural transmission—Mathematical models. I. Title.
II. Series.
QP361.T76 1988 559'.0188 87-25690

British Library Cataloguing in Publication Data

Tuckwell, Henry C.

Introduction to theoretical neurobiology.
– (Cambridge studies in mathematical biology, 8).

Vol. 1, Linear cable theory and dendritic structure
1. Neurobiology—Mathematics
I. Title
591.1'88'0151 QP361

ISBN 0-521-35096-4 (v. 1)
ISBN 0-521-35217-7 (v. 2)

To My Mother, Laura

CONTENTS

PREFACE

This is the first of two volumes dealing with theories of the dynamical behavior of neurons. It is intended to be useful to graduate students and research workers in both applied mathematics and neurobiology. It would be suitable for a one-quarter or one-semester course in quantitative methods in neurobiology.

The book essentially contains descriptions and analyses of the principal mathematical models that have been developed for neurons in the last 30 years. Chapter 1, however, contains a brief review of the basic neuroanatomical and neurophysiological facts that will form the focus of the mathematical development. A number of suggestions are made for further reading for the reader whose training has been primarily mathematical.

The remainder of the book is a mathematical treatment of nerve-cell properties and responses. From Chapter 2 onward, there is a steady increase in mathematical level. An attempt has been made to explain some of the essential mathematics as it is needed, although some familiarity with differential equations and linear algebra is desirable. It is hoped that physiologists will benefit from this method of presentation. Biophysicists, engineers, physicists, and psychologists who are interested in theoretical descriptions of neurons should also find this book useful.

From Chapter 2 onward, the theme is the systematic development of mathematical theories of the dynamical behavior of neurons. The fundamental observation is of the resting membrane potential. Hence Chapter 2 is mainly concerned with the passive properties of cells and is an exposition of the classical theory of membrane potentials (i.e., the Nernst–Planck and Poisson equations). The culmination of this body of work is the highly successful Goldman–Hodgkin–Katz formula, which is much used despite the existence of more basic modern theories. The latter are not included simply because there was not enough space. Glial cells are included in the present discussion as well as the effects of active transport (pumps).

Chapter 3 focuses on a simple neural model, which I have called the Lapicque model after the French physiologist who used it in 1907. The subthreshold potential satisfies a first-order linear differential equation but the introduction of a threshold makes the system nonlinear and nontrivial to analyze except perhaps for simple input patterns. Many basic nerve-cell properties can be illustrated with this simple model, including postsynaptic potentials, strength–duration curves, and phase locking. Scharstein's method of analysis is given, which enables the spike train to be found graphically or numerically for an arbitrary input current. There is a brief discussion of Laplace transform theory, which is used extensively in subsequent chapters.

Chapter 4 begins with a derivation of the usual (linear) cable equation from elementary considerations and gives the various basic physiological boundary conditions. However, throughout the rest of the chapter, the applications are confined to the steady state, which means that only linear ordinary second-order differential equations are involved. The Green's function method of solution is introduced for single cables. The second half of this chapter is concerned with finding the steady-state distribution of potential throughout dendritic trees. A m[illegible] solution is developed that involves solving linear systems o[illegible] be achieved exactly in some cases or by machi[illegible]eral.

Ch[illegible] time-dependent solutions of the cable [illegible]tion method of solution is construed [illegible]bles the solution to be written down [illegible]put-current density. Two methods of [illegible]e developed – separation of variables [illegible] illustrated with examples. Standard [illegible]ay then be considered theoretically, [illegible]r current waves at the soma, and so [illegible]various forms of synaptic input. A [illegible]or various boundary conditions is [illegible] of finding time-dependent poten-[illegible] then addressed and various ap-pr[illegible]de Laplace transforming the sys-tem[illegible]ivalent cylinder, and a mapping proc[illegible]which enables the potential to be foun[illegible]quence of two-point boundary-value proble[illegible]oncludes with some perhaps surprising results on the [illegible] of dendritic branching on neuronal integration. These are called the "principle of independence of response on geometry" and the extended such principle.

The last chapter in this volume, Chapter 6, contains a description of Rall's model nerve cell in which a cylinder representing one or more dendritic trees is attached to a lumped soma. For an infinite cylinder solutions are found in closed form. For finite cables the boundary data is not of standard type and the nonorthogonality of eigenfunctions means that alternative methods must be employed. Solutions suitable for calculation at small and large times are obtained and applied to standard physiological experiments. The problem of parameter estimation is addressed at length and an attempt made at an exhaustive account of components contributing to subthreshold neuronal response within the framework of the present chapter. Standard motoneuron parameters are catalogued and the response of the standard cell to current injection computed and compared with experiment.

This material has been written over a period of several years in several locations. Chapters 1 and 2 were written at UCLA in 1980 and 1982, respectively. Chapters 3, 4, and some of 5 were written at Monash University between 1981 and 1983. The rest of Chapter 5 and Chapter 6 were written at the University of British Columbia in 1983. I am grateful to the institutions I have visited for their partial support. It will be seen from the text and perusal of the references that some of this material has been a consequence of joint work. I am grateful for the contributions of co-workers and students to the present volume.

Los Angeles, March 1987 Henry C. Tuckwell

1

Introductory neuroanatomy and neurophysiology: the properties of motoneurons

1.1 The central nervous system

Our brains and spinal cords contain specialized cells called *nerve cells* or *neurons*, which are collectively referred to as the *central nervous system* (CNS). At one time it was thought that the nervous system was continuous, but it is now firmly established that the neuron is the fundamental discrete unit of the CNS. The nervous system is extremely complex and estimates of the number of nerve cells in the human brain are on the order of 10 billion (i.e., 10^{10}). In addition, there are closely associated cells, as numerous or more so, called *glial cells* or *glia*, that seem to play an important regulatory role. They have several properties in common with neurons but are nevertheless quite distinguishable from them.

From location to location in the CNS, nerve cells differ in their properties and functions. It is convenient, however, to envisage a *paradigm*, or typical nerve cell, with four basic components (see Figure 1.1). The components and their usual roles are as follows.

Cell body or soma

This is the focal part from which branching structures emanate. It roughly delineates the *input* or *information-gathering* parts of the cell from the *output* or *information-transmitting* parts.

Dendrites

There are usually several dendrites that may branch several times to form treelike structures – the *dendritic trees*. Over the dendrites occur many contacts from other cells at specialized sites called *synapses*, though these are also often found on the cell body. The dendrites are the chief information-*gathering* component although they sometimes perform integration and processing of synaptic inputs.

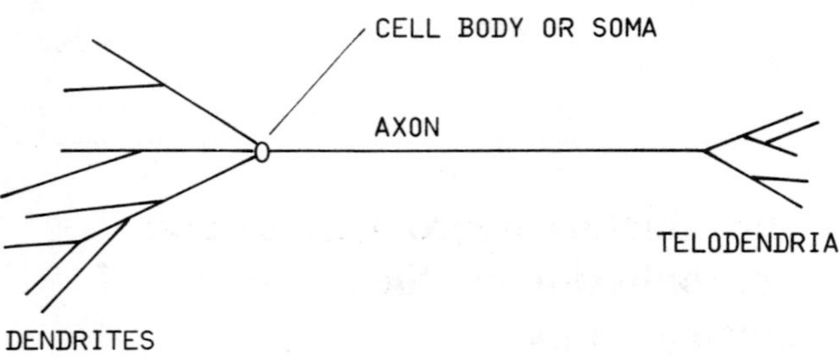

Figure 1.1. Schematic representation of a typical nerve cell.

Axon

The axon is the component along which output electrical signals propagate. The output signal is called an *action potential* and may travel at speeds ranging from 2 to about 120 m/s to eventually reach *target cells* through synaptic contacts. Larger axons are usually covered with a whitish insulating material called *myelin*, which is broken at intervals called *nodes*.

Telodendria

When the axon nears its target cells it branches, possibly several times, to form telodendria. At the ends of these, contacts are made with other cells again by means of synapses.

In order to give some idea of the diversity of neuron types, three kinds of nerve cell are shown in Figure 1.2. These are pictures obtained by a light microscope using the silver staining technique called the *Golgi method*.

The cell on the upper left is a *pyramidal cell*, so called because of the shape of its cell body. Such cells are numerous in the *cerebral cortex* and other parts of the brain. Large pyramidal cells are called *Betz cells*. Their axons (not shown fully in Figure 1.2) project down into the spinal cord to make contact with motoneurons that occur in groups or *pools*. Each pool is associated with a particular muscle. A *motoneuron* is also shown in Figure 1.2. Its shape is quite different from that of the pyramidal cell. The axons of motoneurons leave the spinal cord to make contact with muscle fibers and signals that pass along these axons result in muscular contraction. Most of the introductory anatomy and physiology in this chapter is with reference to motoneurons.

The third kind of neuron shown in Figure 1.2 is a *Purkinje cell*. Such cells are found in the *cerebellum*, which is a specialized structure at the back of the brain involved in the timing of muscle movements.

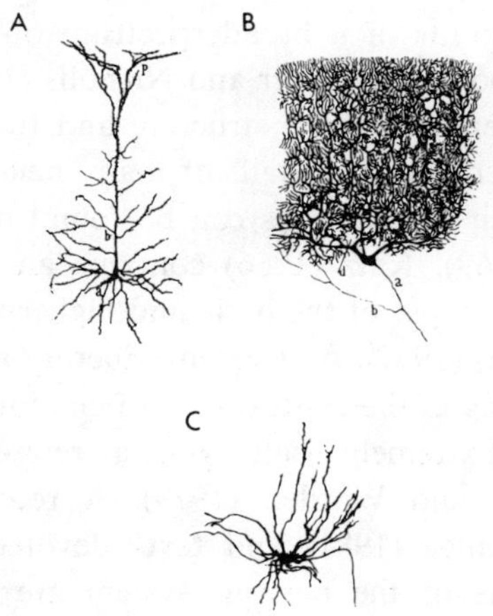

Figure 1.2. Light-microscope pictures of three different kinds of nerve cell. A – Pyramidal cell; B – Purkinje cell; and C – spinal motoneuron. [From drawings by Ramón y Cajal (1909, 1911).]

These can be very simple or they can be associated with very complex skills (e.g., athletic performance, playing of musical instruments, and speech). Estimates vary, but the number of cells in the cerebellum may be as large as in the brain itself (Shepherd 1979). The most striking feature of the Purkinje cell is its massively branched dendritic tree, which is ideally suited to the reception of thousands of inputs from other cells. Estimates of the number of inputs (some of which occur on dendritic *spines*) to Purkinje cells range from 60,000 to 100,000.

Throughout most of this book attention is focused mainly on the properties of single nerve cells. A wealth of neurophysiological and neuroanatomical data has been collected and it is desirable to have an underlying cogent mathematical theory to act as a unifying thread through the myriads of experimental data. It will be seen that even single nerve cells require quite complicated mathematics to describe their behavior.

Supplementary reading

In the rest of this chapter we will briefly examine some anatomical and neurophysiological properties of spinal motoneurons. This represents a very limited, though important, sample of cells of the CNS. The reader should be aware that not all neurons have the same properties. It would be extremely beneficial to become familiar not only with the properties of many other kinds of nerve cell but also how cells interact within populations.

Very lucid and useful accounts of a broad spectrum of neurobiology are contained in the books by Kuffler and Nicholls (1976) and Shepherd (1983). A good reference on the structure and function of the nervous system is Gardner (1963). Details of many neurophysiological investigations and their implications can be found in Eccles' books (1953, 1957, 1964, 1969). Katz (1966) contains an excellent introduction. An interesting glimpse of the basic and higher functions of the CNS are given in Eccles (1977). A clear introductory article on nerve cells and neural modeling in the context of artificial intelligence is given by Stevens (1985). Extremely useful general references are Kandel (1977) and Schmidt and Worden (1979). A recent more theoretical text is that of Junge (1981) and texts devoted almost entirely to theoretical aspects of the nervous system are those of Plonsey (1969), Griffith (1971), Leibovic (1972), Jack, Noble, and Tsien (1985), MacGregor and Lewis (1977), Scott (1977), Stein (1980), Reichardt and Poggio (1981), and Hoppensteadt (1986).

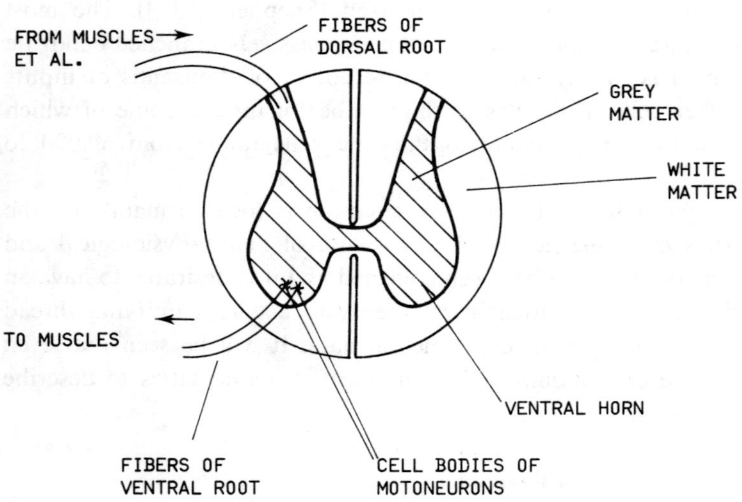

Figure 1.3. Anatomical arrangement showing the location of a spinal motoneuron. A cross section of the spinal cord is depicted. The cell bodies of the motoneurons are located in the ventral horn. The arrows by the dorsal and ventral roots indicate the direction of impulse propagation.

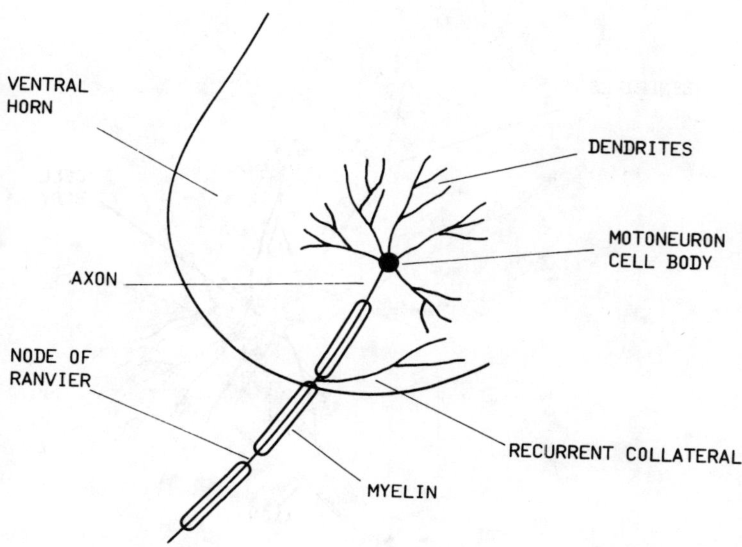

Figure 1.4. An enlarged drawing of the part of a motoneuron within the ventral horn showing the myelin sheath and the collateral branch of the axon that terminates in the gray matter of the cord. [Adapted from Eccles (1957) and Shepherd (1979).]

1.2 Introductory anatomy of motoneurons

We begin our study of the nervous system by examining the anatomical and physiological properties of the most studied mammalian central nervous system cell. This is the *spinal motoneuron*, which has its cell body in the gray matter of the spinal cord and whose axon leaves the cord in the *ventral root* to eventually, after losing its myelin covering, innervate a muscle fiber. This anatomical arrangement is shown in Figure 1.3. Note that in the spinal cord the gray matter is on the inside, whereas in the cerebral cortex the gray matter is on the outside. The gray matter consists of a collection of nerve-cell bodies, dendrites, glial cells, and unmyelinated axons. The white matter consists of tracts of axons covered by myelin sheaths.

Some details not discernible from Figure 1.3 are shown in Figure 1.4. The myelin covering of the axon of the motoneuron begins within the gray matter. The myelin has breaks in it at the *nodes of Ranvier*. Emerging from the first node (but this may not always be the case), we may find a branch from the axon called a *collateral branch*. This branch heads back into the gray matter and is called a *recurrent collateral*. It is usually related to another cell called a *Renshaw cell*.

To get an accurate picture of the cell body and dendrites of the spinal motoneuron, we have in Figure 1.5 the excellent result of a

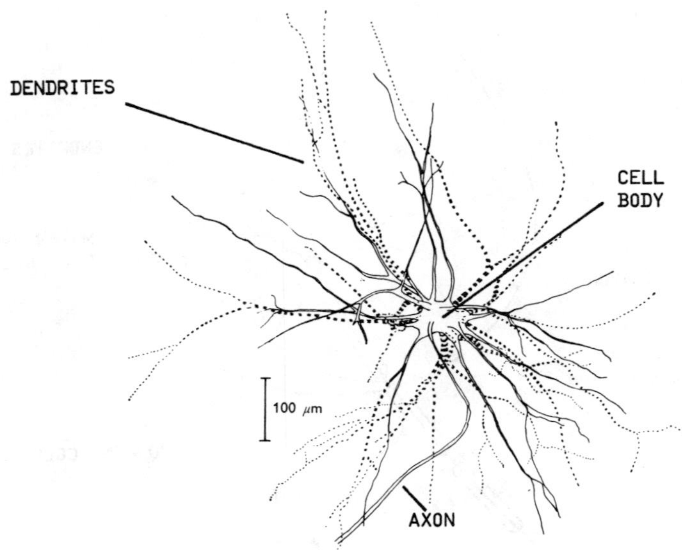

Figure 1.5. Light-microscope picture of a motoneuron in the lumbosacral region of the cat spinal cord. This is "three-dimensional." The dashed portions represent dendrites that are not "in" the plane of the paper. [From Barrett and Crill (1974). Reproduced with the permission of The Physiological Society and the authors.]

Procion dye-injected motoneuron of the cat spinal cord by Barrett and Crill (1974). Note that the cell body is about 80 μm in diameter and the diameters of the dendrites, where they leave the cell body, are about 10 μm. The total "sphere of influence" of the dendritic tree is about 1 mm (1000 μm) in diameter. In this particular study it was found that the average number of primary dendrites was 12 with a range of 7 to 21. The total surface area of the soma and dendrites ranged from 79,000 to 250,000 μm^2 with an average value of 145,000. On average, the surface area of the dendrites is much larger than that of the soma. The average ratio of these quantities is 12.9.

John Barrett has been kind enough to supply some actual numbers on the measurements of the diameters and lengths of some dendrites. Data on one such primary dendrite are given in Figure 1.6. The quantities of interest here are the amount of tapering from proximal (near the soma) dendrites to distal (at the ends of the tree) dendrites and the ratios of the diameters where branching occurs. In the figure we see a total of five branchings. At this point, note that in mathematics a *process* is usually a time-dependent (random) function; to a neurobiologist, the component parts of the neuron are called *processes*.

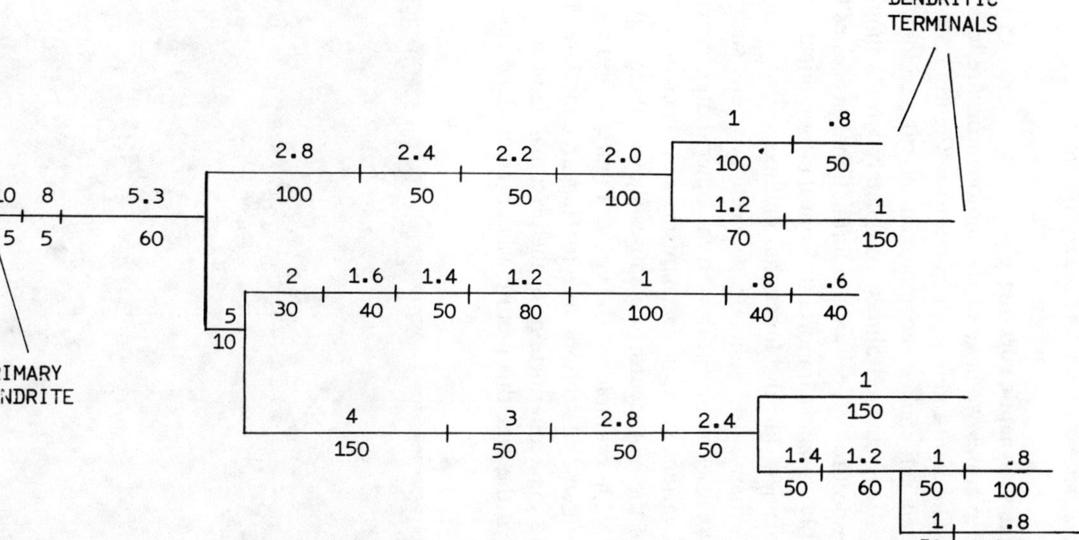

Figure 1.6. Measurements of diameters (upper figures) and lengths of segments (lower figures) of a primary dendrite and its branches from a cat spinal motoneuron. Vertical lines in the upper figure do not represent cell structure. Lengths are not to scale. (Courtesy of J. Barrett.)

1.3 Fine structure: synaptic covering

So far the pictures we have of the motoneuron are in a way highly deceptive. The light-microscope picture in Figure 1.5 of the Procion dye-injected cell makes the motoneuron surface look very "clean." In reality, it is covered almost completely with a jungle of functional (with respect to that cell) and nonfunctional processes from other cells. Probably the most important, from the theoretical neurobiologist's point of view, are the functional endings called *synapses*. We will define a synapse as a junction between two cells such that activity in one cell causes a current flow in the other. Later we will go into the details of various synaptic mechanisms and types in full.

Early pictures of the motoneuron's synapses showed them clustered on the cell body. An example is shown in Figure 1.7. This is a reconstruction from several sections due to Haggar and Barr (1950) and shows only the soma and portions of the proximal dendrites. [It is worth noting that at the time this picture was obtained dendrites were

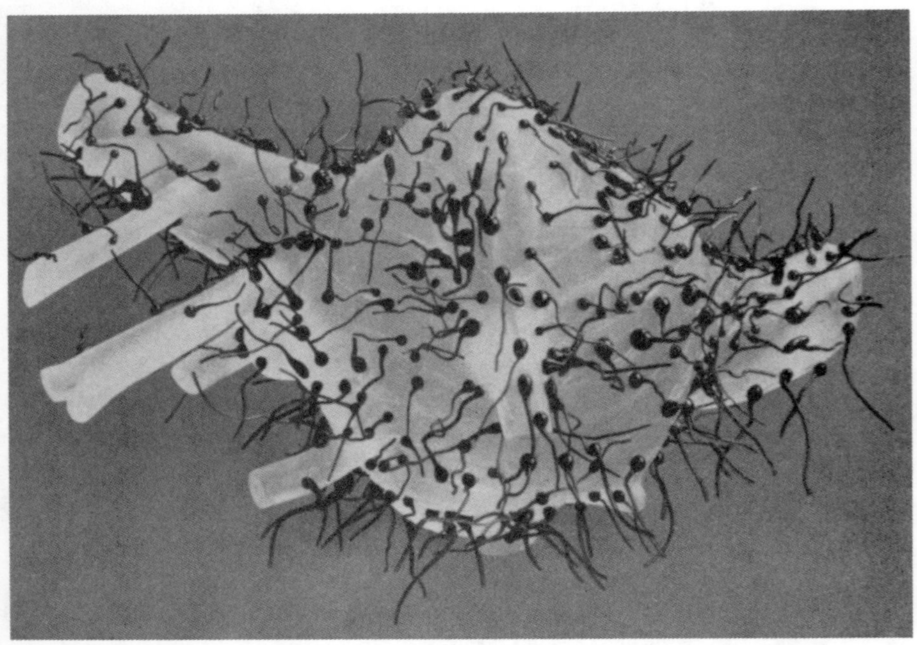

Figure 1.7. An early reconstruction of the spinal motoneuron cell body and proximal dendrites showing the covering with synaptic knobs. [From Haggar and Barr (1950). Reproduced with the permission of Alan R. Liss, Inc.]

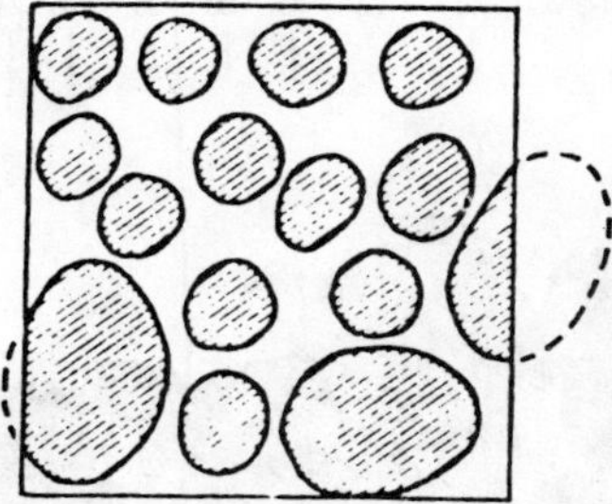

Figure 1.8. Schematic representation of the covering of 100 μm^2 of a motoneuron cell surface with synaptic boutons. [From Illis (1964). Reproduced with the permission of Oxford University Press.]

not thought to play a very important role (if any) in neuronal function.]

Estimates of the number of synaptic endings on spinal motoneurons have tended to increase through the years. A summary of data prior to 1964 has been compiled by Illis (1964). The earliest figure given was between 300 and 350 boutons (end buttons) per cell (Hoff 1932). Wycoff and Young (1956) claimed that there were not less than 2000, and Aitken and Bridger (1961) increased this substantially to 16,000 for a cell with a surface of 80,000 μm^2. The latest estimate based on Barrett and Crill's (1974) light-microscope study and Conradi's (1969) electron-microscope study places the total number of synapses at an average of 22,600 (Koziol and Tuckwell 1978).

The distribution of synapses over the cell surface is also important, as will be seen later, when we consider how input location affects the neuron response. Illis (1964) estimated that between 50 and 60% of the cell surface was covered by boutons with a density of 11.6 synapses/100 μm^2 on the cell body and 7.16 synapses/100 μm^2 on the dendrites. His picture of a typical covering is given in Figure 1.8. For a recent theoretical study of synaptic densities see DeSantis and Limwongse (1983).

The most graphic details of the covering of the spinal motoneuron with synapses has come from Conradi's (1969) mammoth study with the electron microscope. The details of the synaptic covering of a portion of a proximal dendrite are shown in Figure 1.9. The cylindrical surface has been unfolded to form a plane. The various boutons are categorized according to details of their junctional contacts. Conradi's data result in higher estimates of the packing density of boutons on the dendrites than had previously been suspected. The results are shown in Figure 1.10, which shows that there are about

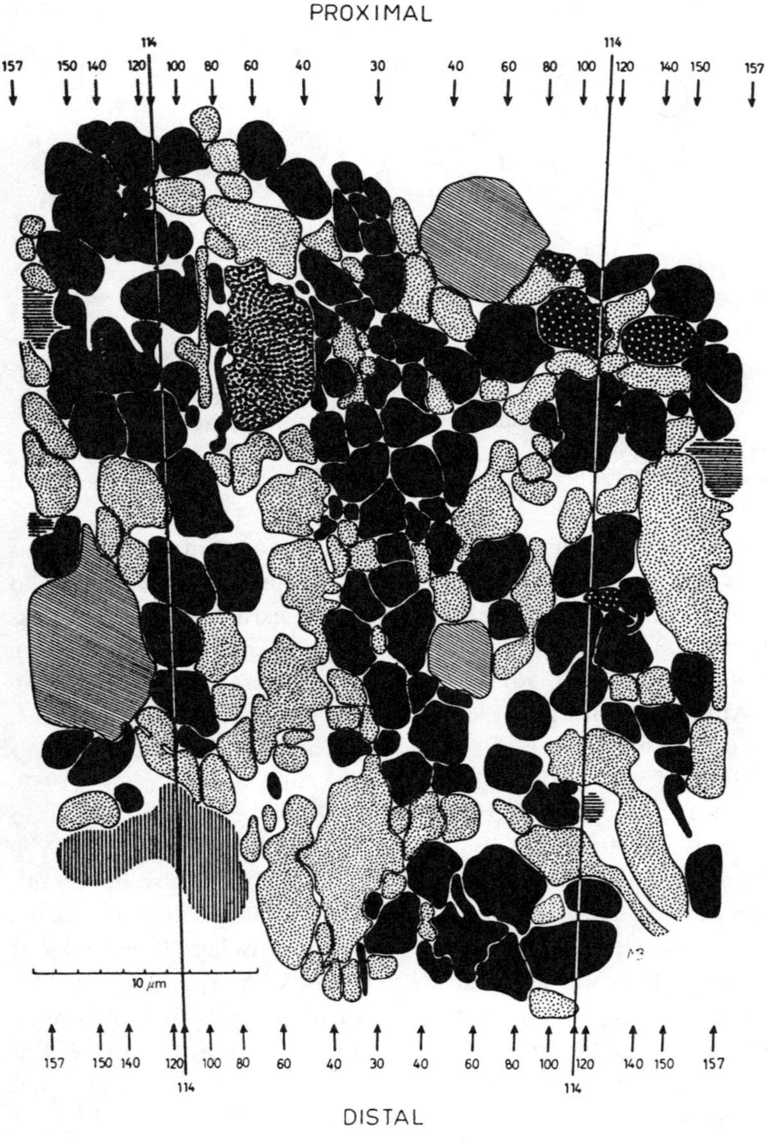

PROXIMAL

157 150 140 120 114 100 80 60 40 30 40 60 80 100 114 120 140 150 157

10 μm

157 150 140 120 114 100 80 60 40 30 40 60 80 100 114 120 140 150 157

DISTAL

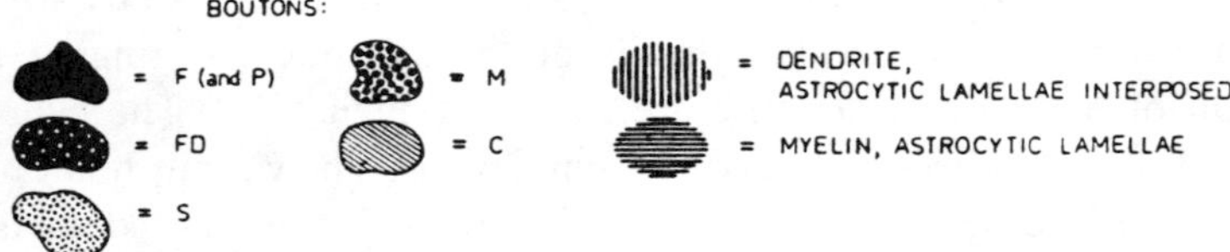

Figure 1.9. The dense packing of the dendrites of a motoneuron with synaptic boutons. This is portion of the surface of a "cylinder." The numbers at the top and bottom are plus and minus degrees from a reference line. The numbers do not start at 0 and end at 180 because the top and bottom slices of the cylinder are missing. The key (F, P, etc.) refers to Conradi's classification of synaptic endings.

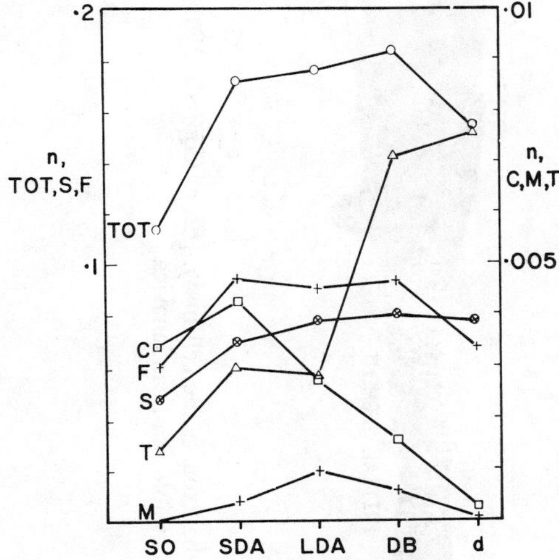

Figure 1.10. Number of synapses per square micrometer versus distance from the cell body. The left-hand scale is for the total number (TOT) and the more numerous bouton types (S and F), whereas the right-hand scale is for the C, M, and T bouton types. [From Koziol and Tuckwell (1978). Reproduced with the permission of Elsevier Biomedical Press, B.V. and the authors.]

11.4 synapses/100 μm^2 on the cell body, rising to 17 to 18 synapses/100 μm^2 on the proximal dendrites, and falling slightly to about 15 synapses/100 μm^2 on the distal dendrites. [The notation for distance are as follows: SO is the soma; DA is the proximal dendrite so that counting from the cell body, 20 boutons are encountered on one side (this was divided into SDA and LDA, the former having membrane lengths less than 100 μm); DB is the distal dendrite; and d is the dendritic fragments, presumed to belong to motoneurons.]

There is one final aspect of the covering of the cell with synaptic boutons as indicated in Figure 1.11. Here the "join" of the axon with the motoneuron cell body is indicated in detail. The thickened part of the axon at the cell body is called the *axon hillock* and we see that it has some bouton covering. Just distal, however, is a region that is devoid of boutons and extends for about 15 μm until the first part of the myelin covering begins. This is the all important *initial segment* of the axon and is believed to be crucial in the instigation of motoneuron action potentials.

1.4 Inhibition and excitation: synaptic potentials

We have indicated that the synaptic boutons are associated with "input" currents to the cell. Here we examine those inputs in

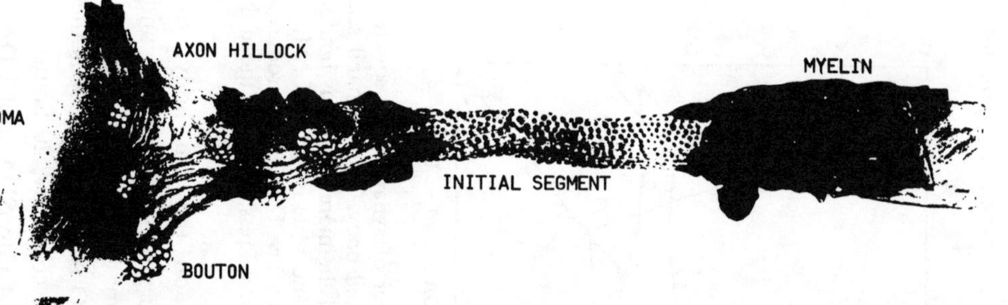

Figure 1.11. Details of the beginning of the axon at the cell body. [From Conradi (1969). Reproduced with the permission of Acta Physiologica Scandanavica.]

detail. Our study begins with the first observations on potential differences across the cat spinal motoneuron membrane, which were made possible with the introduction of glass microelectrodes that tapered to diameters of 0.5 to 1 μm at the tips. A successful penetration of the cell implies a small puncture that does not cause "'injury currents." When such a penetration is made, it is found that the potential suddenly drops by between 60 and 80 mV. In the absence of activity, the potential will remain at this level and this is called the *resting potential* of the nerve cell. The theory of the resting potential is given in Chapter 2.

When a synapse or group of synapses is activated, we see a response in the cell on which the synapses are located called a *synaptic potential*. Sometimes it is called a *postsynaptic potential*. The cell receiving the input is called the *postsynaptic cell* and the cell from which the synapse is being activated is called the *presynaptic cell*. (These are, of course, relative concepts as cells may be simultaneously postsynaptic and presynaptic even relative to each other as in the case of *reciprocal synapses*.) The scheme is sketched in Figure 1.12. The synaptic bouton, or ending, is sometimes called the *presynaptic terminal* and when isolated from its cell processes, it is called a *synaptosome*. Stevens (1968) and Uchizono (1975) contain excellent introductions to the mechanisms involved in synaptic transmission.

We have noted that the resting membrane potential is about 70 mV, inside negative. When a synaptic event occurs that *decreases* the potential difference across the cell membrane, the postsynaptic cell is

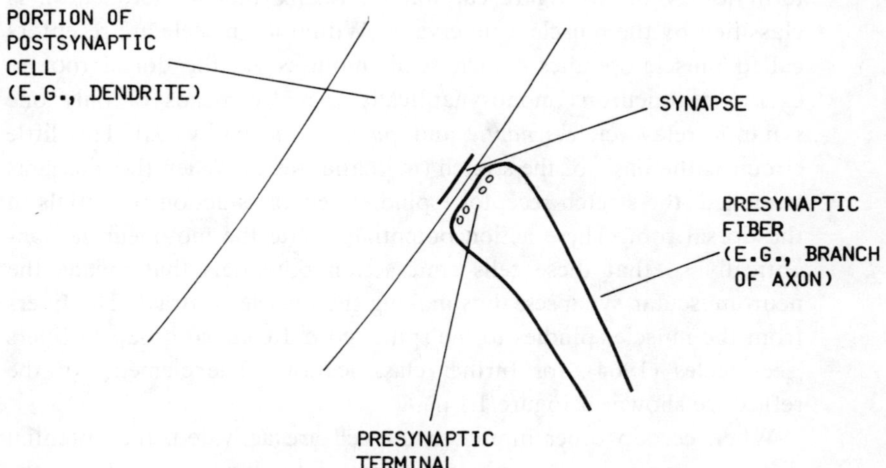

Figure 1.12. The basic terminology for postsynaptic and presynaptic cells.

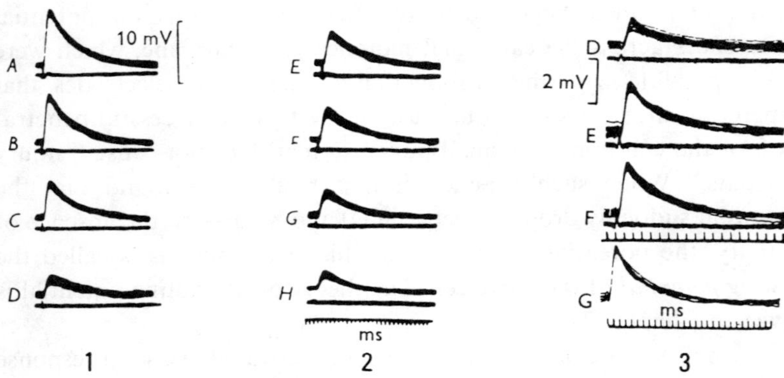

Figure 1.13. EPSPs in cat spinal motoneurons. Time and voltage scales are the same in columns 1 and 2. *Column* 1. Responses of a lateral gastrocnemius neuron to group Ia volleys in lateral gastrocnemius nerve. *Column* 2. Responses in the same cell to group Ia volleys in medial gastrocnemius nerve. *Column* 3. Responses of a posterior biceps semitendinosus motoneuron to group Ia volleys in the biceps semitendinosus nerve. The data of columns 1 and 2 are from Coombs, Curtis, and Eccles (1959) and those of column 3 are from Curtis and Eccles (1959). (Reproduced with the permission of The Physiological Society and the authors.)

said to be *excited* or *depolarized*. The time course of the potential in the postsynaptic cell is called an *excitatory postsynaptic potential* (EPSP). Figure 1.13 shows some EPSPs set up in cat spinal motoneurons by inputs of varying strengths. To understand the terminology in the figure caption we realize that a motoneuron is classified by the muscle it innervates. Within the muscle are receptors called muscle spindles, which send impulses via the dorsal root to excite motoneurons monosynaptically (i.e., by means of only one synaptic relay – cf. *disynaptic* and *polysynaptic* pathways). This little circuit is the basis of the stretch (myotatic) reflex. When the muscle is stretched, the stretch receptor (spindle) generates action potentials in the dorsal root. These action potentials excite the motoneurons synaptically so that these cells emit action potentials that invade the neuromuscular synapses, thus making the muscle contract. The fibers from the muscle spindles to the spinal cord are called group Ia fibers [see Eccles (1964), for further classification]. The elements of the reflex are shown in Figure 1.14.

When certain other inputs to the cell are activated, the potential difference across the membrane increases (or hyperpolarizes) as the inside becomes more negative electrically. Such inputs are called

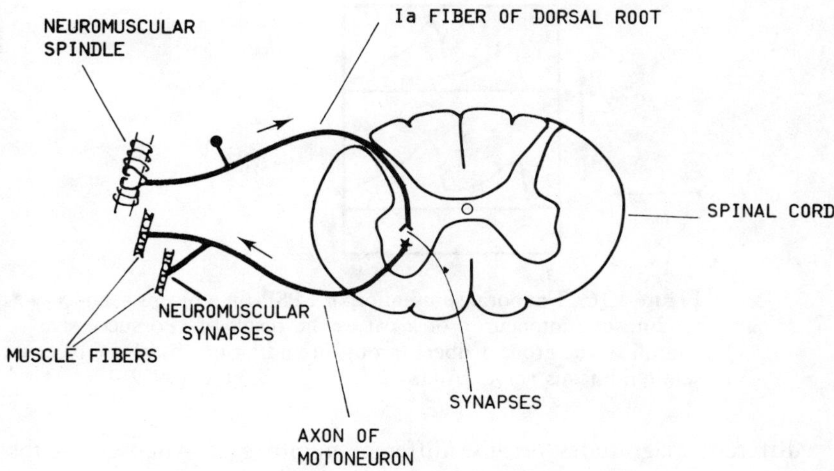

Figure 1.14. The Ia monosynaptic excitation of spinal motoneurons involved in the simple reflex arc. [Adapted from Gardner (1963).]

inhibitory and the time course of the response is called an *inhibitory postsynaptic potential* (IPSP). Examples are shown in Figure 1.15.

There are two important properties of nerve cells and their inputs that we can observe from the pictures of the EPSPs. The first is that after the potential rises or falls fairly abruptly on delivering the input, there is a fairly long-lasting *exponential-like decay* of the potential back towards the resting level. The second property is that responses of various magnitudes are obtained depending on how strongly the input fibers are stimulated. We infer that these responses are of

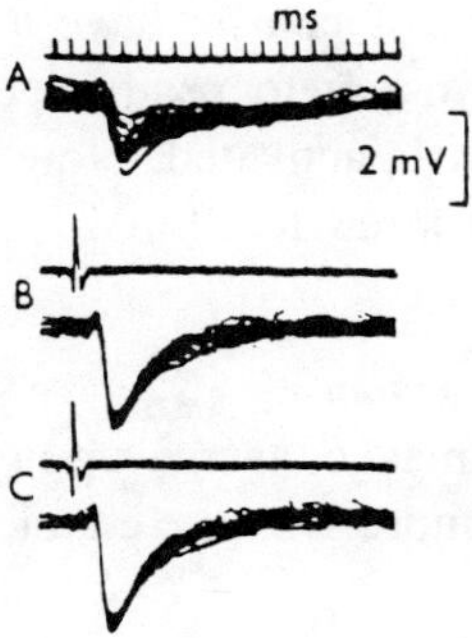

Figure 1.15. Intracellularly recorded IPSPs in a posterior biceps semitendinosus motoneuron as a result of stimulation of the Ia fibers of the quadriceps nerve. [From Curtis and Eccles (1959). Reproduced with the permission of The Physiological Society and the authors.]

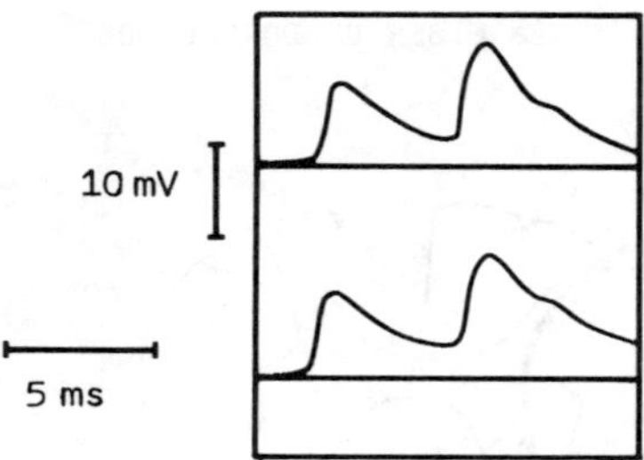

Figure 1.16. Temporal summation of EPSPs in a biceps semitendinosus motoneuron of a cat as the result of two successive stimuli to the group I fibers (group Ia and group Ib) of the biceps semitendinosus nerve. [Adapted from Brock et al. (1952).]

different magnitudes because different numbers of synapses (of the same kind) are being simultaneously (more or less) activated. The addition of the individual responses, which are due to inputs at different locations, is called *spatial summation*.

There is another important summation property that relates to the exponential decay of the EPSPs and IPSPs. This is referred to as *temporal summation*. A graphic example is shown in Figure 1.16. It is clear that the effects of the first stimulus have not died away by the time that the second one is delivered. The observer will notice a kink in the tail of the second EPSP. This is probably an aborted action potential.

Given that the EPSPs and IPSPs of Figures 1.13 and 1.15 are the result of the almost simultaneous activation of many synapses of the same kind, we are naturally led to ask what the contribution from a single synapse is. This question will not, of course, have a definite answer as we would not expect each synapse to produce the same response in the postsynaptic cell when it is activated. Note that the EPSPs in Figure 1.13 are on the order of a few to 10 mV.

1.5 Synaptic potentials from a single synapse

Because several afferent fibers may converge upon a given neuron and each fiber may make one or more synaptic contacts, there are a variety of EPSPs that may occur.

(i) A *single-termininal* EPSP is elicited when a single synapse is activated.

(ii) A *single-fiber* EPSP is elicited when one afferent fiber is stimulated so that all the synaptic endings belonging to that fiber may be activated.

(iii) A *multi-fiber* EPSP is elicited when a bundle of two or more afferent fibers are stimulated. This is sometimes called a *composite* EPSP.

(In Section 9.3 we will encounter another more fundamental EPSP called a *quantal* EPSP.)

Neurophysiologists have succeeded in studying EPSPs in spinal motoneurons, which result from impulses in single Ia fibers. One such study was that of Kuno and Miyahara (1969), who examined responses in 73 motoneurons after stimulation of 36 separate input fibers. Figure 1.17 shows two sets of results. In one motoneuron,

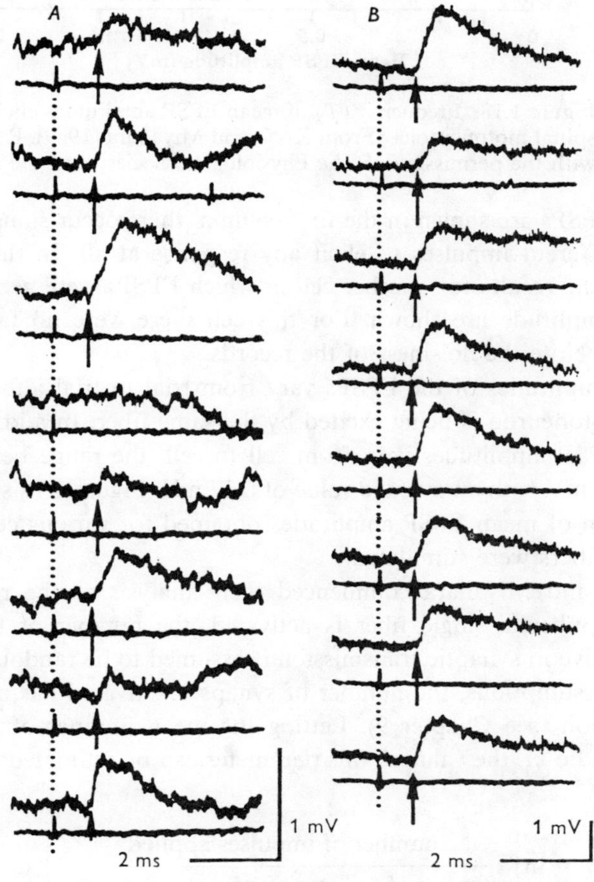

Figure 1.17. EPSPs elicited in two triceps surae motoneurons by stimulation of single fibers of the triceps surae nerve. Columns A and B are the results for two different cells in different trials with the same stimulus. [From Kuno and Miyahara (1969). Reproduced with the permission of The Physiological Society and the authors.]

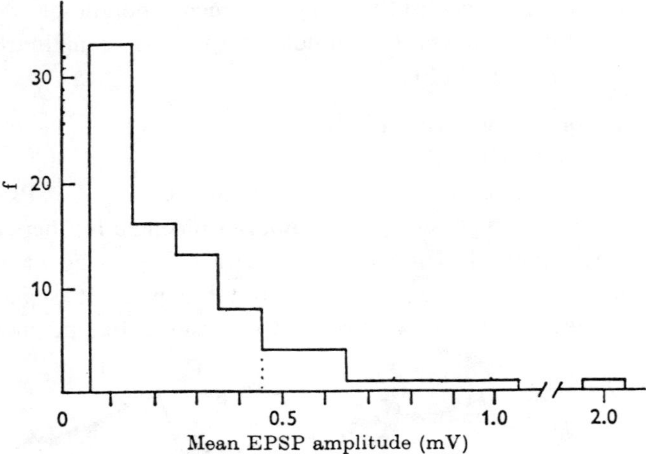

Figure 1.18. Frequency (f) of mean EPSP amplitudes elicited in cat spinal motoneurons. [From Kuno and Miyahara (1969). Reproduced with the permission of The Physiological Society and the authors.]

whose EPSPs are shown in the first column, there occur some failures of the afferent impulses to elicit any response at all. In the second column the results for another cell in which EPSPs were generally of larger amplitude are shown. For this cell there were no failures to respond. Note the noisiness of the records.

The amplitudes of the EPSPs vary from trial to trial although the same motoneuron is being excited by the same fiber. In addition, the *mean* EPSP amplitudes vary from cell to cell, the range being from 0.06 to 2 mV with an average value of 0.27 mV. Figure 1.18 shows the histogram of mean EPSP amplitudes obtained for various cells when various fibers were stimulated.

Kuno and Miyahara commenced their analysis of the results as follows. When a single fiber is activated, the number of terminals made active in synaptic transmission is assumed to be random. Under certain assumptions, the number of synapses activated has a Poisson distribution (see Chapter 9). Letting the mean number of synapses activated be m, the value of this parameter can be estimated from the formula

$$m = \ln\left[\frac{\text{number of impulses applied}}{\text{number of failures of synaptic response}}\right].$$

Values of m were found to lie approximately between 1 and 5 with an average between 2 and 3. For some cells, however, there were no failures to respond and their m values were estimated in another way.

It was pointed out by Kuno and Miyahara that neither the Poisson distribution nor the binomial distribution gives an excellent fit to their data. One major reason for the discrepancy is that many Ia EPSPs originate on distal dendrites and significant nonlinear summation may occur under such circumstances (see Chapter 7). More recently, Jack, Redman, and Wong (1981) have succeeded in separating noise and EPSP. They found that for most EPSPs, the peak EPSP amplitude had values that were integral multiples of a given voltage and that these values occurred with certain probabilities.

1.6 Anatomical connections: wiring diagrams

Given that there are about 20,000 synapses on the surface of an alpha motoneuron (i.e., one that directly innervates a skeletal muscle fiber), it is natural to ask the following. From which cells do the synaptic contacts come? How do these inputs determine the activity of the motoneuron?

By considering the results of various anatomical and physiological investigations, it is possible to provide a partial answer to the question of which cells are presynaptic to the motoneuron. The way in which the results are presented is by means of a *wiring diagram*.

In their simplest form in neurobiology, wiring diagrams incorporate the following elements.

(i) The *anatomical arrangement*, in a very loose sense, of the cells. A single cell, or more usually a population of similar cells, appears as follows:

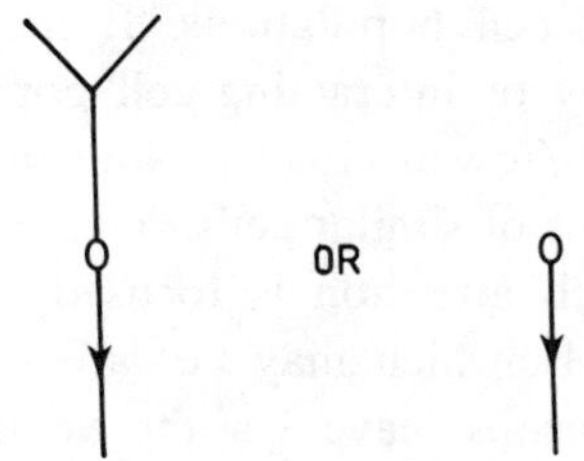

where the circle represents the cell body; the line with the arrow represents the axon, the arrow indicating the direction of normal or *orthodromic* (cf. *antidromic* in the opposite direction) impulse propagation; and the upper branches represent the dendrites. The latter can be omitted unless, as will not be done here, the relative positions of the various inputs are indicated.

(ii) The *synaptic inputs*, which are marked to indicate whether they are excitatory or inhibitory. Here the symbols "+" and "−" are used to denote excitatory or inhibitory inputs. For example, the diagram

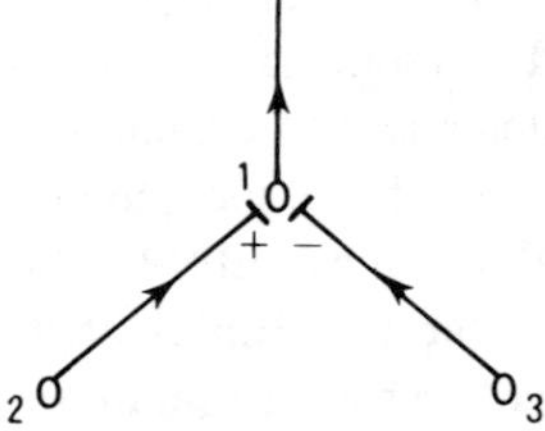

indicates that cells of type 1 receive excitatory inputs from cells of type 2 and inhibitory inputs from cells of type 3.

Using the available evidence, it is possible to construct a tentative wiring diagram showing the origins of the synaptic inputs to an alpha motoneuron. Cells in different segments of the spinal cord will have somewhat different input patterns. The wiring diagram of Figure 1.19 has been compiled for motoneurons of the lumbosacral region of the cord and has been synthesized from Eccles (1957), Burke and Rudomin (1977), and Shepherd (1979).

It will be observed that in Figure 1.19, some cell bodies have been filled in with black, whereas others have not. The black cells make only inhibitory connections with other cells, whereas the others make only excitatory connections. This is an often-used convention and illustrates an important principle known as *Dale's law*. This states that a neuron is *either* excitatory *or* inhibitory in *all* of its connections with other cells. This law is not an absolute one but it provides a useful classification of cell populations. It has been employed by modelers of the activity of interacting cell populations [e.g., Beurle (1956) and Wilson and Cowan (1973)]. Note that each cell depicted may represent thousands of similar cells.

The neuron on which attention is focused in Figure 1.19 is the motoneuron (labeled M), which may be called the *primary* neuron. The axons of motoneurons leave via the ventral roots to directly innervate muscle fibers that receive excitatory synaptic inputs at the *neuromuscular junction*. On the arrival of sufficient excitation the muscle fibers contract.

Pathways from one cell to another are called *mono-*, *di-*, *trisynaptic*, and so forth, according to how many synapses occur along the pathway. The motoneurons receive monosynaptic excitation from:

(i) Ia fibers from muscle spindles that monitor the degree of muscle stretch;

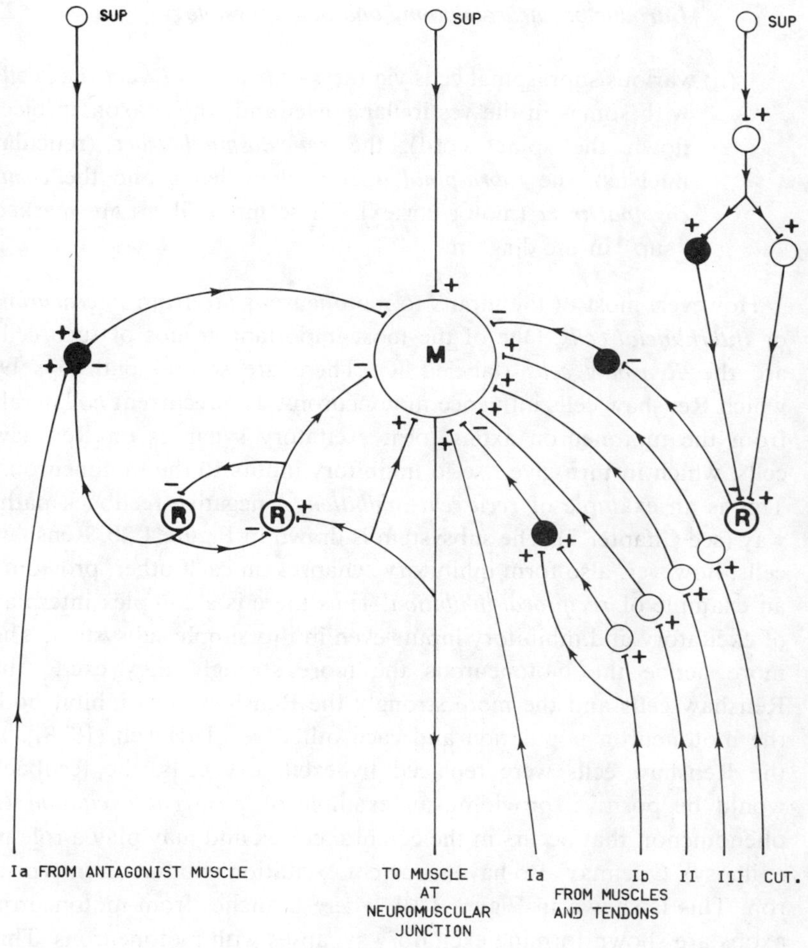

Figure 1.19. Wiring diagram showing some of the known connections to cat spinal alpha motoneurons. The motoneuron is marked M, cells marked R are Renshaw cells, whereas other interneurons are not labeled. The layout does not represent the true anatomical arrangement and the fact that different Renshaw cells are represented in different circuits does not imply that a given Renshaw cell is not involved in all such circuits. The key to the input fibers is as follows. Ia – fibers from the annulospiral endings of muscle spindles; Ib – fibers from Golgi tendon organs of muscle; II – fibers from the secondary endings of muscle fibers; III – small medullated fibers from muscle; sup – descending fibers from the cerebral cortex and other higher centers; and cut – afferent fibers from cutaneous (skin) receptors.

(ii) various supraspinal cells via the *vestibulospinal tract* (i.e., cells with somas in the vestibular nuclei and whose axons project down the spinal cord), the *reticulospinal tract* (reticular nucleus), the *rubrospinal tract* (red nucleus), and the *corticospinal tract* (motor cortex). These input fibers are marked "sup" in the diagram.

However, most of the inputs to motoneurons are from *interneurons* or *internuncial cells*. One of the most important groups of such cells are the *Renshaw cells* (labeled R). There are several pathways by which Renshaw cells influence motoneurons. The recurrent collaterals from the motoneuron axons form excitatory synapses on Renshaw cells, which in turn give rise to inhibitory inputs to the motoneurons. This is an example of *recurrent inhibition*, a negative feedback pathway (see Chapter 7). The subsystem is drawn in Figure 1.20. Renshaw cells, however, also form inhibitory synapses on each other, providing an example of *reciprocal inhibition*. Thus there is a complex interplay of excitatory and inhibitory inputs even in this simple subsystem: The more active the motoneurons the more strongly they excite the Renshaw cells and the more strongly the Renshaw cells inhibit both the motoneuron population and each other [see Tuckwell (1978)]. If the Renshaw cells were replaced by excitatory cells, the feedback would be positive, providing an example of *recurrent excitation* (a phenomenon that occurs in the cerebral cortex and may play a role in epilepsy). One may also have recurrent excitation without an interneuron. This is shown in Figure 1.19, where branches from motoneuron axons are shown forming excitatory synapses with motoneurons. This connection has been claimed to be very likely (Gogan et al. 1977).

In Figure 1.19 is also shown another frequently found subsystem, which is drawn on Figure 1.21. The Renshaw cells inhibit another

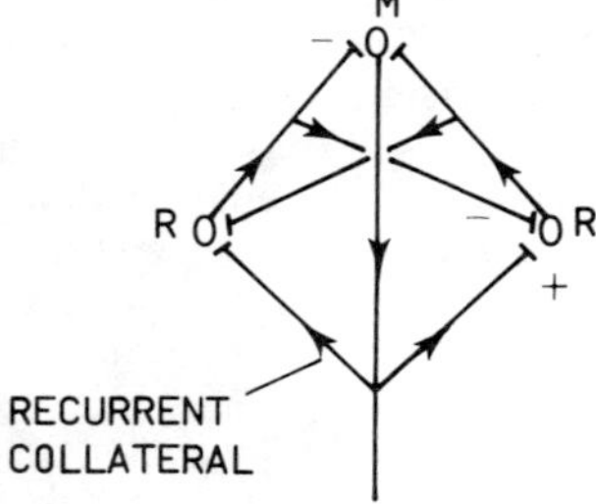

Figure 1.20. The subsystem consisting of motoneurons and Renshaw cells illustrating the phenomena of recurrent inhibition and reciprocal inhibition.

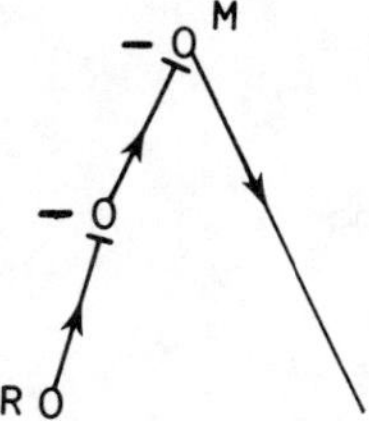

Figure 1.21. The phenomenon of disinhibition.

population of interneurons, which in turn inhibit motoneurons. By means of this pathway Renshaw cells actually have an excitatory effect on motoneurons because their activity decreases the amount of inhibition from the other interneurons. This phenomenon is called *disinhibition*. In addition, motoneurons recurrently excite themselves via this pathway.

A disynaptic inhibitory pathway, which originates from the group Ia fibers of the antagonist muscles, is also shown in Figure 1.19. This provides a means of maintaining certain body postures [see Eccles

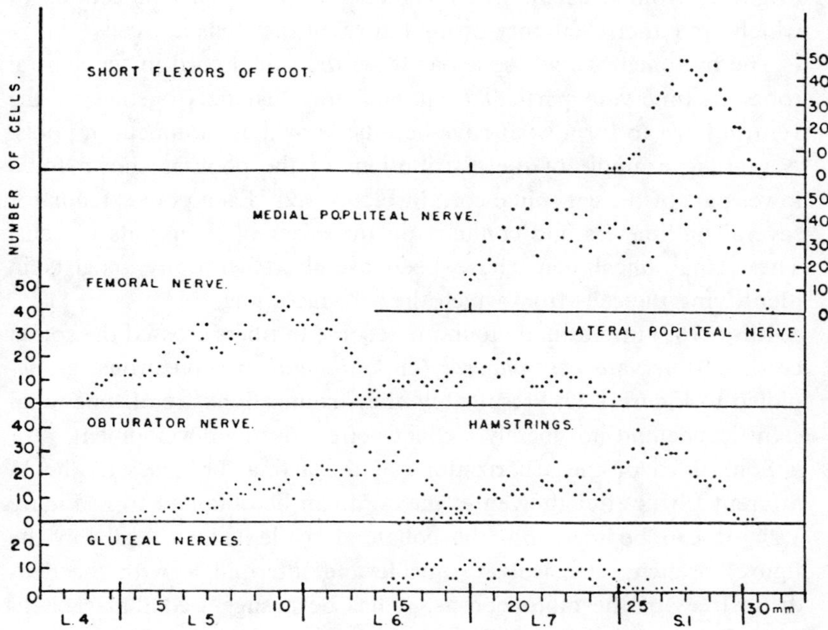

Figure 1.22. Graphs of motoneurons in some of the lumbar (L) and sacral (S) regions of the cat spinal cord. Distance down the cord (in millimeters) goes from left to right. [From Romanes (1951). Reproduced with the permission of Alan R. Liss, Inc.]

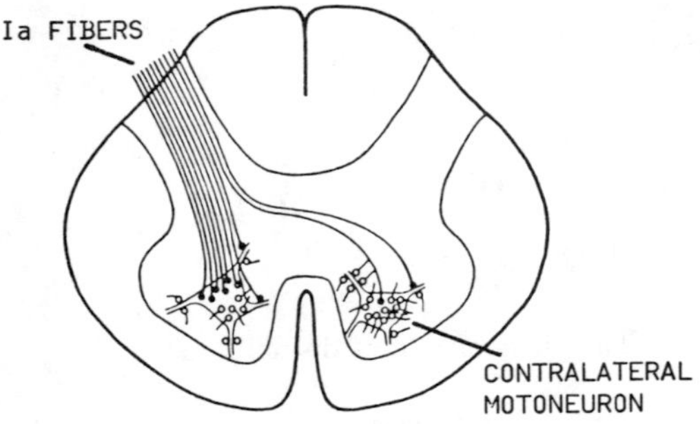

Figure 1.23. How some of the Ia fibers cross over in the spinal cord to innervate motoneurons on the contralateral side. Numbers of fibers are relative. [From Illis (1967). Reproduced with the permission of Elsevier Biomedical Press, B.V. and the author.]

(1977), page 72]. A model for the spinal control of antagonist muscles has been developed by Oğuztöreli and Stein (1983). There are also disynaptic and trisynaptic pathways mediated by the Ib fibers, disynaptic pathways from group II fibers, group III fibers, and afferent fibers from the skin. Inhibitory pathways with more interneurons originate from cells in higher centers, the synaptic connections of which are either excitatory or inhibitory on the Renshaw cells.

The motoneurons whose axons leave the spinal cord in the ventral roots to innervate particular muscles are clustered together in the ventral horn to form what have become known as motoneuron pools. A graphic example of the distributions of the pools is shown for a lower part of the cat spinal cord in Figure 1.22. Each pool extends for several millimeters and contains on the order of thousands of cells. These anatomical maps have been useful to neurophysiologists in identifying the cells from which they are recording.

Illis (1967) interestingly found that some Ia fibers crossed the spinal cord to innervate *contralateral* (cf. *ipsilateral*) motoneurons, as depicted in Figure 1.23. Presumably, these connections are of functional significance and not merely a chance effect during development.

Some idea of the arborizations of the axonal branches of the Ia afferent fibers after they enter the cord can be obtained from Figure 1.24. It can be seen that the collaterals (oriented vertically in the figure) branch extensively to invade and intermingle with the dendritic trees of the motoneurons. It has been suggested that each Ia fiber, by means of these extensive branchings, can make as many as 540 synaptic contacts with various motoneurons (Dodge 1979).

In fact, Mendell and Henneman (1971) suggested that each Ia fiber may sometimes make synaptic contact with *every* motoneuron in the corresponding pool.

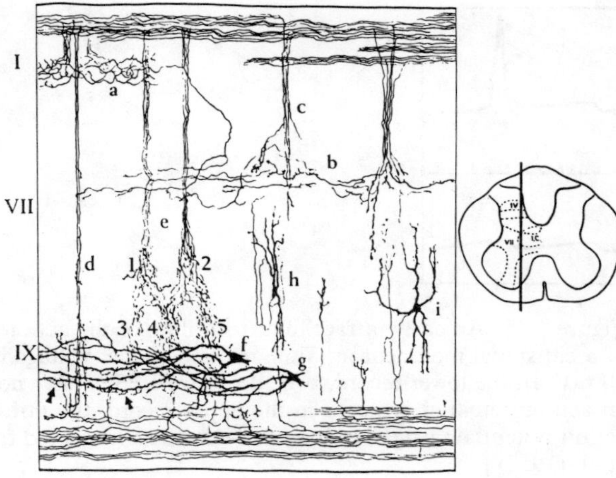

Figure 1.24. How the Ia fibers from muscle spindles form collateral branches that invade, with profuse arborizations, the region occupied by the motoneuron cell bodies (f, g) and dendritic trees. [From Scheibel and Scheibel (1969). Reproduced with the permission of Elsevier Biomedical Press, B.V. and the authors.]

1.7 Action potential of the motoneuron

The responses of the motoneuron that we have considered thus far, the EPSP and IPSP, are essentially local responses. If an excitatory synapse is activated at a point on the soma–dendritic surface, then the depolarization is greatest near that point and diminishes as we move away from that point. In contrast to this local response, there is a depolarization of the cell that propagates usually from near the cell body throughout the entire length of the axon. This propagating wave of depolarization is called an *action potential*.

What is generally found is that an action potential is elicited when the level of depolarization in the cell reaches some critical level called the *threshold* for action-potential generation. An example is shown in Figure 1.25, which shows the time course of the action potential at a recording electrode implanted in the cell body. This figure shows well the concept of a threshold. As in the lower recording, two successive EPSPs did not cause sufficient depolarization through temporal summation to produce an action potential. However, in the upper recording the threshold is attained.

We obtain the following quantitative information from these results. First, the depolarization is about 12 to 15 mV at the threshold when the action potential appeared. Second, the total depolarization during the action potential is on the order of 100 mV. Since the resting membrane potential is about 70 mV, inside negative, the

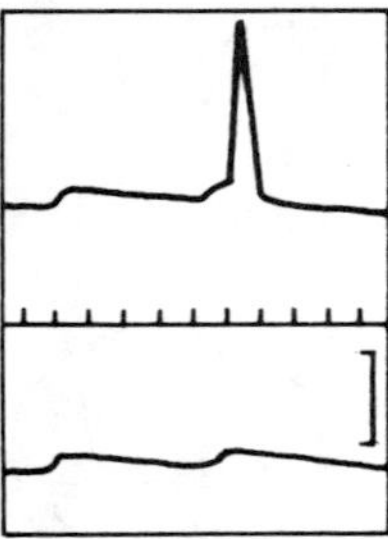

Figure 1.25. An early intracellular recording of the action potential of a cat spinal motoneuron. Time markers, 1 ms; voltage indicated, 50 mV. In the lower record, two successive EPSPs have not elicited an action potential, whereas in the upper record, the threshold for action-potential instigation has been exceeded. [Adapted from Brock et al. (1952).]

polarity of the cell is briefly reversed during the action potential. The inside actually attains a positive value relative to the outside. Note that the main event, this drastic depolarization, is over in a few milliseconds. Indeed, the rate of change of the membrane potential is between 300 and 500 mV/ms for the rising phase of the action potential and between 200 and 250 mV/ms in the falling phase.

Sometimes we use the word *spike* to describe an action potential. We say that when a cell emits a spike it "spikes"; that is, a cell that is "spiking" is emitting action potentials. Another commonly employed synonym is *firing*. Thus we say a cell "fires" an action potential, which refers to the suddenness of the emission, as with a bullet leaving a rifle or revolver. We call the number of action potentials emitted per unit time (usually taken as a second in this context), the *frequency of firing* or the *firing rate*. A typical firing rate for cells in the central nervous system is on the order of 10 to 20 per second. Many cells, however, can fire much more rapidly. Muscle spindles for example, can fire up to 1000 impulses (action potentials) per second, as can Renshaw cells in the early part of their response to a sudden stimulus. Of course, the firing rates of nerve cells vary tremendously depending on the conditions of their stimulation. Some nerve cells do not emit action potentials at all (e.g., retinal amacrine cells).

There are several ways to elicit on action potential in a spinal motoneuron. One is what we could call "natural," in which the dorsal root or other input fibers to the motoneuron are stimulated giving rise to synaptic input over the soma–dendritic surface of the cell. The action potential of Figure 1.25 was obtained by this method, where the nerve stimulated is the dorsal root. Of course, this method is not completely natural because of the way in which the dorsal root is stimulated (electric shock). A second method, which we will focus on in Section 1.9, is impaling the motoneuron with an electrode and passing current into the cell. This is the *current-injection method*. Third, an electric shock can be delivered to the *ventral root* and an action potential, if elicited, will now travel "backward" into the initial segment and into the cell body. The *normal manner* in which an action potential propagates in a spinal motoneuron is away from the cell body and such an action potential (as in the first two methods of elicitation) is said to propagate *orthodromically*, or in the orthodromic direction. When the action potential propagates in the opposite direction, as in ventral root stimulation, the term *antidromic* is employed. Since the distribution in time and space of the potential across the motoneuron surface will depend on the method of elicitation of an action potential, care must be taken when comparing experimental and theoretical results.

According to Brock, Coombs, and Eccles (1952), however, the action potentials recorded (presumably at the cell body) either as a result of orthodromic or antidromic propagation are "virtually identical." Action potentials recorded antidromically reveal some details not apparent in Figure 1.25. These are sketched in Figure 1.26. We see that the "spike" is just the sharp part of the action potential and lasts on average for 1.0 ms. Following the steeply declining falling edge is a slowing down in the rate of decline called the *delayed depolarization* (DD), which commences at about 8 mV from resting level and takes about 2.5 ms. Then a hyperpolarizing phase occurs, called the after-

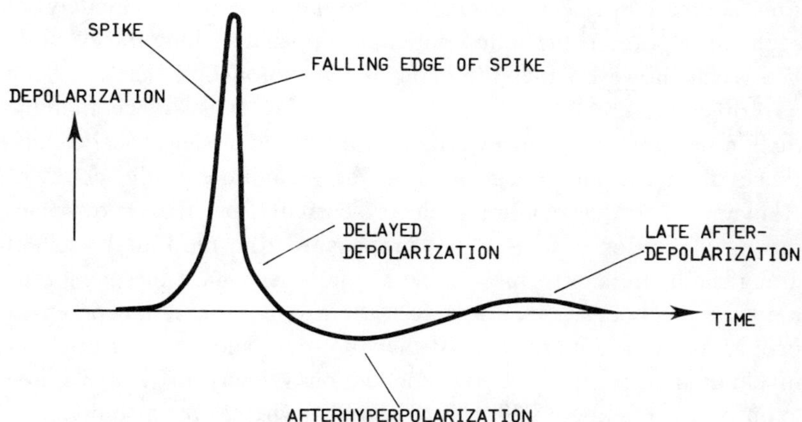

Figure 1.26. Details of the action potential of the motoneuron. [Adapted from Brock et al. (1952).]

hyperpolarization (AHP), which attains its maximum value of about 4.5 mV after 15 ms, but may persist for up to 100 ms. Both the delayed depolarization and afterhyperpolarization have been taken into account in explaining the repetitive firing activity of motoneurons under current injection.

1.8 Anatomical factors in action-potential propagation

Though we usually regard the action potential as a wave of depolarization, we realize that its various components will change in a nonhomogeneous environment. We recall, in fact, that the motoneuron has distinct anatomical regions – dendrites, soma, axon hillock, initial segment, and myelinated axon that eventually arborizes itself after losing its myelin, forming synaptic contact with the muscle fibers. These anatomical differences play an important role in the observations on antidromically propagating action potentials. When an action potential propagates along the motoneuron axon, it sees a different geometry and membrane structure at the initial segment and again when it reaches the hillock, soma, and eventually the dendrites. Sometimes the action potential fails to invade these latter regions successfully, and, since the recording electrode is usually (believed to be) implanted in the soma, the response actually seen depends on the success of the invasion. Thus neurophysiologists talk about *three different components of the "spike"* – the "M" spike found in the myelinated axon; the "IS" (initial segment) spike found in the initial segment and axon hillock; and the "SD" spike found in the soma and dendrites.

In a uniform cylindrical-shaped axon without myelin, we expect the time course of the depolarization to be the same (approximately) at each space point as the action potential propagates along the cylinder. We recall, however, that the axon of the motoneuron has a myelin covering that breaks at the nodes of Ranvier. It is known that the myelin has a much greater resistance (and much less electrical capacitance) than the initial segment and the membrane at the nodes of Ranvier. The transmembrane electric-current flow that takes place during the action potential as it propagates from the initial segment along the axon is only appreciable at the nodes. Since the membrane between the nodes does not have to be excited, the action potential propagates much more quickly than it would along a uniform unmyelinated axon. In fact, the conduction velocity in a myelinated axon of a cat is about 120 m/s (270 mph), whereas for a squid axon, the conduction velocity is about 20 m/s. According to Gardner (1963) a rule of thumb is that myelin speeds up the conduction rate by a

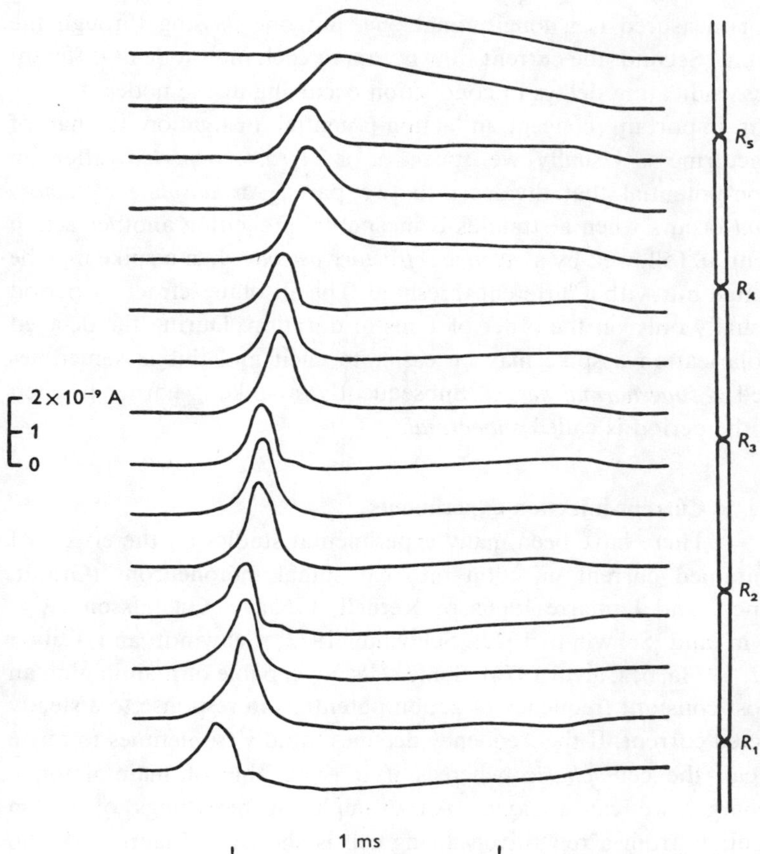

Figure 1.27. Saltatory conduction in a myelinated nerve fiber of a frog. The records show the current flow at three positions between the nodes of Ranvier (marked R_1, etc.). [From Huxley and Stampfli (1949). Reproduced with the permission of The Physiological Society and the authors.]

factor of about six. Some unmyelinated fibers have conduction velocities as low as 2 m/s.

The conduction in myelinated fibers is called *saltatory conduction*. A beautiful illustration of this is shown in Figure 1.27. The current flow during action-potential propagation along a myelinated nerve fiber of a frog is shown at various positions. There are three recordings from each of the *internodes*, as they are called, between R_1 and R_2, R_2 and R_3, R_3 and R_4, and R_4 and R_5. The evidence for the saltatory nature of the conduction is twofold in nature. First, the current flow at various positions within any internode is almost simultaneous and identical in time course. (Note that the current being measured is a longitudinal one, not one passing through the myelin.) Second, the current flow occurs in each internode at different times, indicating delays in conduction occurring at the nodes.

An important concept in action-potential instigation is that of refractoriness. Usually we speak of a *refractory period* after an action-potential that divides into two parts. An *absolute refractory period* occurs when a stimulus is incapable of eliciting another action potential, followed by a *relative refractory period* when a spike may be initiated but with a different threshold. The absolute refractory period is usually only on the order of 1 ms in duration. During the delayed depolarization a spike may be easier to elicit and this is sometimes called a *supernormal period*. Subsequently, a spike is harder to elicit and the period is called *subnormal*.

1.9 Current-injection experiments

There have been many experimental studies on the effects of maintained current injection into cat spinal motoneurons (Granit, Kernell, and Lamarre 1966a, b; Kernell, 1965a–c; Gustafsson 1974; Calvin and Schwindt 1972; Schwindt 1973; Schwindt and Calvin 1972, 1973a, b; Calvin 1974). Some cells are capable of maintaining an almost constant frequency of action potentials in response to a steady injected current. If the frequency declines rapidly (sometimes to zero), we call the cell *phasic*, whereas if it is capable of maintaining a discharge, we call it *tonic*. An example of recordings of action potentials from a repetitively firing cell is shown in Figure 1.28. An

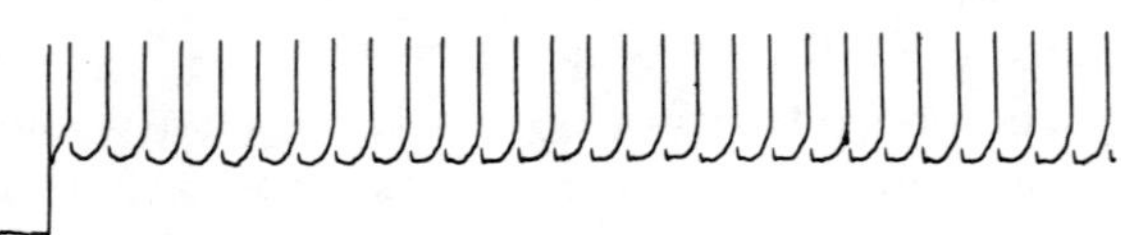

Figure 1.28. Repetitive spiking of a cat spinal motoneuron in response to intracellularly injected current. The upper record shows a rhythmic train of doublet spikes. [From Calvin and Schwindt (1972) and Schwindt and Calvin (1972). Reproduced with the permission of The American Physiological Society and the authors.]

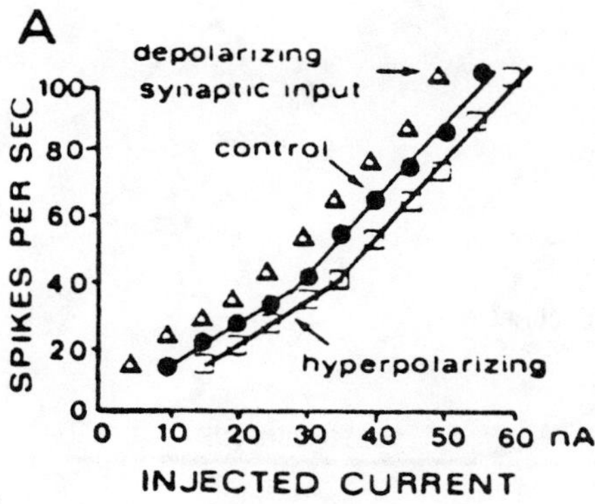

Figure 1.29. An f/I diagram for a cat spinal motoneuron. The solid circles are for current injection alone; the triangles for current injection, together with excitatory synaptic input; and the squares for current injection, together with inhibitory synaptic input. The synaptic inputs are the result of muscle stretch or stimulation of an afferent pathway. [From Schwindt and Calvin (1973a). Reproduced with the permission of Elsevier Biomedical Press, B.V. and the authors.]

interesting example of a cell emitting a train of doublet spikes is also shown in the figure. The second spike of each pair arises out of the delayed depolarization of the first. Such doublets have been implicated in pathological conditions of the nervous system.

When the level of the depolarizing current is raised in these current-injection experiments, the frequency of action potentials increases. There is, of course, a critical level of depolarizing current necessary to elicit rhythmic firing. Once this is exceeded, the frequency f increases linearly with the injected current I in what is called the *primary range*. For many cells the linearity in the primary range is interrupted by a second linear range that has a steeper slope. This is illustrated in Figure 1.29, which shows how the f/I curves are shifted up or down by excitatory and inhibitory synaptic input. This phenomenon is an example of what the neurophysiologists call *algebraic summation* of input currents. One final interesting aspect of these current-injection experiments is the finding of Kernell (1965c), that the minimum repetitive firing rate and the maximum firing rate in the primary range correlate well with the times at which the afterhyperpolarization and late depolarization achieve their maximal values.

When current is injected into a nerve cell, the discharge rate is not necessarily constant as was indicated in the repetitive trains of action potentials in Figure 1.28. Sometimes a cell responds with a burst of spikes and then goes silent. At other times a cell fires rapidly and

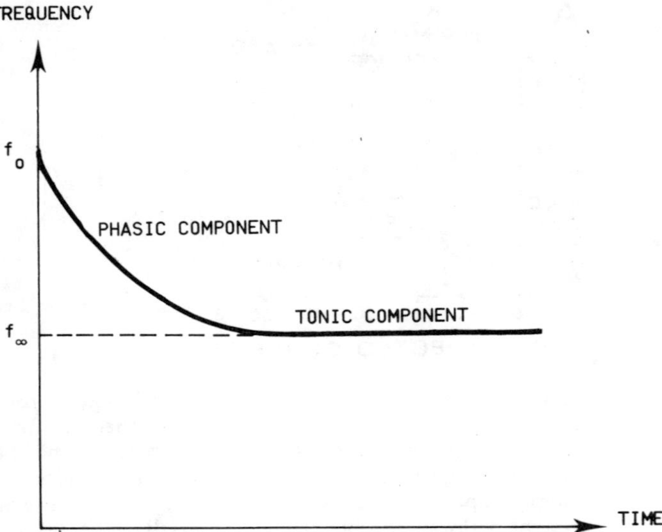

Figure 1.30. The terms "tonic" and "phasic" as applied to the discharge of nerve cells. Here frequency of action potentials is plotted against time.

either maintains that rate or drops to a somewhat lower rate, which remains fairly constant. The drop in firing rate is referred to as *adaptation* or sometimes *accommodation*. A cell that discharges at a uniform rate is called a *tonically* discharging cell. A *phasic* cell is one whose firing rate drops to zero. The terminology has become a little more involved. We talk of the *phasic component* of the discharge, which gives way to a *tonic component*. If we plot the frequency (defined as the reciprocal of the time between impulses) of action potentials versus time, we get the idealized picture shown in Figure 1.30. The initial rate is f_0 and the final rate is f_∞. If $f_0 = f_\infty$ we call it a *purely tonic* cell. If $f_\infty = 0$ we call it a *purely phasic* cell. If $f_0 \neq f_\infty$ and $f_\infty > 0$ we call the initial discharge *phasic*, which dies away to give a *tonic discharge*.

2

The classical theory of membrane potentials

2.1 Introduction

In the previous chapter it was pointed out that there was an electrical-potential difference of about 70 mV across the membrane of the motoneuron. This is true for all nerve cells in their normal, undamaged state, except very briefly during an action potential. When a cell is in an unstimulated steady state, the potential difference across its membrane is called the *resting membrane potential*. This chapter is devoted to the classical theory of such potentials.

We ignore spatial variations and assume that at all interior points the electrical potential is V_i and that at all immediate exterior points the potential is V_0. The membrane potential V_m is by convention defined as interior potential minus exterior potential. That is,

$$V_m = V_i - V_0. \tag{2.1}$$

Usually V_0 is taken to be a zero reference potential so $V_m = V_i$.

The first accurate observation of a nerve-cell membrane potential was on the axon of the *stellar nerve* of the *squid* by Hodgkin and Huxley (1939). A glass microelectrode of 100 μm diameter was inserted in part of a severed axon of diameter 500 μm. Because this diameter is exceptionally large, the preparation is called the *squid giant axon*. The resting and action potential were recorded as shown in Figure 2.1. It is seen that the resting potential was about $V_m = -45$ mV. There is a correction to this due to the *liquid-junction potential*, which develops between the *axoplasm* (the fluid within the axon) and the liquid within the glass microelectrode. This correction, which is elaborated on later in this chapter, leads to the value of about -60 mV for the resting membrane potential of the squid giant axon. Note that in Figure 2.1 the membrane potential remains steady before and

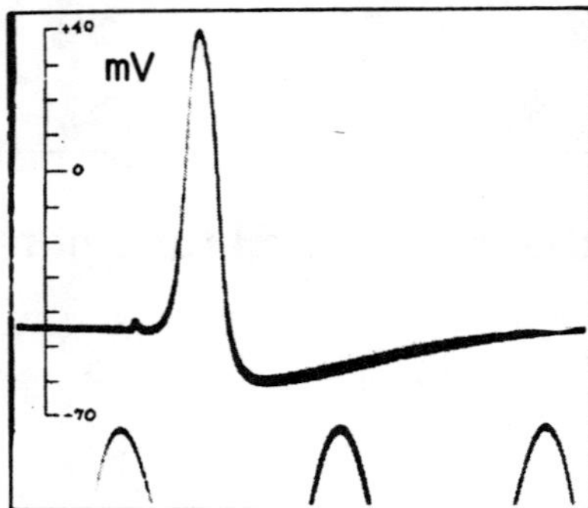

Figure 2.1. The first record of the electrical potential across a nerve-cell membrane in the resting state and during the action potential. Time markers, 500 cycles/s. [From Hodgkin and Huxley (1939), Reprinted by permission from *Nature*, **144**. MacMillan Journals Limited.]

after the action potential. The resting potential is therefore *stable* and is an *equilibrium potential* under normal conditions.

How typical is the resting potential of the squid axon? Since 1939 smaller and smaller microelectrodes have been successfully employed to measure the resting potential of many other nerve cells. Variability is found among cells of a given type in the same species or even in the same animal, and variability is found among nerve cells of different types. Some results are given in Table 2.1. Results for certain glial cells that are ubiquitously enmeshed in nervous systems are also included here. In all cases the interiors of the cells are at a negative electrical potential relative to the exterior, with values between about -40 and -90 mV.

In the classical theory of membrane potentials, the ideas are borrowed for the most part from electrochemistry and date back to the works of Nernst (1888, 1889) and Planck (1890a, b). We begin with the fundamental definitions of flux due to diffussion and motion of ions in electric fields. We incorporate these into an *equation of continuity* to obtain the time-dependent Nernst–Planck equation, which is coupled with Poisson's equation for the electrical potential. The time-dependent problem is difficult and we move quickly to steady-state equations that can be employed in various situations to predict membrane potentials. The *Goldman formula* results from an assumption of a constant electric field in the membrane. Applications of this formula to the calculation of membrane potential and current-density-versus-voltage relations then follow. Finally, the effects of *active transport* are considered.

Table 2.1. *Resting membrane potentials for various neurons and glia*

Type of cell	Resting potential (mV)	Comments
Snail (*Helix aspersa*) abdominal ganglion	−41	Moreton (1968). Average result. External K^+ concentration, 4 m*M*
Squid axon	−60	Hodgkin and Katz (1949), Hodgkin (1951). Corrected for liquid junction potential. Axon in sea water. Range from −65 to −52 mV
Cat pyramidal cell of cerebral cortex	−60	Eccles (1957)
Rabbit pyramidal cell of cerebral cortex	−67	Collewijn and Van Harreveld (1966). Average for 14 cells
Cat spinal motoneuron	−70	Eccles (1957). Range from −60 to −80 mV
Cat dorsal spinocerebellar tract cell	−65	Cited in Burke and Rudomin (1977)
Cat glial cell	−90	Kuffler and Nicholls (1976)
Leech nerve cell	−50	Kuffler and Nicholls (1976)
Leech glial cell	−75	Kuffler and Nicholls (1976)

2.2 Cell-membrane and chemical constituents of the intracellular and extracellular fluids

As revealed by the electron microscope, the limiting membrane of the cell is about 75 Å in thickness. (1 Å $= 10^{-10}$ m.) It has been deduced that the cell membrane is composed of two layers of lipid molecules in which large protein molecules are interspersed irregularly. (The term lipid refers to the "fatty" component of the membrane, soluble in organic solvents.) For views on the structural details, see, for example, White, Handler, and Smith (1973) and Stein (1980).

There are two chief factors that determine the resting membrane potential. First, there is the relative ease with which various ions are able to penetrate the membrane. Some substances, such as ethanol, can pass very easily through the membrane because they have a high lipid solubility. Water, too, readily passes across the cell membrane. Most ions in aqueous solution, however, do not easily penetrate the membrane.

One view (Hille 1977) is the passage of the physiologically relevant ions through the membrane is via sparsely located specialized holes, which are referred to as *channels*. Possibly each kind of ion has its own preferred channel through which it passes more easily than other kinds of ions.

In aqueous solution, ions such as K^+, Na^+, and Cl^- (potassium, sodium, and chloride) are bound to several water molecules. For example, a potassium ion, which unhydrated has a diameter of 2.66 Å, forms a complex with about 3 water molecules. The *potassium channel* is thought to have a minimum diameter of about 3 Å, so that the K^+–water molecule complex (with a total width of about 7 or 8 Å) must be broken down before the potassium ion can pass through it. Selectivity of the potassium channel to potassium rather than sodium ions is possibly as great as 100 : 1.

Though sodium ions have diameters of 1.9 Å and are thus smaller than potassium ions in the unhydrated state, they are bound, on average, to 4.5 water molecules. However, due to geometric and electrical effects, the *sodium channel*, with a minimum diameter of about 5 Å, allows passage of Na^+ much more easily than K^+. It is not as selective as the potassium channel in the sense that selectivity to Na^+ rather than K^+ is probably about 12 : 1 (Hille 1975).

The second key fact in the classical theory of membrane potentials is that the ions that may permeate the cell membrane have differing concentrations inside and outside the cell. Table 2.2 gives the intracellular and extracellular concentrations of various ions for a few

Table 2.2. *Approximate ionic concentrations in intracellular and extracellular fluids for representative nerve cells*

Nerve		Intracellular concentration (mM)	Extracellular concentration (mM)
Squid axon[a]	K^+	410	10
	Na^+	49	460
	Cl^-	40	540
Cat spinal[b] motoneuron	K^+	150	5.5
	Na^+	15	150
	Cl^-	9	125

[a] Hodgkin (1951).
[b] Eccles (1957).

nerve cells.* Only included are the ions that have been found to usually play the most important role in determining membrane potentials. In addition, in both extracellular and intracellular compartments, there are various large organic anions as well as HCO_3^- and cations such as Ca^{2+} and Mg^{2+}. In all cases the K^+ concentration inside the cell is far in excess of that in the external fluid, whereas the concentrations of Na^+ and Cl^- are much greater outside than inside the cell.

2.3 The Nernst–Planck and Poisson equations

We wish to look at equilibrium situations in which the ionic concentrations and potentials are constant in time. Under these conditions, according to the classical theory, the concentrations and potentials satisfy ordinary differential equations. There is a *Nernst–Planck equation* for each kind of ion present, and the *Poisson equation* relates the potential to the spatial distribution of charges.

*One *mole* equals the molecular weight in grams. Thus 1 mol of H_2 contains 2 g. The *mole fraction* is moles of solute/(moles of solute + moles of solvent). The *molality* is the number of moles of solute per kilogram of solvent. The molecular weights of K^+ and Cl^- are 39 and 35.5, respectively. Thus an aqueous solution of KCl of unit molality (abbreviated to 1 M KCl) contains 74.5 g of KCl for each kilogram of water. The *molarity* is the number of moles of solute per liter of solution. For dilute aqueous solutions the molarity and molality are approximately the same. Concentrations are also expressed in terms of *equivalent weights*. The equivalent weight of a substance is "that weight in grams that combines with, displaces, or otherwise plays the part of 1 g of hydrogen." Thus an equivalent weight of K^+ is 39 g, that of Na^+ (whose ionic weight is 23) is 23 g, whereas that of Ca^{2+} (whose ionic weight is 40) is 20 g. Most physiological concentrations are expressed in millimoles per liter, which is the molality divided by 1000 (mM) or in milliequivalents per liter (mEq/l), where a milliequivalent is a thousandth of an equivalent weight. Electrical neutrality requires that the number of equivalents of cations equals the number of equivalents of anions.

The equations governing the equilibrium situation follow naturally by setting time derivatives to zero in the *time-dependent Nernst–Planck equations*, which we will derive first. The time-dependent equations are difficult to solve – in fact, there do not appear to be any solutions available to the coupled Nernst–Planck–Poisson equations in any practical situation.

We begin with the definitions of *concentration*, *flux*, *electric field*, *electrical potential*, and *mobility*. We will restrict our attention to the case of one space dimension. The units we will employ are meters for length and seconds for time.

2.3.1 Basic definitions

Concentration $C(x, t)$

If the ion concentration at x at time t is $C(x, t)$, then in a volume of cross-sectional area A between x and $x + dx$, there are $C(x, t)A\,dx$ mol of the ion. Thus $C(x, t)$ is measured in moles per cubic meter.

Flux $\mathbf{J}(x, t)$

Flux refers to flow that has direction, so flux is a vector quantity. If the magnitude of the flux at x at t is $J(x, t)$, then $|J(x, t)|$ mol of the ion pass through unit area per unit time. Thus the units for $J(x, t)$ are moles per square meter per second (mol m^{-2} s^{-1}). We will forthwith adopt the convention that fluxes are positive in the direction of increasing x (usually to the right) so $\mathbf{J}(x, t) = J(x, t)\mathbf{i}$, where $\mathbf{i}$ is the unit vector in the positive x-direction.

Electric field $\mathbf{E}(x, t)$

Electric field (or electric intensity) is a force and hence is a vector quantity. If the electric field has magnitude $E(x, t)$ at x at t, then this is the magnitude of the force experienced by a unit positive charge (coulomb) at that point in space and time. The electric field is then $\mathbf{E}(x, t) = E(x, t)\mathbf{i}$, since the direction of the field is that in which a positive charge would move.

Electrical potential $\phi(x, t)$

The electrical potential at a point x at time t is related to the electric field by the relation

$$E(x, t) = -\frac{\partial \phi}{\partial x}(x, t). \tag{2.2}$$

It can be seen by integrating both sides of (2.2) with respect to x that the difference in electrical potential between two points is related to the work done in moving unit positive charge from one point to the other. Note that if ϕ is an increasing function of x, then the electric-field vector is in the negative x-direction, since a positive charge tends to move from higher (positive) to lower (negative) potential. The unit of measurement for ϕ is the volt, so electric-field strength is expressible in volts per meter. Also, ϕ has no directional property (i.e., it is a scalar).

Mobility μ

When an ion, in a medium in which it is free to move, is placed in an electric field, it accelerates in accordance with Newton's laws. Usually an opposing force called friction, or drag, impedes the motion in such a way that the force due to the drag is proportional to the velocity of the ion (Stokes' law). Eventually, a constant final, or terminal, velocity is reached and it is left as an exercise to show that the terminal velocity is proportional to the magnitude of the electric-field strength. It therefore makes sense to define an unsigned quantity called the *mobility*, which is the terminal speed in an electric field of unit strength. Thus the speed v is related to mobility μ and the absolute magnitude of the electric field by

$$v = \mu |E|. \tag{2.3}$$

Thus μ is measured in square meters per volt per second.

Total flux $\mathbf{J}(x, t)$

When a solution containing ions is in an electric field there are two contributions to the ionic flux. One is due to *diffusion*, which occurs if concentration differences exist, and is denoted by $\mathbf{J}_D(x, t)$. The other is due to the electric field and is denoted by $\mathbf{J}_E(x, t)$. The total ionic flux is

$$\mathbf{J}(x, t) = \mathbf{J}_D(x, t) + \mathbf{J}_E(x, t). \tag{2.4}$$

Flux due to diffusion

There is a relation called *Fick's principle*, which asserts that the flux due to concentration differences is proportional to the space rate of change of the concentration. Thus

$$\mathbf{J}_D(x, t) = -D\frac{\partial C(x, t)}{\partial x}\mathbf{i}, \tag{2.5}$$

where the constant of proportionality D is called the *diffusion coefficient*. The negative sign appears in (2.5) because if $C(x, t)$ is increasing as x increases so that $\partial C/\partial x$ is positive, then the flow will be in the negative x-direction. Since $J(x, t)$ is measured in moles per square meter per second and $C(x, t)$ is measured in moles per cubic meter, D is measured in square meters per second.

Flux due to the electric field

If the speed with which ions are moving is v, then the ions that pass through one square meter in unit time are those in a rectangular parallelipiped of length v and unit cross-sectional area. The concentration at x at t is $C(x, t)$, so the number of moles passing through unit area per unit time is $vC(x, t)$. If the individual ions carry z positive elementary charges e (so z is positive for cations and negative for anions), then, from (2.3),

$$\mathbf{J}_E(x, t) = \mu E(x, t)\frac{z}{|z|}C(x, t)\mathbf{i}. \tag{2.6}$$

Here $|z|$ is the absolute value of z. The flux $\mathbf{J}_E(x, t)$ is positive if $E(x,t)$ and z have the same sign, otherwise it is negative. From (2.2) we have

$$\mathbf{J}_E(x, t) = -\mu\frac{\partial\phi(x, t)}{\partial x}\frac{z}{|z|}C(x, t)\mathbf{i}. \tag{2.7}$$

Combining (2.5) and (2.7), we obtain the total ionic flux,

$$\mathbf{J}(x, t) = -\left[D\frac{\partial C}{\partial x} + \mu\frac{\partial\phi}{\partial x}\frac{z}{|z|}C\right]\mathbf{i}. \tag{2.8}$$

Einstein's relation between D and μ

Einstein (1905) showed that for spherical particles in a viscous medium in a state of dynamic equilibrium under the influence of a force field, the diffusion coefficient D is related to the speed u per unit force by

$$D = \frac{RTu}{N}. \tag{2.9}$$

R is the gas constant (as in the ideal gas law, $PV = RT$) with value 8.31431 J K^{-1} mol^{-1}. T is the absolute temperature, and N is Avogadro's number (the number of molecules in a gram molecule) with value 6.0225×10^{23} mol^{-1}. The relation (2.9) has been "extrapolated" to apply to ions in solution. If the electric field has magnitude E, then a charge q experiences a force qE. Thus if the terminal speed is v with force qE, we have $v = |qE|u$. For particles each carrying charge $q = ze$, substitution from (2.3) in (2.9) gives

$$D = \frac{RT}{|z|Ne}\mu. \tag{2.10}$$

Also, since $Ne = F$, the faraday (96500 C mol^{-1}),

$$D = \frac{RT}{|z|F}\mu, \tag{2.11}$$

where absolute magnitudes are employed to fit in with our convention that the mobility is an unsigned quantity. We can now write the following expression for the ionic flux in moles per square meter per second,

$$\mathbf{J}(x, t) = -\frac{\mu}{|z|}\left[\frac{RT}{F}\frac{\partial C}{\partial x} + z\frac{\partial \phi}{\partial x}C\right]\mathbf{i}. \tag{2.12}$$

2.3.2 Time-dependent Nernst–Planck equation for a single ion species

The flux given by (2.12) refers to the number of ions, but we are more interested for applications in the *electric-current density*, which is the positive charge crossing unit area per unit time. Since 1 mol of an ion species whose particles carry charge ze contains a total charge of zF, the current density $\mathbf{j}(x, t)$, with the convention that currents are positive in the direction of increasing x, is related to the flux by

$$\mathbf{j}(x, t) = zF\mathbf{J}(x, t). \tag{2.13}$$

Similarly, the charge density is given by

$$\rho(x, t) = zFC(x, t). \tag{2.14}$$

Equation of continuity

The *equation of continuity* expresses conservation of charge. It can be derived by the methods of vector calculus or heuristically as follows. In an infinite medium in which variations only occur in the x-direction, consider a region of unit cross-sectional area between x and $x + \delta x$. We wish to calculate the time rate of change of $\rho(x, t)$,

$$\frac{\partial \rho}{\partial t} = \lim_{\delta t \to 0} \frac{\rho(x, t + \delta t) - \rho(x, t)}{\delta t}. \tag{2.15}$$

Since at time t the charge in the region is $\rho(x, t)\,\delta x$, it follows that the change in the amount of charge in $(t, t + \delta t)$ is

$$\delta Q = [\rho(x, t + \delta t) - \rho(x, t)]\,\delta x. \tag{2.16}$$

Furthermore, the increase in the amount of charge in the region in $(t, t+\delta t)$ is that entering at x minus that leaving at $x+\delta x$. Hence we also have

$$\delta Q = [j(x,t) - j(x+\delta x, t)]\,\delta t. \tag{2.17}$$

Equating the two expressions for δQ gives, on rearranging,

$$\frac{\rho(x, t+\delta t) - \rho(x,t)}{\delta t} = -\frac{[j(x+\delta x, t) - j(x,t)]}{\delta x}. \tag{2.18}$$

Letting δx and δt approach zero, we obtain the *equation of continuity*,

$$\frac{\partial \rho}{\partial t} = -\frac{\partial j}{\partial x} \qquad \text{equation of continuity.} \tag{2.19}$$

The time-dependent Nernst–Planck equation is obtained by substituting the previous expressions for charge density, current density, and flux in this equation. This gives $\partial C/\partial t = -\partial J/\partial x$ or

$$\frac{\partial C}{\partial t} = \frac{\mu}{|z|}\frac{\partial}{\partial x}\left[\frac{RT}{F}\frac{\partial C}{\partial x} + z\frac{\partial \phi}{\partial x}C\right]$$

time-dependent Nernst–Planck equation. (2.20)

2.3.3 Poisson's equation

From Gauss's electric-flux theorem and the divergence theorem of vector calculus one may derive the *Poisson equation*. In one space dimension this takes the form

$$\frac{\partial^2 \phi}{\partial x^2} = -\frac{\rho}{\varepsilon\varepsilon_0}, \tag{2.21}$$

where ε_0 is the permittivity of free space (8.85×10^{-12} C^2 N^{-1} m^{-2}) and ε is the (dimensionless) relative permittivity of the medium under consideration. Note that (2.21) is one of Maxwell's equations, $\nabla . \mathbf{E} = \rho/\varepsilon\varepsilon_0$, when the substitution $\mathbf{E} = -\nabla\phi$ is made.

2.3.4 Steady-state Nernst–Planck and Poisson equations

The coupled system (2.20) and (2.21) is difficult to solve under meaningful assumptions about ρ. The classical theory of membrane potentials usually ignores the temporal evolution of the potential and ion concentrations and is obtained by considering only steady-state solutions. In the steady state, all time derivatives are zero and there is no time dependence for the concentrations. If $\partial C/\partial t = 0$, then $dJ/dx = 0$ or $J(x) = J$, a constant. Thus

$$-\frac{\mu}{|z|}\left[\frac{RT}{F}\frac{dC}{dx} + z\frac{d\phi}{dx}C\right] = J. \tag{2.22}$$

Also, in the steady state the current density is a constant, $j = zFJ$, so we obtain what is usually referred to as the *Nernst–Planck equation*,

$$-\frac{\mu zF}{|z|}\left[\frac{RT}{F}\frac{dC}{dx} + z\frac{d\phi}{dx}C\right] = j$$

steady-state Nernst–Planck equation. (2.23)

For *monovalent cations*, $z = 1$, this becomes

$$-\mu F\left[\frac{RT}{F}\frac{dC}{dx} + \frac{d\phi}{dx}C\right] = j. \tag{2.24A}$$

For *monovalent anions*, $z = -1$, and

$$\mu F\left[\frac{RT}{F}\frac{dC}{dx} - \frac{d\phi}{dx}C\right] = j. \tag{2.24B}$$

Also, in the steady state, $\rho(x, t) = \rho(x)$, so Poisson's equation is

$$\frac{d^2\phi}{dx^2} = -\frac{\rho}{\varepsilon\varepsilon_0} \qquad \text{Poisson's equation.} \tag{2.25}$$

2.4 Solution of first-order linear differential equations

The Nernst–Planck equation is first-order linear and it is useful in order to solve it and equations we will encounter later to have a formula for their solutions.

The general first-order linear differential equation is

$$\frac{dy}{dx} + p(x)y = g(x), \qquad a < x < b. \tag{2.26}$$

If $p(x)$ and $g(x)$ are continuous on the interval (a, b), then this equation is guaranteed to have a unique solution when a value, $y_0 = y(x_0)$, is specified at some point x_0 [see, for example, Boyce and DiPrima (1977), Chapter 2].

The solution of Equation (2.26) with the stated initial condition is

$$y(x) = \frac{1}{\lambda(x)}\left[\int_{x_0}^{x} g(x')\lambda(x')\,dx' + y_0\lambda(x_0)\right], \tag{2.27}$$

where the integrating factor is

$$\lambda(x) = \exp\left[\int^{x} p(x')\,dx'\right]. \tag{2.28}$$

This can be seen as follows. Multiply the differential equation by $\lambda(x)$ to get

$$\lambda \frac{dy}{dx} + \lambda p y = \lambda g. \tag{2.29}$$

Now, by the product rule,

$$\frac{d}{dx}(\lambda y) = \lambda \frac{dy}{dx} + \frac{d\lambda}{dx} y. \tag{2.30}$$

Now, since

$$\frac{d\lambda}{dx} = \frac{d}{dx}\left[\int^{x} p(x')\,dx'\right]\exp\left[\int^{x} p(x')\,dx'\right] = p(x)\lambda, \tag{2.31}$$

we find that the left-hand side of (2.30) is the same as the left-hand side of (2.29). Thus

$$\frac{d}{dx}(\lambda y) = \lambda g. \tag{2.32}$$

Integrating this, we get

$$\lambda y = \int^{x} \lambda(x') g(x')\,dx' + c, \tag{2.33}$$

where c is a constant of integration. This may be written as

$$y(x) = \frac{1}{\lambda(x)}\left[\int^{x} \lambda(x') g(x')\,dx' + c\right]. \tag{2.34}$$

Choosing c such that $y(x_0) = y_0$ gives the solution as in (2.27).

2.5 General solution of the Nernst–Planck and Poisson equations

Consider a single ion species with concentration $C(x)$, mobility μ, current density j, and charge ze on each ion. It is convenient to introduce the constant

$$\gamma = F/RT, \tag{2.35}$$

whereupon the Nernst–Planck equation becomes, in standard form,

$$\frac{dC}{dx} + \gamma z \frac{d\phi}{dx} C = \frac{-zj}{|z| RT \mu}. \tag{2.36}$$

In conjunction with this we will need to consider the Poisson equation,

$$\frac{d^2\phi}{dx^2} = -\frac{\rho(x)}{\varepsilon\varepsilon_0}. \tag{2.37}$$

Assuming $\rho(x)$ is integrable in the region of interest we may integrate the Poisson equation directly to obtain

$$\frac{d\phi}{dx} = -\frac{1}{\varepsilon\varepsilon_0}\int^x \rho(x')\,dx' + k, \tag{2.38}$$

where k is a constant of integration, which can be found if $\phi(x)$ is known at one space point.

Now let

$$\alpha(x) = \gamma z \frac{d\phi}{dx}, \tag{2.39}$$

and

$$\beta = \frac{-zj}{RT|z|\mu}. \tag{2.40}$$

The Nernst–Planck equation now reads simply

$$\frac{dC}{dx} + \alpha(x)C = \beta. \tag{2.41}$$

Thus, using our standard solution (2.27),

$$C(x) = \frac{1}{\lambda(x)}\left[\beta\int_{x_0}^{x}\lambda(x')\,dx' + C(x_0)\lambda(x_0)\right], \tag{2.42}$$

where x_0 is some space point at which $C(x)$ is known and

$$\begin{aligned}\lambda(x) &= \exp\left\{\int^x \alpha(x')\,dx'\right\}\\ &= \exp\left\{\gamma z\int^x \frac{d\phi}{dx'}\,dx'\right\}\\ &= \exp[\gamma z\phi(x)].\end{aligned} \tag{2.43}$$

For future purposes it is convenient to rearrange (2.42) to obtain an expression for the current density,

$$j = \frac{|z|RT\mu}{z}\,\frac{[\lambda(x_0)C(x_0) - \lambda(x)C(x)]}{\int_{x_0}^{x}\lambda(x')\,dx'}. \tag{2.44}$$

In general, it is necessary to specify $\rho(x)$ in order to obtain a solution, but in some cases we may obtain expressions for the membrane potential *without making any assumptions about* $\rho(x)$ (Mullins and Noda 1963). Such cases will be dealt with in the next three sections.

2.6 The membrane is permeable only to ions of the same valence

We will now be specific. Suppose the membrane extends from $x = 0$ to $x = a$ with $a > 0$. *Inward currents will be taken as positive*

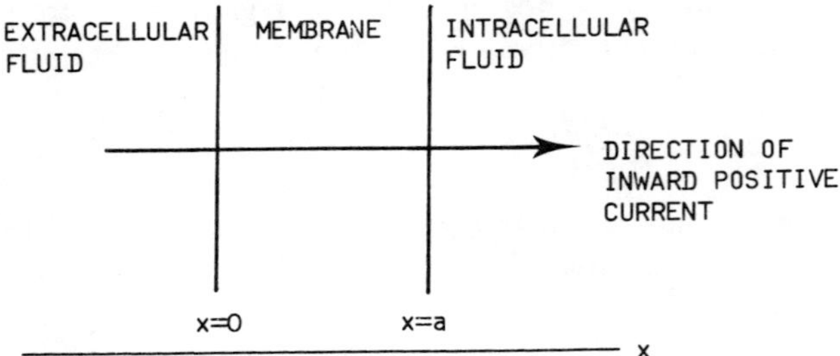

Figure 2.2. The membrane extends from $x = 0$ to $x = a$ and is bounded by extracellular fluid for $x < 0$ and intracellular fluid for $x > a$.

currents so we retain the notation of earlier sections if we take $x = 0$ to be the outside boundary of the membrane and $x = a$ to be the inside. This is sketched in Figure 2.2.

We suppose there are n species of ion present, each of which has the same valence z. We let $C_k(x)$, j_k, and μ_k, $k = 1, \ldots, n$, be the concentrations, current densities, and mobilities, respectively. Then for the kth species the current density is, from (2.44),

$$j_k = \frac{|z| RT\mu_k}{z} \frac{[\lambda(x_0)C_k(x_0) - \lambda(x)C_k(x)]}{\int_{x_0}^{x}\lambda(x')\,dx'}, \qquad 0 \le x \le a. \tag{2.45}$$

At $x = 0$ the potential is set at zero so $\phi(0) = 0$ and at $x = a$ the potential is equal to the membrane potential, so $\phi(a) = V_m$. If we set $x_0 = 0$ and $x = a$ in the last equation, then, since $\lambda(0) = 1$ and $\lambda(a) = \exp(\gamma z V_m)$,

$$j_k = \frac{|z|}{z} RT\mu_k \frac{[C_k(0) - C_k(a)e^{\gamma z V_m}]}{\int_0^a \lambda(x')\,dx'}. \tag{2.46}$$

Boundary values

In the extracellular and intracellular compartments the concentrations are assumed constant, independent of x,

$$C_k(x) = \begin{cases} C_k^0, & x < 0, \\ C_k^i, & x > a. \end{cases} \tag{2.47}$$

The concentrations are not necessarily continuous at the membrane–fluid interfaces, $x = 0$ and $x = a$. In fact, if the ions have difficulty penetrating the membrane and can only do so at sparsely located sites, we expect sudden large decreases in concentrations of ions as we pass from intracellular and extracellular fluid to within the membrane. The drop in concentration of species k at the fluid–membrane interfaces is given in terms of the *partition coefficient* p_k. Thus

$$C_k(0) = p_k C_k^0, \tag{2.48A}$$

$$C_k(a) = p_k C_k^i, \tag{2.48B}$$

where $0 \le p_k \le 1$.

A measure of how easily an ion species may pass through the membrane is the *permeability coefficient* P_k. This is defined as the absolute magnitude of the flux when there is a *unit concentration difference* between external and internal fluids and the concentration is a linear function of distance within the membrane. With these assumptions and the definition of partition coefficient, $|dC_k/dx| = p_k/a$, and from eq. (2.5) the flux is $|J_k| = D_k p_k/a = P_k$. Substituting for D from Einstein's formula (2.11), we find

$$P_k = \frac{R}{Fa}\frac{\mu_k p_k}{|z_k|}. \tag{2.49}$$

The dimensions of the permeability coefficients are the same as for velocity (in meters per second).

Incorporating the boundary values and the definition of permeabilities in (2.46), we obtain

$$j_k = zaFP_k \frac{\left[C_k^0 - C_k^i e^{\gamma z V_m}\right]}{\int_0^a \lambda(x')\,dx'}. \tag{2.50}$$

Electrical equilibrium

If the potential difference across the membrane is stable, a necessary condition is that there be no net transfer of charge. Thus we assume that *the net electric-current density is zero*,

$$\sum_{k=1}^{n} j_k = 0. \tag{2.51}$$

From this equation and (2.50) we get, assuming that the integral in the denominator is nonzero,

$$\sum_{k=1}^{n} P_k\left[C_k^0 - C_k^i e^{\gamma z V_m}\right] = 0. \tag{2.52}$$

This leads to the following expression for the membrane potential:

$$V_m = \frac{RT}{zF}\ln\left[\frac{\sum_{k=1}^n P_k C_k^0}{\sum_{k=1}^n P_k C_k^i}\right]. \tag{2.53}$$

In the case when there is only one permeable ion species, the kth, we obtain the formula for the *Nernst potential* V_k for that species,

$$V_k = \frac{RT}{zF}\ln\left[\frac{C_k^0}{C_k^i}\right] \qquad \text{Nernst potential.} \tag{2.54}$$

We may state therefore that when the membrane is permeable only to species k, regardless of the charge density in the membrane, the condition of electrical equilibrium implies that the membrane potential is given by the Nernst formula (2.54). In this case a true equilibrium situation exists because not only is the electric-current density zero, but also the total ionic flux is zero. Thus when $n = 1$, formula (2.53) is a true equilibrium potential, whereas, in contrast, when $n \geq 2$, there is electrical equilibrium (zero current density) but there is a flux of ions across the membrane, so a true equilibrium does not exist.

When the permeable ions are sodium and potassium, both with $z = 1$, formula (2.53) becomes

$$V_m = \frac{RT}{F}\ln\left[\frac{P_{\mathrm{K}}C_{\mathrm{K}}^0 + P_{\mathrm{Na}}C_{\mathrm{Na}}^0}{P_{\mathrm{K}}C_k^i + P_{\mathrm{Na}}C_{\mathrm{Na}}^i}\right]. \tag{2.55}$$

This formula has been widely used with the more familiar notation

$$V_m = \frac{RT}{F}\ln\left[\frac{P_{\mathrm{K}}[\mathrm{K}]_0 + P_{\mathrm{Na}}[\mathrm{Na}]_0}{P_{\mathrm{K}}[\mathrm{K}]_i + P_{\mathrm{Na}}[\mathrm{Na}]_i}\right], \tag{2.56}$$

where $[\ ]_i$ and $[\ ]_0$ denote internal and external ion concentrations. Formula (2.53) is also a special case of the *Goldman formula*.

Examples

Application of formulas (2.53) and (2.54) is straightforward and the formulas are widely used despite the nature of the assumptions under which they have been derived. The value of RT/F depends on temperature and values at a few temperatures in the physiological range are shown in Table 2.3. These values were obtained with $R = 8.31431$ J K^{-1} mol^{-1}, $F = 96500$ C mol^{-1}, and with $0°\mathrm{C} = 273$ K. (Note that 1 V $=$ 1 J C^{-1}.)

(i) Glial cell-membrane potential

When a cell membrane is permeable to only one ion species (X, say) so that we might expect the Nernst formula to have its greatest predictive power, the cell is said to act like an X-electrode. There is experimental evidence that over wide ranges of external

Table 2.3. *Values of RT/F for various temperatures*

T (°C)	RT/F (mV)
5	24.0
10	24.4
15	24.8
20	25.2
25	25.7
30	26.1
35	26.5
40	27.0

potassium concentrations, many glial cells act like *potassium electrodes*.

Figure 2.3 shows data points of membrane potential at various external potassium ion concentrations for a glial cell in the optic nerve of the mudpuppy (*Necturus*). The external K^+ concentration was known and the intracellular K^+ concentration was estimated as the value of the external concentration at which the cell had zero membrane potential. This gave $[K]_i = 99$ mEq/l. Substituting in the Nernst formula for $T = 24°C$ gives

$$V_m = (25.6\ln[K]_0 - 117.6)\text{ mV}. \tag{2.57}$$

Values from this formula are shown as the straight line in the figure,

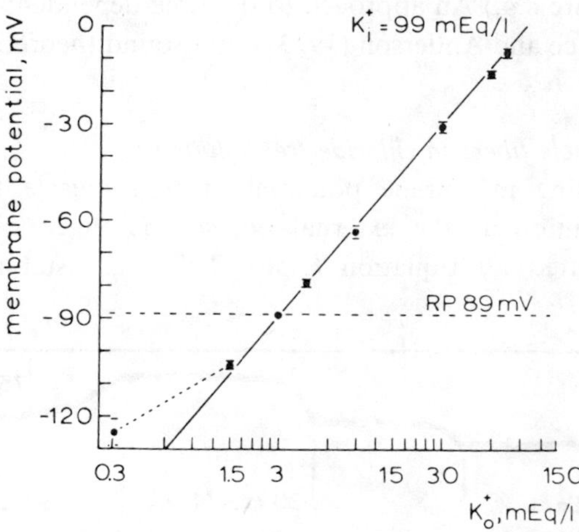

Figure 2.3. Membrane potential of a glial cell versus logarithm of the external K^+ concentration. The straight line is the potential using the Nernst formula. [From Kuffler, Nicholls, and Orkand (1966). Reprinted with the permission of The American Physiological Society and the authors.]

where V_m is plotted against the logarithm of the external K^+ concentration.

For values of $[K]_0$ greater than about 1.5 mEq/l, formula (2.57) fits the data well. When $[K]_0 = 0.3$ mEq/l, the Nernst formula fails as it predicts $V_m = -148.4$ mV, whereas the observed membrane potential is about -125 mV. Other ions, such as Cl^- and Na^+, had negligible effects on the membrane potential in the physiological range of K^+ concentrations ($[K]_0 \geq 3$ mEq/l).

Transients

Suppose the extracellular potassium ion concentration is suddenly changed. The membrane potential does not suddenly change to the new value predicted by the Nernst formula. Instead it achieves the new value with an exponential-like time course. This is seen in Figure 2.4, which is also taken from the above study on glial cells of the mudpuppy optic nerve. Until $t = t_1$ the external K^+ concentration was 3 mEq/l, whereupon it was suddenly changed to 0.3 mEq/l. At $t = t_2$, $[K]_0$ was changed back to 3 mEq/l and, subsequently, at $t = t_3$, it was increased to 30 mEq/l. Finally, at $t = t_4$, $[K]_0$ was changed back to its original value.

The theory of steady-state electrodiffusion that we have considered cannot, of course, make predictions concerning the transient changes in membrane potential of the kind seen in Figure 2.4. (Note that the time constant for rises in membrane potential differs from that for decreases in Figure 2.4.) An approach to the time-dependent problem was made by Brace and Anderson (1973) but a sound theoretical basis is lacking.

(ii) Muscle fibers in chloride-free solution

The resting membrane potential of frog muscle fibers in chloride-free solution as the external potassium ion concentration varies is well fitted by Equation (2.56). This was established by

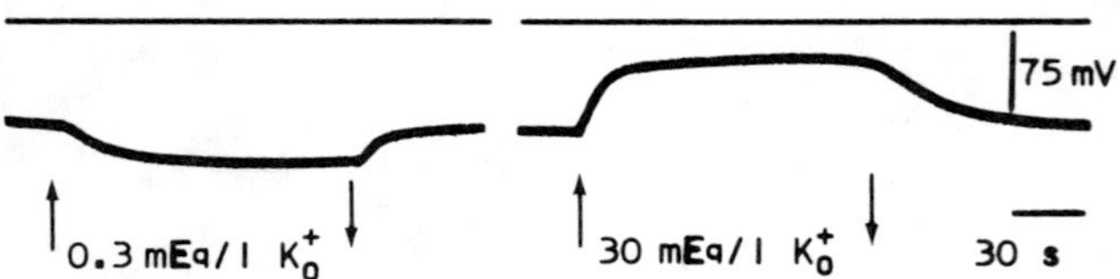

Figure 2.4. Time course of the glial cell-membrane potential when the external potassium ion concentration is changed suddenly at the times t_1, t_2, t_3, and t_4. [From Kuffler et al. (1966). Reproduced with the permission of The American Physiological Society and the authors.]

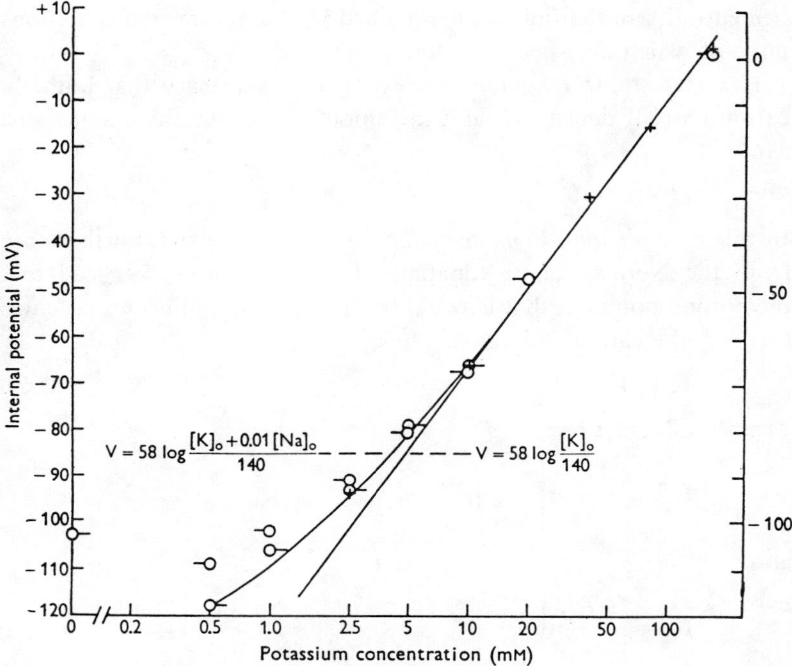

Figure 2.5. Membrane potential of frog muscle fibers at various external potassium concentrations. The straight line is the Nernst potential for K^+. The other continuous line is the prediction by formula (2.58). Crosses are points for measurement made between 10 and 60 min after change in concentration. Circles represent measurements made under a minute. The short horizontal line attached to circle is to the right (left) if the measurement was made after a decrease (increase) in $[K]_0$. [From Hodgkin and Horowicz (1959). Reproduced with the permission of The Physiological Society and the authors.]

Hodgkin and Horowicz (1959) who wrote

$$V_m = \frac{RT}{F} \ln \left[\frac{[K]_0 + \alpha [Na]_0}{[K]_i + \alpha [Na]_i} \right], \tag{2.58}$$

where $\alpha = P_{Na}/P_K$. The value of α was 0.01 and the term $\alpha[Na]_i$ is neglected in the calculated curve shown in Figure 2.5. Here $[K]_i = 140$ mM and $[Na]_0$ was different at different data points.

2.7 Gibbs–Donnan equilibrium

First, we are interested in the following kind of situation. A membrane is permeable to a certain cation species and a certain anion species, both contained in intracellular and extracellular fluids. Electroneutrality in the fluids is maintained by the presence of other ions, none of which may penetrate the membrane.

If a state of true equilibrium exists, it is necessary that both the cation current density j^+ and the anion current density j^- are zero. Thus

$$j^+ = j^- = 0. \tag{2.59}$$

In this case a *Gibbs–Donnan equilibrium* is said to exist. Furthermore, from the Nernst–Planck equations, the conditions (2.59) lead to a membrane potential that is equal to the Nernst equilibrium potential for *both* the cation and anion. Thus

$$V_m = V^+ = V^-. \tag{2.60}$$

From the Nernst formulas,

$$V^+ = \frac{RT}{z^+ F} \ln\left[\frac{C_0^+}{C_i^+}\right], \tag{2.61}$$

and

$$V^- = \frac{RT}{z^- F} \ln\left[\frac{C_0^-}{C_i^-}\right], \tag{2.62}$$

we obtain, from (2.60),

$$\left[\frac{C_0^+}{C_i^+}\right]^{1/z^+} = \left[\frac{C_0^-}{C_i^-}\right]^{1/z^-} \tag{2.63}$$

In the case where both the anion and cation are univalent, this becomes

$$C_i^+ C_i^- = C_0^+ C_0^-, \tag{2.64}$$

so that the product of the concentrations of the anion and cation are the same for both sides of the membrane.

The Gibbs–Donnan equilibrium, like the Nernst equilibrium, is a true equilibrium because there is zero electric current and zero ion flux. The establishment of the equilibrium is usually presented as follows [see, for example, Aidley (1978)].

Suppose to begin with there are equal concentrations of a salt (KCl, say) on both sides of the membrane. Under these conditions there is zero flux and zero electrical potential. Suppose now an amount of a salt KA is introduced to the interior compartment, where A^- is an anion that cannot permeate the membrane. The concentration of K^+ is now greater inside than outside so K^+ tends to diffuse outward. To maintain electrical neutrality Cl^- also flows outward. This movement is promoted by the excess of positive (K^+) ions there. The electrodiffusion continues until the Gibbs–Donnan relation holds,

$$[\mathrm{K}]_i[\mathrm{Cl}]_i = [\mathrm{K}]_0[\mathrm{Cl}]_0. \tag{2.65}$$

Gibbs–Donnan systems

Boyle and Conway (1941) found that the frog sartorious muscle membrane was practically impermeable to Na^+ but freely permeable to K^+ and Cl^- when the external K^+ concentration was raised above 12 mEq/l. Thus if the external K^+ level is raised, K^+ passes inward against a concentration gradient. Similarly, if the external Cl^- concentration is raised, Cl^- travel inward. For various values of the concentrations of sodium, potassium, and chloride in the extracellular fluid, the relation at equilibrium between potassium and chloride concentrations was very well described by Equation (2.65) when $[\mathrm{K}]_0$ was above 12 mEq/l. Further experiments are reported in Hodgkin and Horowicz (1959). For the applicability of the Gibbs–Donnan relation to other preparations including some nerve cells, see Ussing (1949). For the squid axon in laboratory situations the ratio $[\mathrm{K}]_i[\mathrm{Cl}]_i/[\mathrm{K}]_0[\mathrm{Cl}]_0$ is about 5.6, which is far from the Gibbs–Donnan value of unity (Hodgkin, 1951).

2.7.1 Extended Gibbs–Donnan equilibrium

It is sometimes found that the membrane potential is accurately predicted by formula (2.58) and that the membrane potential is also very close to the chloride Nernst potential. An example is the muscle of the fruit fly (*Drosophila melanogaster*) larva (Jan and Jan 1976). Suppose that sodium, chloride, and potassium ions are the only relevant ions. If the resting membrane potential is independent of chloride concentration, then there are two possibilities in the classical framework. Either the membrane is impermeable to Cl^-, or chloride is *distributed passively*, which means there is zero chloride current. The latter was found to be the case in the study of Jan and Jan.

In the context of the Nernst–Planck electrodiffusion picture, this situation arises if there is zero net cation current density

$$j_{\mathrm{K}} + j_{\mathrm{Na}} = 0, \tag{2.66}$$

and zero net anion current density

$$j_{\mathrm{Cl}} = 0, \tag{2.67}$$

where the subscripts refer to the ion species. Without making any assumptions about the charge density in the membrane, (2.66) and

(2.67) lead to

$$V_m = \frac{RT}{F}\ln\left[\frac{[\mathrm{K}]_0 + (P_{\mathrm{Na}}/P_{\mathrm{K}})[\mathrm{Na}]_0}{[\mathrm{K}]_i + (P_{\mathrm{Na}}/P_{\mathrm{K}})[\mathrm{Na}]_i}\right]$$

$$= V_{\mathrm{Cl}} = \frac{RT}{F}\ln\left[\frac{[\mathrm{Cl}]_i}{[\mathrm{Cl}]_0}\right], \tag{2.68}$$

or

$$\{[\mathrm{K}]_0 + (P_{\mathrm{Na}}/P_{\mathrm{K}})[\mathrm{Na}]_0\}[\mathrm{Cl}]_0 = \{[\mathrm{K}]_i + (P_{\mathrm{Na}}/P_{\mathrm{K}})[\mathrm{Na}]_i\}[\mathrm{Cl}]_i. \tag{2.69}$$

Since (2.66) implies that neither j_{K} nor j_{Na} is zero, it is clear that this situation is not a true equilibrium and that there must be some means of countering the continual fluxes of sodium and potassium ions in order to maintain an equilibrium state.

2.8 Diffusion potential

Consider the following situation. A "porous structure" of thickness a separates two solutions of a salt whose cation and anion may both permeate the structure. Electroneutrality is maintained everywhere by virtue of equal concentrations of the cation and anion, $C^+(x) = C^-(x) = C(x)$, for all x.

Let the mobilities of the anion and cation be μ^- and μ^+, respectively, and assume there is zero net electric-current density

$$j^+ + j^- = 0. \tag{2.70}$$

To simplify matters, let both the anion and cation be *univalent*. Then the steady-state Nernst–Planck equations are

$$-\mu^+ F\left[\frac{RT}{F}\frac{dC}{dx} + \frac{d\phi}{dx}C\right] = j^+, \tag{2.71}$$

$$\mu^- F\left[\frac{RT}{F}\frac{dC}{dx} - \frac{d\phi}{dx}C\right] = j^-. \tag{2.72}$$

Under condition (2.70) we have

$$\frac{RT}{F}[\mu^+ - \mu^-]\frac{dC}{dx} + (\mu^+ + \mu^-)\frac{d\phi}{dx}C = 0, \tag{2.73}$$

and it is left as an exercise to show that on integration, with $\phi(0) = 0$ and $\phi(a) = V_m$, this yields

$$V_m = \frac{RT}{F}\left(\frac{\mu^+ - \mu^-}{\mu^+ + \mu^-}\right)\ln\left[\frac{C^0}{C^i}\right], \tag{2.74}$$

where C^i and C^0 are the "internal" and "external" concentrations, respectively. Again, no assumption has been made concerning charge density, though the equality of the cation and anion concentrations in the membrane forces the charge density to be zero. Condition (2.70) implies electrical equilibrium but does not give rise to a true equilibrium since there are continual cation and anion fluxes. Note that if either μ^+ or μ^- is zero, formula (2.74) reduces to the Nernst formula.

The potential given by (2.74) is commonly referred to as a *diffusion potential* [see, for example, Katz (1966)]. It is of interest in relation to the *liquid-junction potentials* that we mentioned in Section 2.1. When two solutions are brought into contact, the integrity of each may be preserved by surface tension. This might occur, for example, when a glass pipette containing a salt solution is employed as a recording electrode and inserted in the intracellular fluid. The analysis, which leads to formula (2.74) or its generalization, remains valid for extremely small values of a, and, in fact, remains valid mathematically when the limit is taken as a approaches zero. Verification of this is left as an exercise.

In the situation described above, if the mobilities of the monovalent cation and anion are equal, then the diffusion potential is zero. This will rarely be the case. The mobilities of Na^+ and Cl^- in aqueous solution are given by Conway (1952) as $\mu^+ = 5.2 \times 10^{-8}$ and $\mu^- = 7.91 \times 10^{-8}\ m^2\ s^{-1}\ V^{-1}$. Thus if a solution of 0.1 *M* NaCl is brought into contact with a solution of strength 0.01 *M*, formula (2.74) gives, at 20°C, $V_m = 12.0$ mV. On the other hand, if solutions of KCl of these strengths are in contact, a junction potential of only 1.0 mV is obtained, using the value 7.64×10^{-8} for the mobility of K^+. Microelectrodes are often filled with KCl at concentrations far greater than that in the intracellular fluid in order to reduce the magnitude of the liquid-junction potential.

2.9 The constant-field assumption

When there is a net current density of zero involving both cations and anions or involving cations or anions of different valency, we cannot obtain expressions for V_m unless we make assumptions about the charge density $\rho(x)$.

The simplest and most frequent assumption is that the charge density in the membrane is identically zero. Apart from leading to predictions, which are often in reasonable agreement with the experiment, this assumption has the advantage of leading to a simple mathematical analysis. Of course, it is impossible in any particular patch of membrane to have zero charge density except at a genuine hole in the membrane. Perhaps because movement of ions is through such holes, the zero charge density assumption performs so well.

If we set

$$\rho(x) = 0, \tag{2.75}$$

then from Poisson's equation we must have

$$d^2\phi/dx^2 = 0. \tag{2.76}$$

Integrating gives

$$d\phi/dx = k, \tag{2.77}$$

where k is a constant. If we set $E = -k$, then this is the same as the assumption of a constant electric field in the membrane,

$$E(x) = E, \qquad 0 \le x \le a \qquad \text{constant field.} \tag{2.78}$$

Now, from (2.77),

$$\phi(x) = kx + k_1, \tag{2.79}$$

where k_1 is a further integration constant. When we impose the values

$$\phi(0) = 0, \tag{2.80}$$

$$\phi(a) = V_m, \tag{2.81}$$

we obtain

$$\phi(x) = V_m x/a, \qquad 0 \le x \le a. \tag{2.82}$$

Inserting this expression for $\phi(x)$ in (2.43), we get, for ion species k,

$$\lambda_k(x) = e^{\gamma z_k V_m x/a}, \tag{2.83}$$

so now

$$\int_0^a \lambda_k(x')\,dx' = \frac{a}{\gamma z_k V_m}\left[e^{\gamma z_k V_m} - 1\right]. \tag{2.84}$$

Using this in expression (2.50) for the current density, we obtain, for the kth ion species,

$$j_k = \gamma F z_k^2 P_k V_m \frac{\left[C_k^0 - C_k^i e^{\gamma z_k V_m}\right]}{\left(e^{\gamma z_k V_m} - 1\right)} \qquad \text{constant-field current density.} \tag{2.85}$$

The constant-field assumption will be employed in the rest of this chapter.

2.10 Goldman's formula for V_m

The *Goldman formula* is widely used to predict resting membrane potentials for nerve cells. It was first obtained by Goldman (1943), rederived in a special case by Hodgkin and Katz (1949), and is commonly referred to as the *Goldman–Hodgkin–Katz* (GHK) *formula*.

To obtain this formula we suppose that the membrane separates solutions containing l permeable cation species and m anion species, with $n = l + m$ total ion species. Let $C_1, \ldots, C_l$ be the cation concentrations and let $C_{l+1}, \ldots, C_n$ be the anion concentrations. Corresponding subscripts will be employed for mobilities, valencies, and current densities. The constant-field assumption is made so that $E(x) = E$ and $\phi(x) = V_m x/a$.

Under the assumption of zero net electric-current density,

$$\sum_{k=1}^{n} j_k = 0, \tag{2.86}$$

we have, from (2.85),

$$\sum_{k=1}^{n} z_k^2 P_k \frac{\left[C_k^0 - C_k^i e^{\gamma z_k V_m}\right]}{\left[e^{\gamma z_k V_m} - 1\right]} = 0. \tag{2.87}$$

It is convenient to separate terms for anions and cations. The same subscript will be employed, being omitted from the summations, but a "+" and "−" will indicate sums over cation and anion species, respectively.

With this notation (2.87) splits into

$$\sum_{+} z_k^2 P_k \frac{\left[C_k^0 - C_k^i e^{\gamma |z_k| V_m}\right]}{\left(e^{\gamma |z_k| V_m} - 1\right)} + \sum_{-} z_k^2 P_k \frac{\left[C_k^0 - C_k^i e^{-\gamma |z_k| V_m}\right]}{\left(e^{-\gamma |z_k| V_m} - 1\right)} = 0. \tag{2.88}$$

Multiplying the numerators and denominators of the terms in the second sum by $e^{\gamma |z_k| V_m}$ gives

$$\sum_{+} z_k^2 P_k \frac{\left[C_k^0 - C_k^i e^{\gamma |z_k| V_m}\right]}{\left(e^{\gamma |z_k| V_m} - 1\right)} + \sum_{-} z_k^2 P_k \frac{\left[C_k^i - C_k^0 e^{\gamma |z_k| V_m}\right]}{\left(e^{\gamma |z_k| V_m} - 1\right)} = 0. \tag{2.89}$$

In this equation everything is "known" except V_m.

2.10.1 Monovalent ions only

When all the permeant ions are monovalent, $|z_k| = 1$ for all species. Solving (2.89) for V_m gives

$$V_m = \frac{RT}{F} \ln \left[\frac{\Sigma_+ P_k C_k^0 + \Sigma_- P_k C_k^i}{\Sigma_+ P_k C_k^i + \Sigma_- P_k C_k^0}\right] \quad \text{Goldman formula.} \tag{2.90}$$

When the only permeable ions are sodium, potassium, and chloride, this becomes, in the usual notation,

$$V_m = \frac{RT}{F}\ln\left[\frac{P_K[K]_0 + P_{Na}[Na]_0 + P_{Cl}[Cl]_i}{P_K[K]_i + P_{Na}[Na]_i + P_{Cl}[Cl]_0}\right], \tag{2.91}$$

or

$$V_m = \frac{RT}{F}\ln\left[\frac{[K]_0 + P'_{Na}[Na]_0 + P'_{Cl}[Cl]_i}{[K]_i + P'_{Na}[Na]_i + P'_{Cl}[Cl]_0}\right], \tag{2.92}$$

where

$$P'_{Na} = P_{Na}/ = P_K, \qquad P'_{Cl} = P_{Cl}/P_K. \tag{2.93}$$

Note that (2.91) reduces to (2.55) when chloride is absent.

2.10.2 Monovalent and divalent ions

We introduce the following notation: $\Sigma_{\pm,1}$ means sum over monovalent cations (+) and anions (−); $\Sigma_{\pm,2}$ means sum over divalent cations (+) and anions (−). It is left as an exercise to show that the membrane potential is then given explicitly by

$$V_m = \frac{RT}{F}\ln\left(\frac{-\bar{\beta} + \sqrt{\bar{\beta}^2 - 4\bar{\alpha}\bar{\gamma}}}{2\bar{\alpha}}\right), \tag{2.94}$$

where

$$\bar{\alpha} = \sum_{+,1} P_k C_k^i + \sum_{-,1} P_k C_k^0 + 4\left\{\sum_{+,2} P_k C_k^i + \sum_{-,2} P_k C_k^0\right\}, \tag{2.95}$$

$$\bar{\beta} = \sum_{+,1} P_k\left[C_k^i - C_k^0\right] + \sum_{-,1} P_k\left[C_k^0 - C_k^i\right], \tag{2.96}$$

$$\bar{\gamma} = -\left[\sum_{+,1} P_k C_k^0 + \sum_{-,1} P_k C_k^i + 4\left\{\sum_{+,2} P_k C_k^0 + \sum_{-,2} P_k C_k^i\right\}\right]. \tag{2.97}$$

Note that in the literature factors of 2 may appear in some terms due to a different definition of mobility [e.g., Piek (1975) and Attwell and Jack (1978)].

2.11 Applications of Goldman's formula to the calculation of membrane potentials

Squid axon membrane potential

The Goldman formula was first applied to squid axon data by Hodgkin and Katz (1949). The internal concentrations were estimated from previous studies as $[K]_i = 345$ mM, $[Na]_i = 72$ mM, and

Table 2.4. *Measured and Goldman formula resting membrane potentials for the squid axon in solutions of various ionic strengths*

Name of external fluid	$[K]_0$	$[Na]_0$	$[Cl]_0$	Measured RP (mV)	Calculated RP (mV)	Difference (mV)
Artificial sea water	10	455	587	−60	−60	0
A	0	465	587	−63	−65	+2
B	15	450	587	−58	−58	0
C	20	445	587	−56	−56	0
Sea water	10	455	540	−59	−59	0
D	7	324	384	−59	−60	+1
E	5	227	270	−61	−61	0
F	3	152	180	−61	−61	0
G	2	91	108	−63	−62	−1
H	10	573	658	−60	−59	−1
I	10	711	796	−57	−59	+2

Adapted from Hodgkin and Katz (1949). Concentrations in mM.

$[Cl]_i = 61$ mM. Table 2.4 contains the results and predictions from the Goldman formula for various external solutions. All of the measured resting potentials have been adjusted by adding −11 mV to allow for the liquid-junction potential between the electrode and axoplasm. In the calculations a temperature of 18°C has been employed ($RT/F = 25$ mV).

For the calculated resting potentials in Table 2.4 the Goldman formula (2.92) is applied with $P'_{Na} = 0.04$ and $P'_{Cl} = 0.45$. From the last column in the table it can be seen that agreement of measured and calculated potentials is very good despite the fact that the values of P'_{Na} and P'_{Cl} were obtained by "trial and error."

Snail neurons

Moreton (1968) examined resting membrane potentials of a large number of snail (*Helix aspersa*) neurons. He did not apply the Goldman formula (2.92) but rather the formula (2.55), which omits the chloride terms. Furthermore, the term involving the internal sodium concentration was neglected. Moreton obtained values of 0.074, 0.127, and 0.217 for P'_{Na} for various cells, values considerably larger than the above value for the squid axon.

2.12 Estimation of parameters in the Goldman formula

The Goldman formula can only be applied if the relative permeabilities and concentrations are known. Usually the external concentrations are more easily obtained than the internal ones, so the relative permeabilities and internal ion concentrations need to be estimated. Also, interest often focuses on the estimation of the relative permeabilities themselves rather than the confirmation or otherwise of the Goldman formula. We give some methods for estimating the relative permeabilities when the internal concentrations are either known or not known. Finally, a method of numerically determining all the unknown parameters is described.

Notation

It is convenient to introduce the following notation in which we restrict our attention to only potassium, sodium, and chloride ions. Let the membrane potential at the jth measurement be $V_{m,j}$ and let the corresponding external ion concentrations be

$$x_j = [\mathrm{K}]_{0,j}, \qquad y_j = [\mathrm{Na}]_{0,j}, \qquad z_j = [\mathrm{Cl}]_{0,j}, \tag{2.98}$$

whereas the internal concentrations are

$$x = [\mathrm{K}]_i, \qquad y = [\mathrm{Na}]_i, \qquad z = [\mathrm{Cl}]_i. \tag{2.99}$$

Let the relative permeabilities be

$$u = P'_{\mathrm{Na}}, \qquad v = P'_{\mathrm{Cl}}. \tag{2.100}$$

Define

$$U_j = \exp(FV_{m,j}/RT). \tag{2.101}$$

Knowledge of the membrane potentials $V_{m,j}$ implies a knowledge of the U_j, which have the advantage of giving rise to equations that are linear in u and v,

$$U_j(x + uy + vz_j) = x_j + uy_j + vz. \tag{2.102}$$

2.12.1 Estimation of relative permeabilities when the internal concentrations are known

Internal ion concentration measurements are possible with ion-selective microelectrodes. If these are accurate the only parameters to estimate are the relative permeabilities.

Exact solution

Theoretically, only two sets of measurements are required to estimate u and v. If these measurements are for $j = 1$ and $j = 2$, we obtain two equations for u and v,

$$u(y_j - U_j y) + v(z - U_j z_j) = U_j x - x_j, \qquad j = 1, 2. \tag{2.103}$$

Using Cramer's rule this gives, with

$$D = \begin{vmatrix} y_1 - U_1 y & z - U_1 z_1 \\ y_2 - U_2 y & z - U_2 z_2 \end{vmatrix}, \tag{2.104}$$

values of the relative permeabilities as

$$u = \frac{1}{D}\begin{vmatrix} U_1x - x_1 & z - U_1z_1 \\ U_2x - x_2 & z - U_2z_2 \end{vmatrix}, \tag{2.105}$$

and

$$v = \frac{1}{D}\begin{vmatrix} y_1 - U_1y & U_1x - x_1 \\ y_2 - U_2y & U_2x - x_2 \end{vmatrix}. \tag{2.106}$$

Least-squares estimates

Consider the simple equation (where x and y are not related to our problem)

$$ax + by + c = 0. \tag{2.107}$$

If x and y are unknown and we try a solution (x^*, y^*), then the *residual* $r(x^*, y^*)$ is defined as

$$r = ax^* + by^* + c, \tag{2.108}$$

which is zero for the exact solution. We seek estimates of u and v, which, in the *least-squares sense*, minimize the residuals for the equations

$$u(y_j - U_jy) + v(z - U_jz_j) + x_j - U_jx = 0, \qquad j = 1, 2, \ldots, n. \tag{2.109}$$

To this end, define

$$a_j = y_j - U_jy, \qquad b_j = z - U_jz_j, \qquad c_j = x_j - U_jx, \tag{2.110}$$

so (2.109) reads

$$ua_j + vb_j + c_j = 0. \tag{2.111}$$

As u and v vary we obtain different values for the residuals

$$r_j = ua_j + vb_j + c_j, \qquad j = 1, 2, \ldots, n. \tag{2.112A}$$

The sum of the squares of the residuals is

$$S = \sum_{j=1}^{n} r_j^2 = \sum_{j=1}^{n} (ua_j + vb_j + c_j)^2. \tag{2.112B}$$

The required values of u and v are those that minimize S. We therefore find the partial derivatives of S with respect to u and v and equate them to zero

$$\frac{\partial S}{\partial u} = 2\sum_j (ua_j + vb_j + c_j)a_j = 0, \tag{2.113}$$

$$\frac{\partial S}{\partial v} = 2\sum_j (ua_j + vb_j + c_j)b_j = 0. \tag{2.114}$$

Solving these equations, we obtain

$$u = \frac{1}{D'}\begin{vmatrix} -\sum c_j a_j & \sum b_j a_j \\ -\sum c_j b_j & \sum b_j^2 \end{vmatrix}, \tag{2.115}$$

$$v = \frac{1}{D'}\begin{vmatrix} \sum a_j^2 & -\sum c_j a_j \\ \sum a_j b_j & -\sum c_j b_j \end{vmatrix}, \tag{2.116}$$

where

$$D' = \begin{vmatrix} \sum a_j^2 & \sum b_j a_j \\ \sum a_j b_j & \sum b_j^2 \end{vmatrix}. \tag{2.117}$$

2.12.2 Estimation of the relative permeabilities when the internal concentrations are unknown

We return to Equations (2.102). Consider three of these with $j = i,\ i+1$, and $i+2$,

$$U_i(x + uy + vz_i) = x_i + uy_i + vz, \tag{2.118}$$

$$U_{i+1}(x + uy + vz_{i+1}) = x_{i+1} + uy_{i+1} + vz, \tag{2.119}$$

$$U_{i+2}(x + uy + vz_{i+2}) = x_{i+2} + uy_{i+2} + vz. \tag{2.120}$$

Eliminating vz from the first two and then from the second two of these equations gives

$$U_i(x + uy + vz_i) - (x_i + uy_i) = U_{i+1}(x + uy + vz_{i+1}) - (x_{i+1} - uy_{i+1}), \tag{2.121}$$

$$U_{i+1}(x + uy + vz_{i+1}) - (x_{i+1} + uy_{i+1}) = U_{i+2}(x + uy + vz_{i+2}) - (x_{i+2} - uy_{i+2}). \tag{2.122}$$

We now eliminate uy from this pair of equations, using the notation

$$\delta U_i = U_{i+1} - U_i, \tag{2.123}$$

$$\delta x_i = x_{i+1} - x_i, \tag{2.124}$$

$$\delta y_i = y_{i+1} - y_i. \tag{2.125}$$

This gives an equation in u and v only

$$u(\delta U_i \delta y_{i+1} - \delta U_{i+1}\delta y_i) + v(\delta U_{i+1}(U_i z_i - U_{i+1} z_{i+1}) - \delta U_i(U_{i+1} z_{i+1} - U_{i+2} z_{i+2})) = \delta U_i \delta x_{i+1} - \delta U_{i+1}\delta x_i. \tag{2.126}$$

The above procedure can be performed for equations $i+1$, $i+2$, and

$i+3$ to yield

$$u(\delta U_{i+1}\delta y_{i+2}-\delta U_{i+2}\delta y_{i+1})+v(\delta U_{i+2}(U_{i+1}z_{i+1}-U_{i+2}z_{i+2}) \\ -\delta U_{i+1}(U_{i+2}z_{i+2}-U_{i+3}z_{i+3})) \\ =\delta U_{i+1}\delta x_{i+2}-\delta U_{i+2}\delta x_{i+1}. \tag{2.127}$$

Equations (2.126) and (2.127) may be solved for u and v. Note, however, that, by judicious choices of the external ion concentrations, values of u and v can be obtained immediately. If $y_i = y_{i+1} = y_{i+2}$ (i.e., the three solutions all have the same sodium concentration), then the coefficient of u in (2.126) is zero and v is found immediately. Similarly, if the three solutions have zero chloride concentration so that $z_i = z_{i+1} = z_{i+2} = 0$, then the coefficient of v is zero and u can be found immediately.

Once the values of u and v are obtained, we may now turn to three of the original equations (2.102) as equations in the unknowns x, y, and z and solve them for the internal ion concentrations.

Alternatively, the five unknown parameters u, v, x, y, and z in the original form of the Goldman formula can be estimated using *nonlinear least-squares algorithms*, such as those of Levenberg (1944) and Marquardt (1963). The theory was outlined in a simpler case of the present problem by Gedulgig (1968). Routines for nonlinear least-squares estimation are found in many software packages. When one of these was applied to the data of Table 2.4, the estimates of P'_{Na} and P'_{Cl} were 0.04316 and 0.4323, respectively, which are very close to Hodgkin and Katz's estimates.

2.13 Current–voltage relationships with constant permeabilities

In addition to providing a simple expression for resting membrane potential, the constant-field assumption enables us to obtain some insight into current-density-versus-voltage relationships for the nerve-cell membrane. Such relationships are important for understanding conductance changes during synaptic transmission and also the generation of action potentials.

If the membrane were an *ohmic resistance*, the magnitude of the current density would be proportional to the potential difference across it. Letting the resistance of a unit area of membrane to current flowing perpendicularly to it be R_m Ω m^2 (recall that resistance is inversely proportional to area), then the current density when the membrane potential is V_m would be

$$j = -V_m/R_m \\ = -G_m V_m \tag{2.128}$$

(in amperes per square meter), where $G_m = R_m^{-1}$ is the conductance per unit area (in mho per square meter). The negative sign appears in (2.128) because of our convention that inward current is positive and the definition of V_m. Thus the current-density-versus-voltage relation is linear with graph a straight line through the origin with slope $-G_m$.

Definition

A resistance is said to possess *rectification* properties if it passes current more easily in one direction than another.

An ohmic resistance has no rectification properties. A membrane is said to possess *inward* (*outward*) *rectification* if it passes *inward* (*outward*) current more easily than *outward* (*inward*) current. We will return to this shortly.

Let the resting membrane potential be V_R. Then the constant-field expression for the current density of the kth ion species at *resting level* is

$$j_{k,R} = \gamma F z_k^2 P_k V_R \frac{\left[C_k^0 - C_k^i e^{\gamma z_k V_R}\right]}{\left(e^{\gamma z_k V_R} - 1\right)}. \tag{2.129}$$

Effect of an applied field

Suppose by some means we are able to hold the interior of the cell at a potential V_m different from the resting potential. Let

$$V_m = V_R + V. \tag{2.130}$$

If $V > 0$ the membrane is *depolarized*, whereas if $V < 0$ the membrane is *hyperpolarized* relative to the resting state. Assuming a constant field, the new electric field in the membrane is

$$\begin{aligned} E_m &= -(V_R + V)/a \\ &= E_R - E, \end{aligned} \tag{2.131}$$

where $E = V/a$. Thus $E > 0$ corresponds to a reduction in field strength. This is illustrated in Figure 2.6. The membrane is said to be *clamped* at the new level V_m.

If we solve the Nernst–Planck equations we obtain, using the same boundary conditions as previously, the same expressions (2.85) for the current densities of the kth ion species at the membrane potential V_m. In terms of the Nernst equilibrium potentials V_k, these may be written, showing the explicit dependence on V_m,

$$j_k(V_m) = \gamma F z_k^2 P_k V_m \frac{C_k^0\left[1 - e^{\gamma z_k (V_m - V_k)}\right]}{\left(e^{\gamma z_k V_m} - 1\right)}. \tag{2.132A}$$

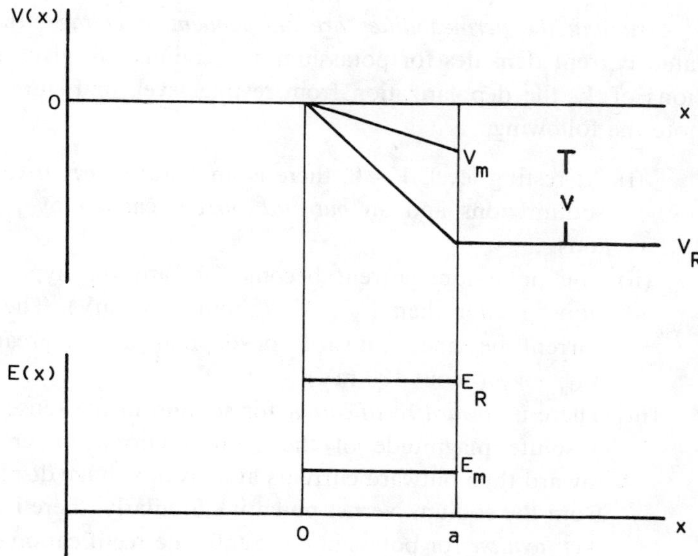

Figure 2.6. Effect of an applied field on the distribution of electrical potential and field strength in the membrane. Here the membrane is depolarized.

In this form we see immediately that

$$j_k(V_k) = 0, \tag{2.132B}$$

so that *the current density for each ion species vanishes at its Nernst potential.* Furthermore, $j_k(V_m)$ is a strictly monotonic (i.e., always increasing or always decreasing) function of V_m, so *the current density for species k undergoes a reversal of direction as V_m passes through V_k.*

In order to obtain current-density-versus-voltage relations in a practical situation, it is necessary to have values for the permeabilities. A rough estimate of the potassium permeability coefficient for the squid axon can be obtained from the measurement of fluxes of radioactive ions.

From Table 2.4 for the squid axon in sea water, $C_K^i = 345$ mM, $C_K^0 = 10$ mM, and $V_R = -59$ mV. The outward K^+ flux for the squid axon at rest is approximately 5×10^{-11} mol cm^{-2} s^{-1} (Caldwell and Keynes 1960). Inserting these numbers in Equation (2.129) with $z_k = 1$ gives a value for the potassium permeability coefficient $P_K = 9 \times 10^{-9}$ m s^{-1}. We may use the relative permeabilities for sodium and chloride, $P'_{Na} = 0.04$ and $P'_{Cl} = 0.45$, to obtain P_{Na} and P_{Cl} for the resting squid membrane.

Assuming the permeabilities are independent of voltage, the calculated current densities for potassium and sodium are given as functions of V, the depolarization from resting level, in Figure 2.7. We note the following.

(i) At resting level, $V = 0$, there is an *inward current* carried by sodium ions and an *outward current* carried by potassium ions.

(ii) The potassium current becomes inward for hyperpolarizations greater than $V_K - V_R$ (about -30 mV). The sodium current becomes outward for depolarizations greater than $V_{Na} - V_R$ (about 100 mV).

(iii) There is *inward rectification* for sodium in the sense that the absolute magnitude of the sodium current is greater for inward than outward currents at a given voltage displacement from the sodium Nernst potential. Similarly, there is *outward rectification* for potassium current. The rectification is in the direction expected from the concentration gradients.

Since at resting level there is a steady influx of Na^+ and a steady efflux of K^+, the concentration gradients for these ions would slowly disappear. If these fluxes persisted, the membrane potential would approach zero. Hence, to maintain the resting potential, a means of countering these fluxes must exist. Evidence for such means, among which are *active transport mechanisms or pumps*, has been convincing and is discussed later in this chapter.

2.14 Reversal potentials and permeability changes

In Chapter 1 we described the excitatory and inhibitory postsynaptic potentials (EPSP and IPSP) observed in spinal motoneurons. The EPSPs and IPSPs we looked at were elicited when the postsynaptic cell was in the resting state with $V_m = V_R$.

We noted in the last section that it is possible to hold the membrane potential at some level other than V_R. One method of doing this is to inject current by means of an intracellular microelectrode. When the steady-state membrane potential is not equal to V_R and a postsynaptic potential is elicited, it is found that the response depends on the holding potential. This is illustrated in Figure 2.8.

EPSPs in a biceps semitendinosus motoneuron when the membrane potential is held at various levels are shown in Figure 2.8A. At the resting potential, $V_m = V_R = -66$ mV, the EPSP amplitude is about 9 mV. When the cell is hyperpolarized to $V_m = -84$ and -102 mV, the EPSP is about the same as that at rest. At the depolarized values

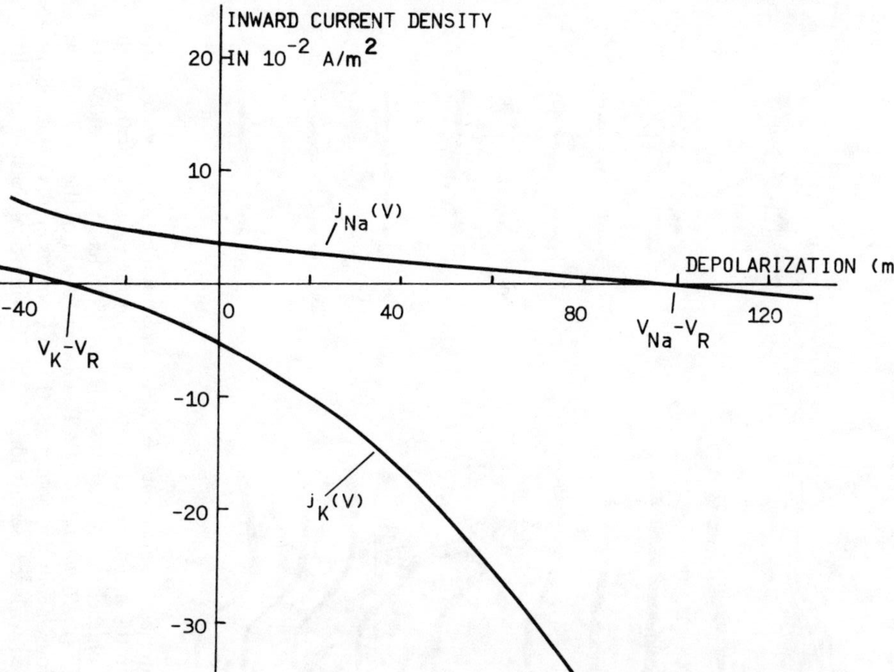

Figure 2.7. Current density versus depolarization from resting level for potassium and sodium ions under the constant-field assumption and the assumption that the permeabilities do not depend on voltage. Inward current is positive and parameter values were estimated from data on the squid axon.

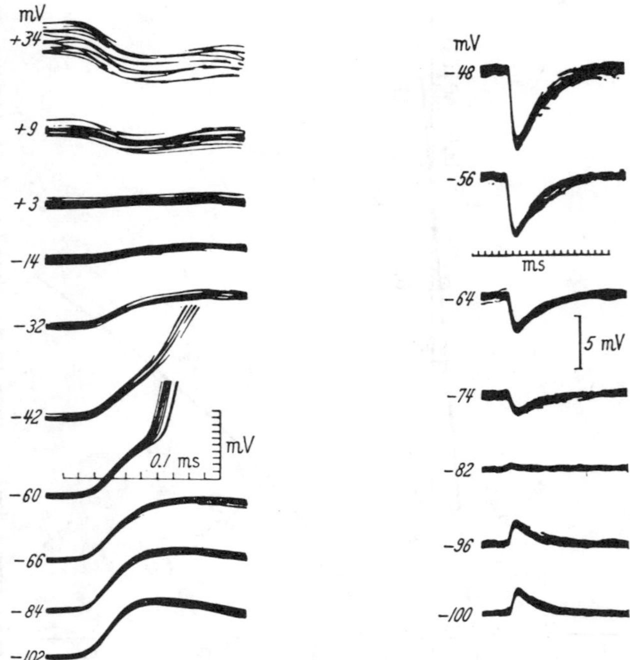

Figure 2.8. *Left column*. EPSPs evoked in a cat spinal motoneuron when the membrane potential is held at the values indicated to the left of each set of responses. [From Coombs, Eccles, and Fatt (1955a).] *Right column*. IPSPs in another cell of the same type at various membrane potentials. [From Coombs et al. (1955b). Reproduced with the permission of The Physiological Society and the authors.]

$V_m = -60$ and -42 mV, an action potential is elicited by the EPSP. Further depolarization ($V_m = -32$ and -14 mV) results in a reduction in the EPSP amplitude and at a depolarization to $V_m = +3$ mV the EPSP has "vanished." Still further depolarization ($V_m = +9$ and $+34$ mV) causes the EPSP to be "reversed," and have the shape of an IPSP.

In Figure 2.8B a set of IPSPs is shown when the membrane potential was maintained at the levels indicated. The resting potential was $V_R = -74$ mV in this cell. Depolarizing the membrane made the IPSP amplitude larger ($V_m = -64$, -56, and -48 mV), a hyperpolarization made it smaller ($V_m = -82$ mV), and further hyperpolarizations ($V_m = -96$ and -100 mV) made the IPSP reverse and have the shape of an EPSP.

To understand these phenomena it is necessary to consider the events that are likely to occur during synaptic transmission (Katz 1969; Krnjevic 1974). The neurotransmitter (e.g., acetylcholine, glutamic acid, gamma-amino-butyric acid etc.) released from presynaptic terminals binds to receptor molecules on the postsynaptic membrane. This alteration in local structure causes changes in the permeabilities (or conductances) of the postsynaptic membrane to various ion species.

Constant-field theory provides one method of obtaining expressions for the postsynaptic reversal (or equilibrium) potentials in terms of permeabilities, concentrations, and so forth. [Many references to this and other approaches are contained in Lewis (1979).] We assume the total membrane current density at a membrane potential V_m is given by

$$j(V_m) = \sum_k j_k(V_m), \tag{2.133}$$

where each term $j_k(V_m)$ is given by the constant-field expression (2.132). The *reversal potential* V_m^* for the postsynaptic potential is the value of V_m at which $j(V_m^*)$ is zero, where now the permeabilities P_k are those induced by the action of the transmitter substance. Thus V_m^* is obtained by solving

$$\sum_k j_k(V_m^*) = 0. \tag{2.134}$$

But this equation is the same as that from which the Goldman formula is obtained. *Hence the reversal potential is given by the Goldman formula with the appropriate values of the permeabilities for the postsynaptic membrane under the action of the transmitter substance.*

Motoneuron IPSP

Coombs, Eccles, and Fatt (1955b) assumed that during a motoneuron IPSP (as in Figure 2.8B) the membrane becomes equally permeable to potassium and chloride, with negligible permeabilities for other ions. (Note that spatial effects are ignored.) In this case,

$$V_m^* = \frac{RT}{F}\ln\left[\frac{[\mathrm{K}]_0 + [\mathrm{Cl}]_i}{[\mathrm{K}]_i + [\mathrm{Cl}]_0}\right]. \tag{2.135}$$

The values of $[\mathrm{K}]_0 = 5.5$ mM, $[\mathrm{Cl}]_0 = 125$ mM, $[\mathrm{K}]_i = 151$ mM, and the IPSP reversal potential, $V_m^* = -79$ mV, enabled the estimation of the internal chloride concentration as $[\mathrm{Cl}]_i = 8.9$ mM. From this, V_{Cl}, the chloride Nernst potential, is -71 mV, which is close to the resting membrane potential. Thus chloride is "almost passively" distributed with a very small current density at rest in this cell.

Endplate currents

At many synaptic junctions between nerve and muscle, generally called *neuromuscular junction*, it is thought that the transmitter substance is acetylcholine [see, e.g., Katz (1969)]. Though it was originally thought that acetylcholine caused a nonspecific increase in the permeability of the postsynaptic membrane (i.e., the permeability coefficients of all ion species increased), it is more likely that the permeability of only the cations Na^+, K^+ (Takeuchi and Takeuchi 1960), and also Ca^{2+} (Lewis 1979) increases.

The junction between nerve and muscle is also referred to as the *endplate* and currents set up in the postsynaptic membrane during synaptic transmission are referred to as *endplate currents* (e.p.c.). The reversal potential for such endplate currents has been measured in solutions of various ionic compositions [see Lewis (1979) for references].

If we assume, as did Takeuchi and Takeuchi (1960), that during the e.p.c. only the potassium and sodium permeabilities are significant, then the reversal potential for the e.p.c. should be

$$V_m^* = \frac{RT}{F}\ln\left[\frac{[\mathrm{K}]_0 + P'_{\mathrm{Na}}[\mathrm{Na}]_0}{[\mathrm{K}]_i + P'_{\mathrm{Na}}[\mathrm{Na}]_i}\right]. \tag{2.136}$$

This may be rearranged to give an expression for P'_{Na} in terms of experimentally measurable quantities

$$P'_{\mathrm{Na}} = \frac{[\mathrm{K}]_i e^{FV_m^*/RT} - [\mathrm{K}]_0}{[\mathrm{Na}]_0 - [\mathrm{Na}]_i e^{FV_m^*/RT}}. \tag{2.137}$$

Using the data of Takeuchi and Takeuchi, $[\mathrm{K}]_i = 126$, $[\mathrm{K}]_0 = 2.5$, $[\mathrm{Na}]_0 = 113.6$, $[\mathrm{Na}]_i = 15.5$ (all in mM), and the measured reversal potential of $V_m^* = -15$ mV, gives, with $RT/F = 25.1$ mV, the ratio of sodium to potassium permeability as $P'_{\mathrm{Na}} = 0.64$. Recalling from Section 2.6 that for resting muscle membrane P'_{Na} is about 0.01, we see that during synaptic transmission at the (frog) neuromuscular junction, the transmitter causes over a 50-fold increase in P'_{Na}. Performing the *same calculation* with the more recent data of Lewis (1979) with the *activities* (rather than concentrations) $[\mathrm{Na}]_i = 6.5$, $[\mathrm{K}]_i = 80$, $[\mathrm{Na}]_0 = 88.2$, $[\mathrm{K}]_0 = 1.92$, $RT/F = 24.8$ mV, and the measured e.p.c. reversal potential of -3.8 mV gives $P'_{\mathrm{Na}} = 0.81$. Lewis obtained $P'_{\mathrm{Na}} = 0.83$ and $P'_{\mathrm{Ca}} = 0.85$, which shows that the postsynaptic membrane becomes roughly equally permeable to the three cations Na^+, K^+, and Ca^{2+}.

Ion-channel selectivity

The Goldman formula can also be employed to evaluate the relative permeabilities of channels to various ions, as the following illustrates. Hille (1973) measured reversal potentials for currents through the frog nodal (i.e. at the nodes of Ranvier) membrane when the external solution contained tetrodotoxin (TTX), which is known to block sodium channels.

Suppose we now wish to find the relative permeability of the potassium channels to various cations. Let a particular cation species be labeled X, with permeability P_X. Assuming the only significant internal cation to be K^+, the reversal potential V_1^* with potassium in the external solution is

$$V_1^* = \frac{RT}{F}\ln\left[\frac{[K]_0}{[K]_i}\right]. \tag{2.138}$$

Similarly, if the reversal potential is now obtained with the species X rather than potassium in the external solution, this will be

$$V_2^* = \frac{RT}{F}\ln\left(\frac{P_X[X]_0}{P_K[K]_i}\right). \tag{2.139}$$

Hence the difference in reversal potentials is

$$V_1^* - V_2^* = \frac{RT}{F}\ln\left(\frac{P_K[K]_0}{P_X[X]_0}\right). \tag{2.140}$$

This can be rearranged to give a formula for the permeability of the channel to X relative to that of K^+,

$$P'_X = \frac{P'_X}{P_K} = \frac{[K]_0}{[X]_0}e^{-F(V_1^* - V_2^*)/RT}. \tag{2.141}$$

Applying this to the cations Tl^+ (thallium), Rb^+ (rubidium), Na^+, and NH_4^+ (ammonium) gives $P'_{Tl} = 2.3$, $P'_{Rb} = 0.92$, $P'_{NH_4} = 0.13$, and $P'_{Na} < 0.01$. Thus these are the estimates of the relative permeabilities of the potassium channels at the frog node for various ions. No other ions were found to have appreciable permeability through potassium channels.

2.15 Current–voltage relationships for nodal membrane

The question arises as to how well constant-field theory is able to predict current densities as a function of depolarization. We will examine data obtained for the nodal membrane of the toad sciatic nerve, where constant-field theory has been applied somewhat successfully.

In *voltage-clamp* experiments, the membrane potential is held fixed in order to eliminate *capacitative currents* that arise from the charging and discharging of the membrane capacity. This enables currents due to ionic fluxes to be measured. One set of voltage-clamp data is shown in Figure 2.9. Each curve represents the current flow at the depolarization indicated to the right of each curve. *Here inward currents are downward.*

For depolarizations less than 114 mV there is a transient inward initial current, which gives way to an outward current. The inward current is presumed carried by sodium ions because it disappears at the sodium Nernst potential for various values of the external sodium concentration. Note that the current densities are considerably greater than those for the squid axon.

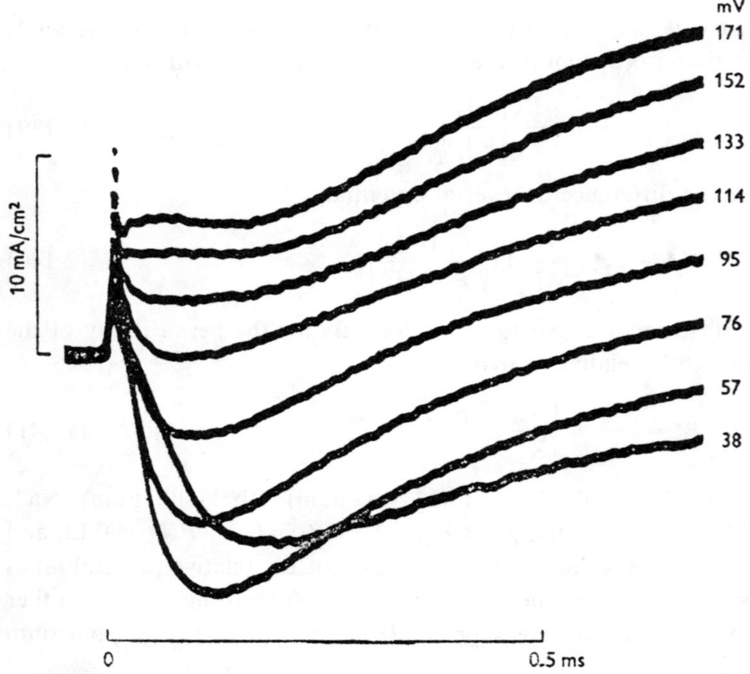

Figure 2.9. Current density versus time under voltage clamp at the nodal membrane of the toad nerve. Inward current is *downward* and depolarizations are marked at the right of each curve. [From Dodge and Frankenhaeuser (1959). Reproduced with the permission of The Physiological Society and the authors.]

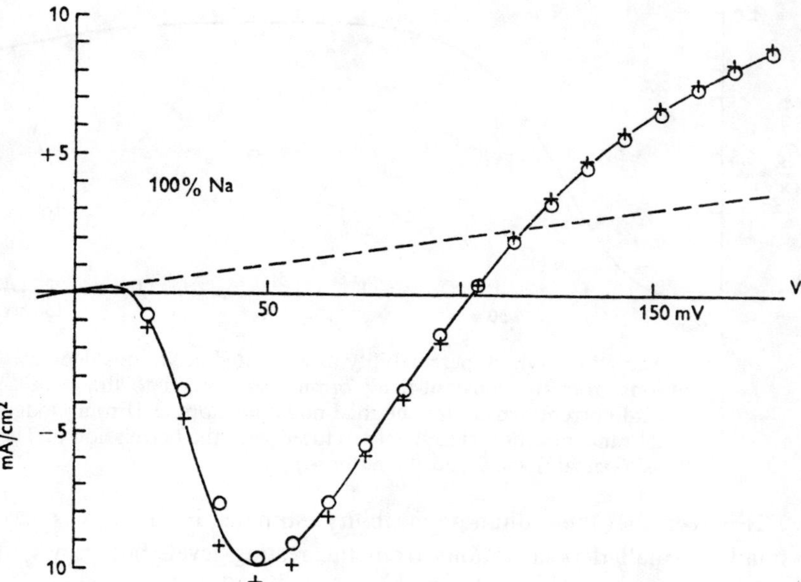

Figure 2.10. Peak inward-current density (inward negative) as a function of depolarization in the toad nodal membrane. [From Dodge and Frankenhaeuser (1959). Reproduced with the permission of The Physiological Society and the authors.]

The dependence of the peak inward current on depolarization is shown in Figure 2.10. This is taken as a measure of the maximal sodium current. The shape of the curve in Figure 2.10 is in contrast to that in Figure 2.7, where the sodium permeability was assumed independent of voltage. In Figure 2.7 $j_{\mathrm{Na}}(V)$ is a monotonic function of V, whereas in Figure 2.10 it has the value zero at three different voltages.

Although the results of Figure 2.10 are *not* for a steady state, whereas those of Figure 2.7 *are* based on a steady state, Dodge and Frankenhaeuser used the constant-field expression for the sodium current density to obtain the sodium permeability as a function of membrane potential

$$P_{\mathrm{Na}}(V_m) = P_{\mathrm{Na}}(V_R + V) = \frac{j_{\mathrm{Na}}(V_m)(e^{\gamma V_m} - 1)}{\gamma F^2 V_m C^0_{\mathrm{Na}}\left[1 - e^{\gamma(V_m - V_{\mathrm{Na}})}\right]}. \tag{2.142}$$

The resulting values of P_{Na} as a function of depolarization are shown as a smooth curve in Figure 2.11.

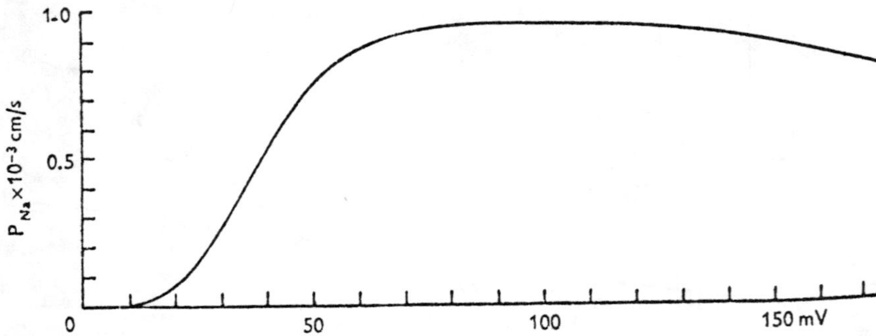

Figure 2.11. Sodium permeability as estimated at various depolarizations from the constant-field formula as applied to the peak inward-current density for the toad nodal membrane. [From Dodge and Frankenhaeuser (1959). Reproduced with the permission of The Physiological Society and the authors.]

It is seen that the sodium permeability estimated in this way is very small at small depolarizations from the resting level, but grows to reach a maximum at depolarizations of about 70 mV. The construction of P_{Na} as a function of depolarization says nothing about the validity or otherwise of constant-field theory.

However, the results of Figure 2.10 and the $P_{\text{Na}}(V_m)$ curve of Figure 2.11 were obtained at an external sodium concentration of 112 mM. When the *same* sodium permeability function is used to calculate the peak inward current density at an external sodium concentration of 41 mM, the calculated values are very close to the experimental values. This is seen in Figure 2.12.

It was deduced that *the constant-field expression for the current density, in conjunction with the appropriate sodium permeability, can be employed to predict sodium current density at the frog or toad node* [see also Dodge and Frankenhaeuser (1958)]. Note that the permeability is a function of membrane potential alone in the above, not of external sodium concentration. The same was found to be true for the potassium current density (Frankenhaeuser 1962). These facts are the basis of the *Frankenhaeuser–Huxley equations* (Frankenhaeuser and Huxley 1964), which are found to quite accurately predict the nodal action potential. In contrast, constant-field theory does not provide a theoretical basis for ionic currents in the squid axon, so a different approach is employed (see Chapter 8).

2.16 Effects of surface charges

The constant-field assumption (i.e., zero charge density) may be modified to accommodate the possibility of *surface charges* on the

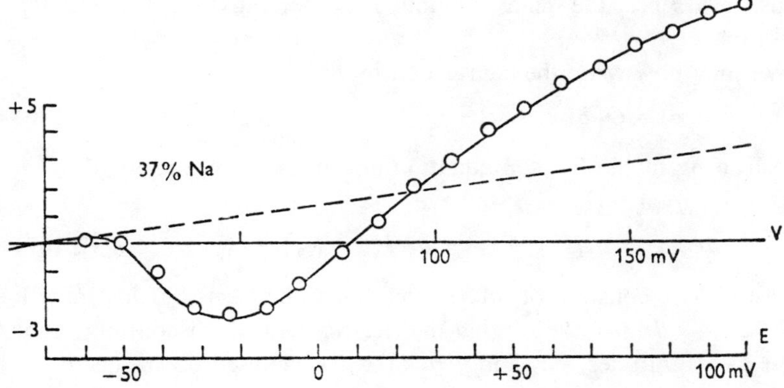

Figure 2.12. Calculated and measured peak inward-current densities when the sodium concentration is reduced to 37% of its original value. [From Dodge and Frankenhaeuser (1959). Reproduced with the permission of The Physiological Society and the authors.]

membrane. References to the experimental evidence for *negative* surface charges on the internal and external surfaces of the squid axon and the external surface of the frog node are given by Brown (1974). The modification to the expressions for the current densities was given, upon the suggestion of Hodgkin, by Frankenhaeuser (1960) and has been employed in the analysis of endplate current reversal potentials (Lewis 1979).

It is convenient to begin with the Poisson equation,

$$\frac{d^2\phi}{dx^2} = -\frac{\rho(x)}{\varepsilon\varepsilon_0}, \tag{2.143}$$

and assume that contributions to the charge density come from charges Q_1 and Q_2 concentrated at the points $x = x_1$ and $x = x_2$, with $0 \le x_1 < x_2 \le a$. This will, in a three-dimensional membrane, represent charge distributed over the planes $x = x_1$ and $x = x_2$.

Mathematically, this situation is described with the use of delta functions (see the end of this section). If a delta function $\delta(x - x_0)$ is "concentrated at the point $x = x_0$," the property of relevance here is

$$\int_{-\infty}^{x} \delta(x' - x_0)\, dx' = H(x - x_0), \tag{2.144}$$

where $H(x)$ is the unit step function,

$$H(x) = \begin{cases} 0, & x < 0, \\ 1, & x \ge 0. \end{cases} \tag{2.145}$$

Thus, in a sense, the delta function is the derivative of the unit step function.

We may now write the charge density as

$$\rho(x) = Q_1 \delta(x - x_1) + Q_2 \delta(x - x_2). \tag{2.146}$$

Substituting in the Poisson equation and integrating once, with $q_1 = -Q_1/\varepsilon\varepsilon_0$, $q_2 = -Q_2/\varepsilon\varepsilon_0$, we have

$$d\phi/dx = q_1 H(x - x_1) + q_2 H(x - x_2) + k_1, \tag{2.147}$$

where k_1 is a constant of integration. Since the electric-field strength is $E(x) = -d\phi/dx$, we see that the electric field has discontinuities or steps of magnitudes $-q_1$ and $-q_2$ at $x = x_1$ and $x = x_2$, respectively.

Now

$$\int_{-\infty}^{x} H(x' - x_0)\, dx' = \begin{cases} 0, & x < x_0, \\ x - x_0, & x \geq x_0, \end{cases}$$
$$= (x - x_0) H(x - x_0). \tag{2.148}$$

Hence the electrical potential is

$$\phi(x) = q_1(x - x_1) H(x - x_1) + q_2(x - x_2) H(x - x_2) + k_1 x + k_2, \tag{2.149}$$

where k_2 is a second integration constant.

Imposing the constraints $\phi(0) = 0$ and $\phi(a) = V_m$ gives

$$\phi(x) = q_1(x - x_1) H(x - x_1) + q_2(x - x_2) H(x - x_2) + (V_m - q_1(a - x_1) - q_2(a - x_2))\frac{x}{a}. \tag{2.150}$$

With negative charges at x_1 and x_2 (so q_1 and q_2 are positive), the electrical potential might appear as in Figure 2.13 (solid line). If x_1 and x_2 represent the edges of the membrane and the points $x = 0$ and $x = a$ are in the extracellular and intracellular fluids, then the potential difference across the membrane will be less than that measured between extracellular and intracellular compartments.

Frankenhaeuser's expression for current density

If x_1 is very small and x_2 is very close to a, then as can be seen from Figure 2.13, the situation will closely resemble that in which the electrical potential has step discontinuities at $x = 0$ and $x = a$. If we assume that the potential at $x = 0^+$ is V' rather than zero, and a constant field exists in the membrane (see Figure 2.14), then we may proceed as follows to obtain Frankenhaeuser's (1960) expression for the effect of surface charges at the external surface.

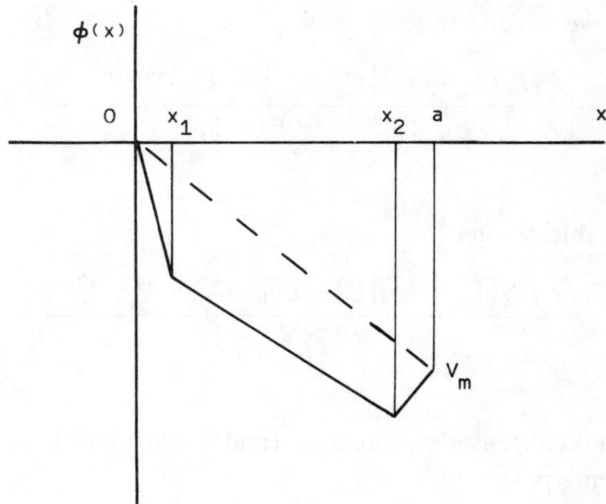

Figure 2.13. Effect of charges at x_1 and x_2 on the electrical potential. The dashed line is the constant-field case with zero charge.

We have

$$\phi(x) = ((V_m - V')/a)x + V'. \tag{2.151}$$

From (2.43), for the kth ion species,

$$\lambda_k(x) = e^{\gamma z_k V'} e^{(V_m - V')\gamma z_k x/a}, \tag{2.152}$$

so that

$$\int_0^a \lambda_k(x')\,dx' = \frac{a e^{\gamma z_k V'}}{(V_m - V')\gamma z_k}\left[e^{(V_m - V')\gamma z_k} - 1\right]. \tag{2.153}$$

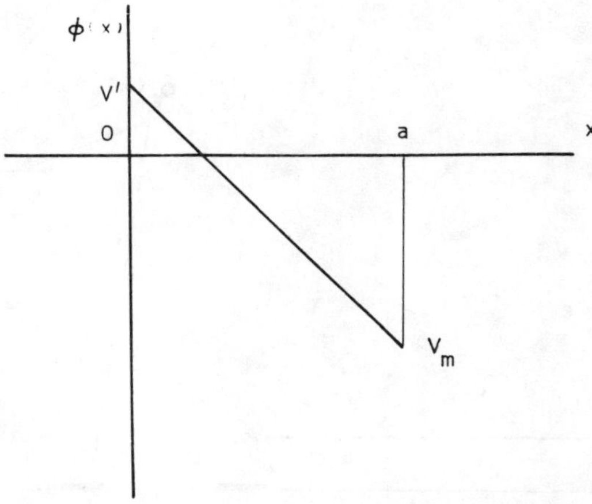

Figure 2.14. Assumed electrical potential in the derivation of Frankenhaeuser's formula for the current density.

From (2.50) we have

$$j_k = \frac{z_k^2 F\gamma P_k (V_m - V')\left[C_k^0 e^{-\gamma z_k V'} - C_k^i e^{-\gamma z_k (V' - V_m)}\right]}{\left(e^{-(V' - V_m)\gamma z_k} - 1\right)}. \tag{2.154}$$

Rearranging this, we get

$$j_k = \frac{z_k^2 F\gamma P_k (V_m - V')\left[C_k^i - C_k^0 e^{-\gamma V' z_k} e^{-(V_m - V')\gamma z_k}\right]}{\left(e^{-\gamma z_k (V_m - V')} - 1\right)}, \tag{2.155}$$

which is Frankenhaeuser's expression (multiplied by -1 as we have inward currents positive).

2.16.1 The delta function

Heuristically, the delta function can be introduced as follows. Consider the functions $\delta_\varepsilon(x)$, defined for $\varepsilon > 0$ by

$$\delta_\varepsilon(x) = \begin{cases} 1/\varepsilon, & -\varepsilon/2 \le x \le \varepsilon/2, \\ 0, & \text{otherwise.} \end{cases}$$

The function $\delta_\varepsilon(x)$ with a large value of ε is sketched in the left part of Figure 2.15, whereas the function with a much smaller value of ε is sketched on the right. As ε gets smaller and smaller, the width of the nonzero part of $\delta_\varepsilon(x)$ gets very small and its height gets very large; nevertheless in all cases the area between $\delta_\varepsilon(x)$ and the x-axis is unity. The delta function $\delta(x)$ can be considered to be the limiting form of $\delta_\varepsilon(x)$ as $\varepsilon \to 0$.

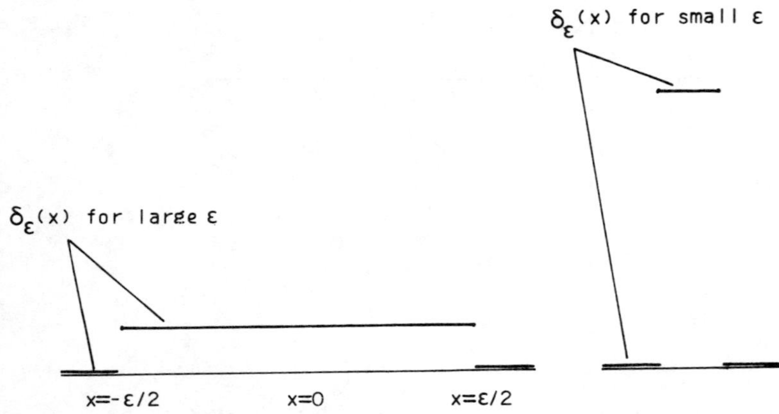

Figure 2.15. The functions $\delta_\varepsilon(x)$ for a large and a small value of ε.

We then see that

$$\int_{-\infty}^{\infty} \delta(x)\, dx = 1.$$

The most important property of the delta function is the following (which may be used as a definition in the generalized function sense):

$$\int_{-\infty}^{\infty} f(x)\, \delta(x)\, dx = f(0).$$

Let us see how this arises when $f(x) = \cos x$. Define

$$I_\varepsilon = \int_{-\infty}^{\infty} \delta_\varepsilon(x) \cos x\, dx.$$

Because $\delta_\varepsilon(x)$ vanishes outside the interval $(-\varepsilon/2, \varepsilon/2)$ and is $1/\varepsilon$ on this interval,

$$\begin{aligned} I_\varepsilon &= \frac{1}{\varepsilon}\int_{-\varepsilon/2}^{\varepsilon/2} \cos x\, dx \\ &= \frac{1}{\varepsilon} \sin x \Big|_{-\varepsilon/2}^{\varepsilon/2} \\ &= \frac{1}{\varepsilon}\left[\sin\frac{\varepsilon}{2} - \sin\left(-\frac{\varepsilon}{2}\right)\right] \\ &= \frac{2}{\varepsilon}\sin\frac{\varepsilon}{2}. \end{aligned}$$

Since

$$\lim_{x\to 0} \frac{\sin x}{x} = 1,$$

we see that as $\varepsilon \to 0$,

$$\begin{aligned} I &= \int_{-\infty}^{\infty} \delta(x) \cos x\, dx \\ &= \lim_{\varepsilon\to 0} \frac{2}{\varepsilon}\sin\frac{\varepsilon}{2} \\ &= 1 = \cos x|_{x=0} \\ &= f(0). \end{aligned}$$

Thus multiplying $\cos x$ by $\delta(x)$ and integrating picks out the value of $\cos x$ at $x = 0$. Similarly, the delta function concentrated at $x = x_0$, has the property

$$\int_{-\infty}^{\infty} \delta(x - x_0) f(x)\, dx = f(x_0).$$

It is left as an exercise to establish property (2.144).

2.17 Effects of active transport

The constant-field approach, or the approach of Section 2.6 when there is more than one ion species present, predicts a steady flux of sodium ions into the nerve cell and a steady efflux of potassium ions. In order to maintain long-term equilibrium there must be means of countering these fluxes.

Among such mechanisms are the *active-transport processes*, which are usually referred to as *pumping*. The term "pump" describes the chemical apparatus that executes it. *Active transport* is contrasted with *passive transport*, which refers to the movement of ions (or other substances) by diffusion or which is due to differences in electrical potential.

The term "active" indicates that the movement is brought about by chemical processes in the membrane, whereby ions form complexes with *carrier molecules* on one side of the membrane and are subsequently disassociated from them to gain passage to the other side. To perform this work, metabolic energy is required and this is derived from the breakdown of *adenosine triphosphate* (ATP) into *adenosine diphosphate* (ADP). For details of the chemistry and a discussion of the variety of transport processes, see Stein (1980) and White et al. (1973).

A pumping process is called *electrogenic* if it involves a net electrical current: otherwise it is called *neutral*. If the pumping of two ions, such as Na^+ and K^+, is performed by the same carrier, which is sometimes assumed to be the case, we talk of a *coupled* Na–K pump. If a coupled Na–K pump transported sodium and potassium ions on a one-to-one basis, it would result in zero net electric-current flow and would be neutral. If, on the other hand, more sodium ions were extruded than potassium ions intruded, the Na–K pump would be electrogenic.

The existence of a sodium pump was first hypothesized by Dean (1941) who realized that if the skeletal muscle membrane were permeable to Na^+, the accumulated internal sodium would have to somehow be extruded. Since then the existence of various ion pumps has been fairly well confirmed in a variety of nerve and muscle cells [see Phillis and Wu (1981) for a review].

An electrogenic pump should, by virtue of its net electric-current flow in one direction, contribute to the membrane potential. A pump that has an associated net influx of positive ions should lead to a resting potential, which is *depolarized* relative to that in the absence of pumping. A net efflux of positive ions should lead to *hyperpolarization* of the membrane.

The Goldman formula predicts that the membrane potential is proportional to the absolute temperature. If the temperature is increased, chemical reactions including those associated with active transport will proceed more rapidly, thus magnifying the contribution of active transport to the membrane potential. Gorman and Marmor (1970) found that when the giant neuron of the marine mollusc, *Anisodoris nobilis*, was warmed, a hyperpolarization occurred that was 10 times that predicted by the Goldman formula! Similar findings, though less dramatic, have been found for other cells. Furthermore, at low temperatures, when active transport is expected to be minimal, the relationship between membrane potential and external K^+ concentration was predicted by the Goldman formula with chloride terms omitted. Above 5°C, the membrane potential could not be predicted by the Goldman formula, except when the bathing solution contained ouabain (at a concentration of 0.1 m*M*), which is known to inhibit active transport. This is shown in Figure 2.16. Gorman and Marmor concluded that "the membrane potential is the sum of ionic and metabolic components, and that the behavior of the ionic component can be predicted by a constant-field-type equation."

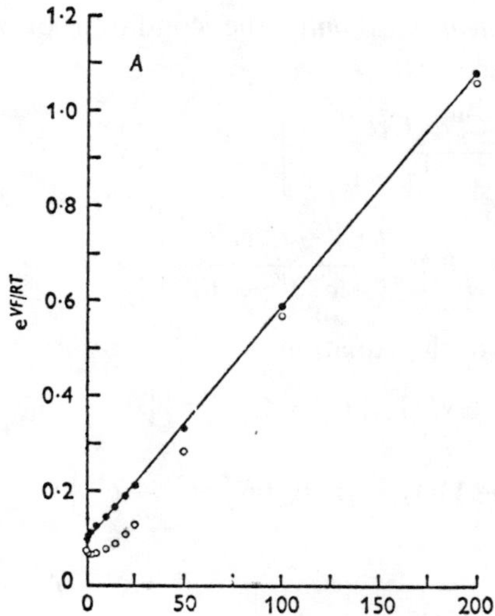

Figure 2.16. Values of $\exp(FV_m/RT)$ plotted against external K^+ concentration for the G-cell of the mollusc *Anisdoris mobilis*. Filled circles are data points when there was 0.5 m*M* ouabain in the bathing solution. Empty circles represent the absence of ouabain. The straight line is the Goldman prediction without chloride terms. [From Gorman and Marmor (1970). Reproduced with the permission of The Physiological Society and the authors.]

There have been many attempts to include the effects of active transport in formulas for the membrane potential. The one we will consider was introduced by Moreton (1969). For references to others see Sjodin (1980).

Modification of the Goldman formula to account for ion pumps

In addition to the contributions to the electric-current density for each ion due to diffusion and the electric field, we assume there is also a contribution from active transport. Though active-transport contributions will depend on such things as concentrations of ions, temperature, the presence of pump inhibitors, and so forth, we will assume they are constants. This is expected to be reasonable if the cell is in its resting state at a fixed temperature.

Let there be n ion species, indexed by k, and let the constant pumping rate for the kth species be A_k (in moles per square meter per second), where A_k is positive if pumping is in the positive x-direction. The total current density for the kth species is obtained by adding $z_k F A_k$ to the previous expression (2.85),

$$j_k = \frac{\gamma F z_k^2 P_k V_m \left[C_k^0 - C_k^i e^{\gamma z_k V_m} \right]}{\left(e^{\gamma z_k V_m} - 1 \right)} + F z_k A_k. \tag{2.156}$$

For the case of *monovalent ions only*, the condition of zero net electric-current density becomes

$$\sum_{+} \left[A_k + \gamma P_k V_m \left\{ \frac{C_k^0 - C_k^i e^{\gamma V_m}}{e^{\gamma V_m} - 1} \right\} \right] + \sum_{-} \left[-A_k + \gamma P_k V_m \left\{ \frac{C_k^0 - C_k^i e^{-\gamma V_m}}{e^{-\gamma V_m} - 1} \right\} \right] = 0. \tag{2.157}$$

Some manipulation leads to the equation

$$\sum_{+} \left[\left(e^{\gamma V_m} - 1 \right) A_k + \gamma P_k V_m \left\{ C_k^0 - C_k^i e^{\gamma V_m} \right\} \right] - \sum_{-} \left[\left(e^{\gamma V_m} - 1 \right) A_k + \gamma P_k V_m \left\{ C_k^0 e^{\gamma V_m} - C_k^i \right\} \right] = 0, \tag{2.158}$$

which may be "solved" for $e^{\gamma V_m}$,

$$e^{\gamma V_m} = \frac{\Sigma_{+} \left(\gamma P_k V_m C_k^0 - A_k \right) + \Sigma_{-} \left(A_k + \gamma P_k V_m C_k^i \right)}{\Sigma_{-} \left(A_k + \gamma P_k V_m C_k^0 \right) - \Sigma_{+} \left(A_k - \gamma P_k V_m C_k^i \right)}. \tag{2.159}$$

This is a transcendental equation for V_m, which we may write as

$$e^V = (AV + C)/(BV + C) = f(V), \tag{2.160}$$

which defines $f(V)$, where

$$V = \gamma V_m, \tag{2.161}$$

$$A = \sum_{+} P_k C_k^0 + \sum_{-} P_k C_k^i \tag{2.162}$$

$$B = \sum_{+} P_k C_k^i + \sum_{-} P_k C_k^0, \tag{2.163}$$

$$C = \sum_{-} A_k - \sum_{+} A_k. \tag{2.164}$$

Thus to find the membrane potential with active transport present we solve (graphically, for example) $e^V = f(V)$.

Properties of $f(V)$

(i) $f(0) = 1$.

(ii) $\lim_{V \to \pm\infty} f(V) = A/B$. Usually $A/B < 1$.

(iii) $f(V) = 0$ when $V = V_1 = -C/A$. Thus $f(V)$ has a zero at a positive value of V if C is negative or at a negative value of V if C is positive.

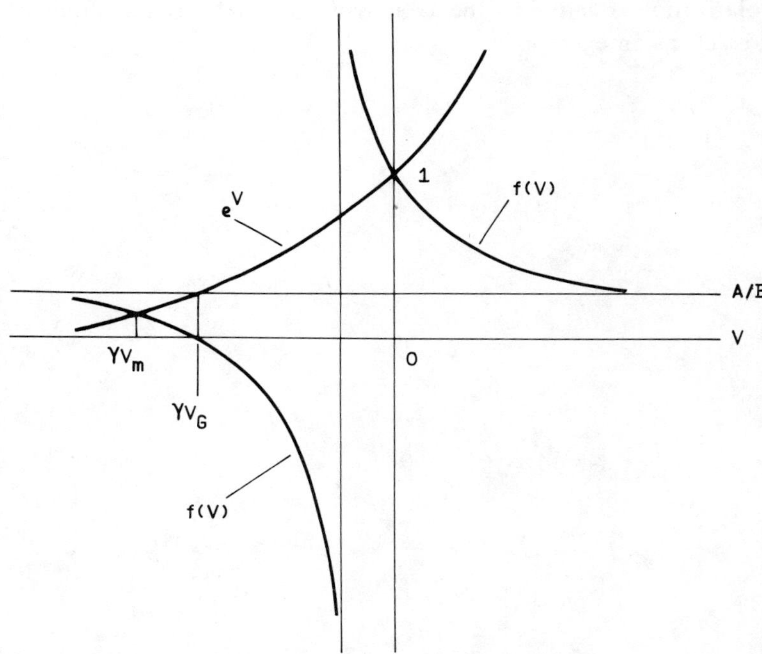

Figure 2.17. Showing that $C > 0$ leads to a hyperpolarization relative to the Goldman formula. The intersection of $f(V)$ with $\exp(V)$ gives the membrane potential γV_m.

(iv) $f(V)$ is infinite when $V = V_2 = -C/B$. Thus V_1 and V_2 have the same sign. Usually $A/B < 1$ so $|V_1| > |V_2|$.

(v) If $C > 0$, then $f(V)$ is increasing through $V = V_2$, whereas if $C < 0$, then $f(V)$ is decreasing through $V = V_2$.

In Figure 2.17 we show graphically that if $C > 0$, a hyperpolarization relative to the Goldman formula without active transport occurs. Here the intersection of e^V with the ordinate A/B gives the value of γV_m in the absence of active transport, $C = 0$ (labeled γV_G). When active transport is included, the intersection of e^V with $f(V)$ occurs at a smaller value (labeled γV_m). Thus a hyperpolarization relative to γV_G occurs.

Assuming only positive ions (e.g., K^+ and Na^+) are being pumped, a value of $C > 0$ implies an excess of positive ions are being pumped out of the cell. This would be the case when there is an excess of sodium over potassium pumping. Indeed, excessive loading of the intracellular compartment, which occurs due to the injection of sodium ions or by repetitive firing at a rapid rate (called *tetanus*), often leads to a hyperpolarization. In the latter case it is referred to as *posttetanic hyperpolarization* (Holmes 1962; Phillis and Wu 1981). If $C < 0$, a depolarization, relative to the Goldman potential, occurs. Proof of this is left as an exercise.

3

The Lapicque model of the nerve cell

3.1 Introduction

One fundamental principle in neural modeling is that one should use the simplest model that is capable of predicting the experimental phenomena of interest. A nerve-cell model must necessarily contain parameters that admit of physical interpretation and measurement, so that it is capable of predicting the different quantitative behaviors of different cells.

The model we will consider in this chapter is very simple and leads only to first-order linear differential equations for the voltage. However, when we employ the model in many situations of neurophysiological interest, we find that the mathematical analysis becomes quite difficult, due mainly to the nonlinearities introduced by the imposition of a firing threshold. This will become even more apparent in Chapter 9, where we consider stochastic versions of this model.

The model will be called the *Lapicque model* after the neurophysiologist who first employed it in the calculation of firing times (Lapicque 1907). Other names for this model, which have recently appeared in the literature are *the leaky integrator* or the *forgetful integrate and fire model*.

According to Eccles (1957) the resting motoneuron membrane can be represented by the circuit shown in Figure 3.1A. A battery with a potential difference equal to that of the resting membrane potential maintains that potential across the membrane circuit elements consisting of a resistor and capacitor in parallel. We call this a *lumped model* or a *point model* to indicate that the whole cell (with attention focused on the soma and dendrites) is lumped together into one representative circuit. Hence with this model we cannot address questions concerning the effects of input position or concerning the interaction between inputs at various locations on the soma–dendritic surface.

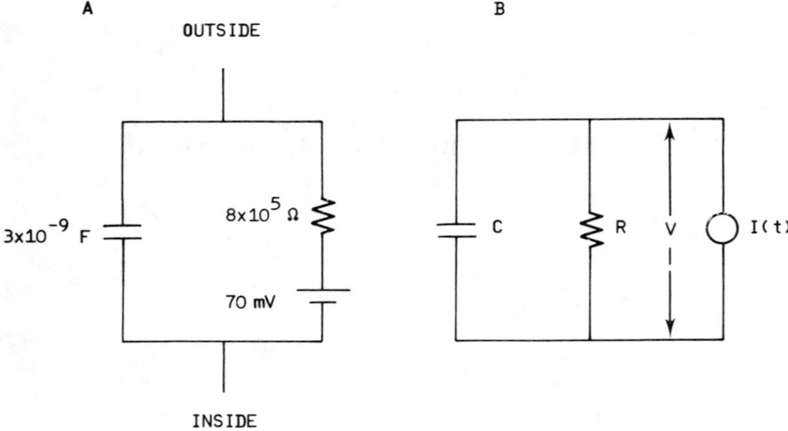

Figure 3.1. A – Electrical circuit employed to represent the resting nerve cell. Values of the resistance, capacitance, and resting potential are average values from Eccles (1957). B – Electrical circuit employed in the Lapicque model for subthreshold depolarizations.

We let $V(t)$ be the potential difference across the cell membrane, minus the resting potential at time t. That is, $V(t)$ is the *depolarization* and in the resting state $V = 0$. We remove the battery from the circuit as in Figure 3.1B, where a resistance R and capacitance C are in parallel. The depolarization varies according to the effects of the input current $I(t)$, which may come from activation of synaptic inputs or other natural means (e.g., due to a sensory input in a receptor) or from current injection. In Chapter 7 a more realistic model for the effects of synaptic inputs will be employed.

Subthreshold behavior

Assume for now that the membrane resistance R does not depend on the voltage and is, in fact, constant. The current through it is, by Ohm's law, V/R. The current through the capacitance is $C\,dV/dt$ so we must have, by conservation of current,

$$C\frac{dV}{dt} + \frac{V}{R} = I(t), \qquad t > 0. \tag{3.1}$$

The solution of this differential equation, given an initial value $V(0)$, will give the time course of the depolarization for subthreshold voltages.

Threshold

Equation (3.1) is only appropriate for subthreshold responses. If the nerve cell under consideration is capable of generating action

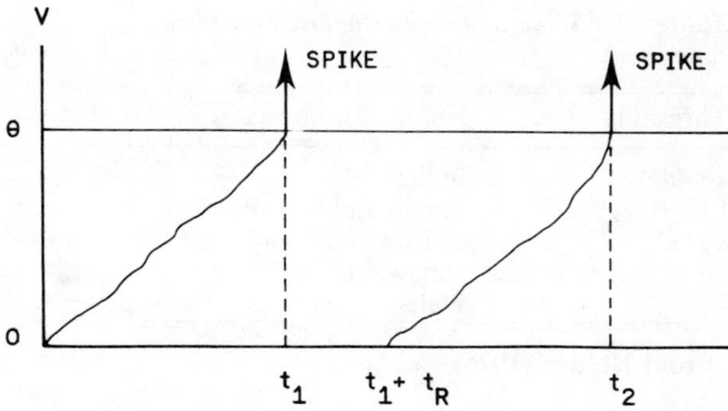

Figure 3.2. Spike generation in the Lapicque model. Whenever $V(t)$ reaches $\theta(t)$, the threshold, an action potential is generated. The threshold function depicted here is the constant threshold (3.2).

potentials, a threshold condition must be superimposed, because (3.1) has no *natural* threshold properties (see Chapter 8). Let the threshold depolarization for action-potential generation be $\theta(t)$, $t \geq 0$. The model nerve cell is completed by imposing the condition that when $V(t)$ reaches $\theta(t)$, an action potential is generated. Following the action potential the depolarization and the threshold are reset, usually to their initial values.

A simple choice for the threshold, which is appropriate for a cell that is not firing rapidly, is to assume that it is *constant* at θ until the generation of a spike, after which it becomes infinite for the duration of an *absolute refractory period* of length t_R. Let the sequence of times at which action potentials occur be $\{t_i,\ i = 1, 2, \dots\}$ as depicted in Figure 3.2. Then

$$\theta(t) = \begin{cases} \infty, & t_i < t < t_i + t_R, \\ \theta, & \text{otherwise.} \end{cases} \tag{3.2}$$

Note that the spikes have no structure in this model. The output train is completely described by the sequence of t_i's. Other threshold functions commonly employed are given in Table 3.1. In the next three sections we study the subthreshold responses of a Lapicque model neuron and then consider the problem of determining the sequence of times of occurrence of spikes.

3.2 Subthreshold response to current steps

By means of intracellular current injection a neuron may be either depolarized or hyperpolarized. The results of applying constant depolarizing and hyperpolarizing currents for a certain length of time are shown in Figure 3.3. Figure 3.3A shows a depolarization reaching

Table 3.1. *Some commonly employed threshold functions*[a]

Threshold	Name of proposer
$k\tau/t$	Buller
$k(1-\tau t)$	Calvin and Stevens
$k/(e^{\tau t}-1)$	Geisler and Goldberg
$ke^{\tau/t}$	Hagiwara
$ke^{-\tau t}$	Weiss

[a] From Holden (1976).

a final value of about 8 mV in response to a current step of 3 nA (1 nA) = 10^{-9} A), whereas Figure 3.3B shows an earlier recording of the response of a cat spinal motoneuron to a depolarizing and hyperpolarizing current step. Indeed, the Lapicque model predicts these responses.

We introduce the unit (Heaviside) step function,

$$H(t) = \begin{cases} 0, & t < 0, \\ 1, & t \geq 0. \end{cases} \tag{3.3}$$

Then with a maintained current of magnitude I = constant,

$$I(t) = IH(t). \tag{3.4}$$

Equation (3.1) for subthreshold voltages is a first-order linear differential equation with integrating factor $\exp(t/RC)$. Thus (see Section 2.4) its general solution is

$$V(t) = \exp\left(-\frac{t}{RC}\right)\left[\int^t \frac{I(t')}{C}\exp\left(\frac{t'}{RC}\right)dt' + k\right], \tag{3.5}$$

where k is a constant to be determined by the initial value of V. If we take the cell to be initially at rest, then $V(0) = 0$, and

$$V(t) = \exp\left(\frac{-t}{RC}\right)\int_0^t \frac{I(t')}{C}\exp\left(\frac{t'}{RC}\right)dt'. \tag{3.6}$$

For a constant current we insert (3.4) in (3.6) to get

$$V(t) = IR(1 - e^{-t/RC}), \qquad t \geq 0. \tag{3.7}$$

If the current were maintained *indefinitely*, then as $t \to \infty$, the depolarization would approach the *steady-state value IR*. However, if the current is switched off at $t = t_1$ so that

$$I(t) = I[H(t) - H(t - t_1)], \tag{3.8}$$

then the depolarization will decay exponentially according to

$$\frac{dV}{dt} = -\frac{V}{RC}, \qquad t \geq t_1, \tag{3.9}$$

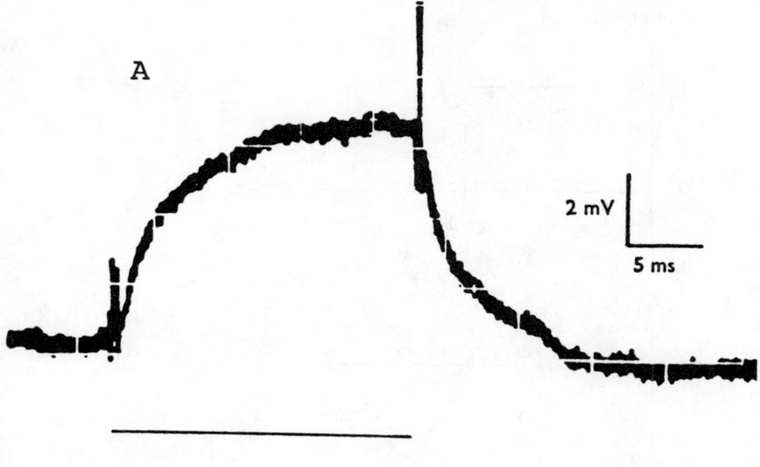

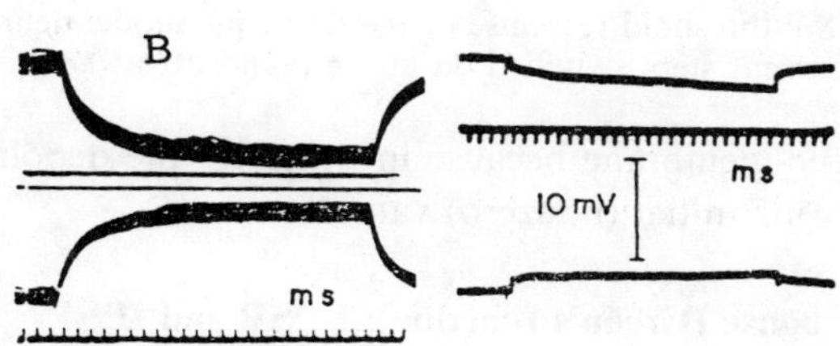

Figure 3.3. Time course of the membrane potential of a cat spinal motoneuron under current steps. A – Response of a cell to a 3-nA current step of duration about 25 ms. [From Barrett and Crill (1974). Reproduced with the permission of The Physiological Society and the authors.] B – *Left column*. Intracellular potential as a result of an 8.5-nA current step that depolarized the cell (lower figure) and another that hyperpolarized the cell (upper figure). *Right column*. Corresponding extracellular potentials illustrating difficulty in measurement techniques. [From Eccles (1957). Reproduced with the permission of Johns Hopkins University Press and the author.]

with "initial" value equal to $V(t_1)$. Thus the results shown in Figure 3.3 will be described by

$$V(t) = \begin{cases} IR(1 - e^{-t/RC}), & 0 \le t \le t_1, \\ IR(1 - e^{-t_1/RC})e^{-(t-t_1)/RC}, & t > t_1. \end{cases} \tag{3.10}$$

This solution is sketched in Figure 3.4 for the case $I > 0$. The simple model performs reasonably well. Furthermore, the results of the current-step experiment enable the parameters R and C to be estimated. The quantity RC has the dimensions of time and is called the

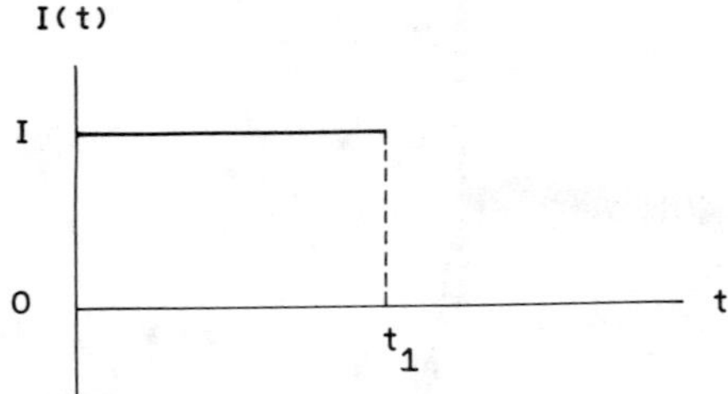

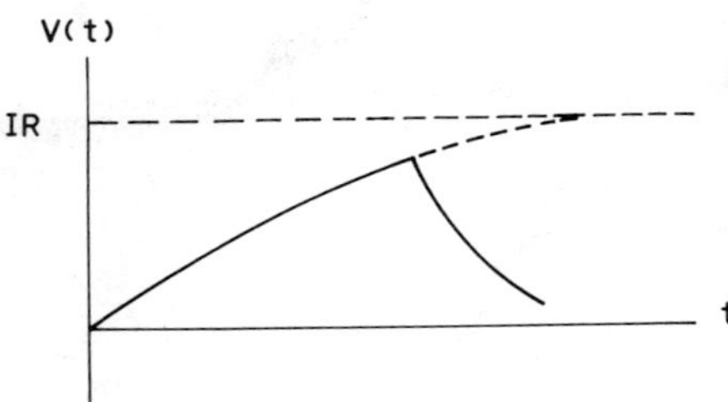

Figure 3.4. Subthreshold response of the Lapicque model neuron to a constant current step, switched on at $t = 0$ and off at $t = t_1$.

time constant τ of the membrane because in time RC the depolarization drops to e^{-1} of its initial (nonzero) value.

3.3 Impulse response (Green's function): EPSP and IPSP

The response of a linear system to an impulsive input is called the *impulse response* or *Green's function* (a term we will come across frequently in the chapters ahead). Suppose a charge C is delivered instantaneously at $t = 0$ to the resting nerve cell. Then, using the delta function introduced in Section 2.16.1, the input current is

$$I(t) = C\delta(t). \tag{3.11}$$

The solution of (3.1) with this input current is defined as the Green's function $G(t)$. Thus G satisfies

$$\frac{dG}{dt} + \frac{G}{\tau} = \delta(t), \qquad G(t) = 0, \quad t < 0. \tag{3.12}$$

Inserting $I(t)$ given by (3.11) in (3.6) gives

$$G(t) = \begin{cases} 0, & t < 0, \\ e^{-t/\tau}, & t \geq 0, \end{cases} \tag{3.13}$$

or

$$G(t) = H(t)e^{-t/\tau}. \tag{3.14}$$

The Green's function is useful to have because the response to an arbitrary input can be expressed in terms of it by means of the integral equation,

$$V(t) = \frac{1}{C}\int_0^t G(t-t')I(t')\,dt', \qquad V(0)=0. \qquad (3.15)$$

The proof of this is left as an exercise.

If Q units of charge are delivered to the nerve cell at $t=0$, the response is $QG(t)/C$. Thus the voltage has a discontinuity of magnitude Q/C at $t=0$. If $Q>0$, an abrupt depolarization occurs at $t=0$, followed by an exponential decay of the potential towards zero (resting level) with time constant τ. This response can be employed as an approximation to an EPSP. Similarly, if $Q<0$, an abrupt hyperpolarization occurs corresponding to an IPSP. Such theoretical EPSPs and IPSPs are sketched in Figure 3.5.

Using the results so far, we can make a rough estimate of the *charge delivered* to a motoneuron during the generation of an EPSP. From Figure 1.13 we see that some EPSPs have amplitudes of about 10 mV. Using the above standard value, $C = 3\times10^{-9}$ F, the charge delivered during this EPSP is about 3×10^{-11} C.

The equation satisfied by $V(t)$ can be rearranged to give

$$I(t) = C\left[\frac{dV}{dt}+\frac{V}{\tau}\right]. \qquad (3.16)$$

Hence if $V(t)$ and dV/dt are known, we can obtain the input current, within the framework of the present model. Experimentalists can, in fact, find dV/dt directly with electronic circuitry. This was utilized by

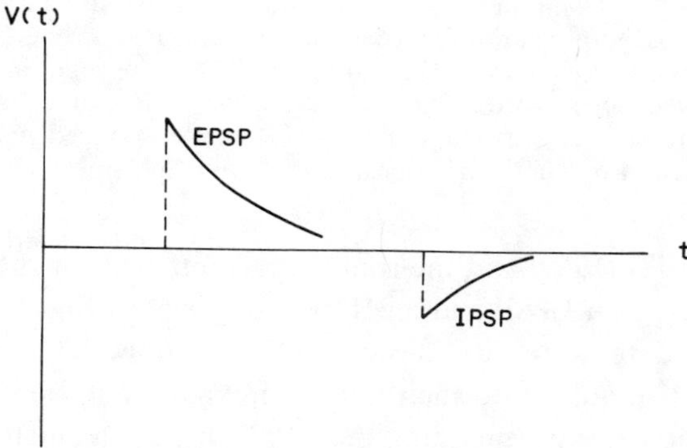

Figure 3.5. Approximations to EPSP and IPSP in the Lapicque model when impulsive currents are applied.

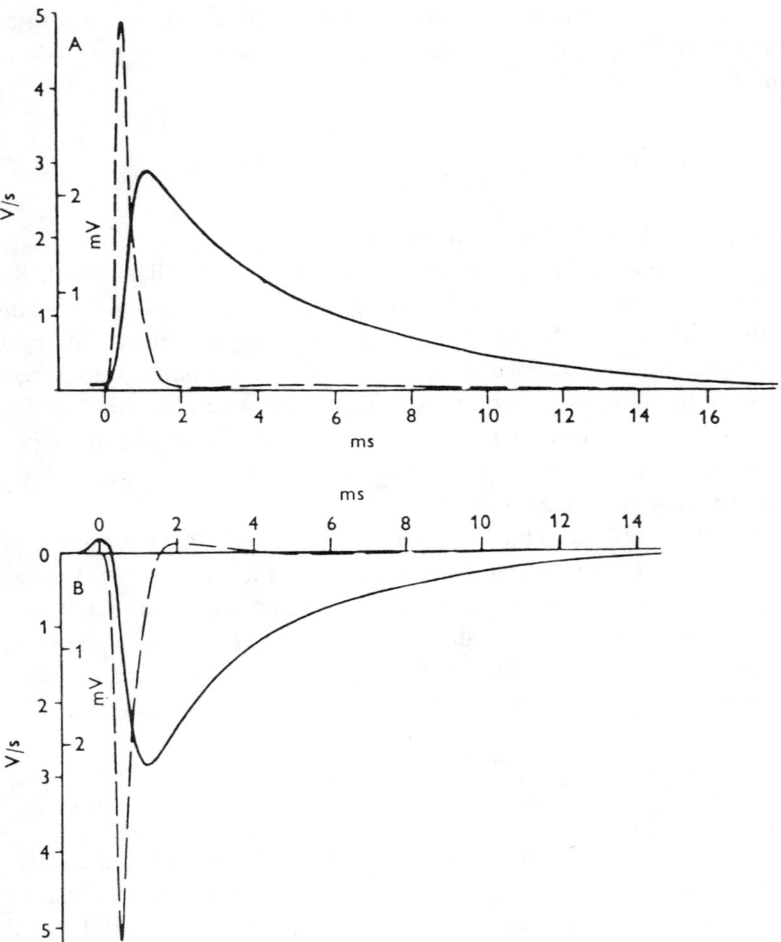

Figure 3.6. Computed current flow during the EPSP and IPSP of a cat spinal motoneuron. The dashed lines are the currents deduced from the Lapicque model via Equation (3.16). The solid lines are the observed postsynaptic potentials. Note the scales for voltage and current. [From Curtis and Eccles (1959). Reproduced with the permission of The Physiological Society and the authors.]

Curtis and Eccles (1959) and the results they obtained for the current that flows during an EPSP and an IPSP are shown in Figure 3.6.

According to these results, the currents that flow, while synaptic inputs occur, rise quickly to their maximum values and decline more slowly. For the current generating the EPSP there is a small residual depolarizing current and for that generating the IPSP there is an overshoot past zero. It was further noted by Curtis and Eccles (1959) that the time constant of decay of the EPSP, the time constant of decay of the IPSP, and the time constant of decay after a current step is terminated were all different. Such discrepancies are accounted for in a natural way with models that incorporate the spatial extent of the nerve cell (see Chapter 5).

Impulsive currents lead to discontinuous voltage trajectories, whereas the EPSPs and IPSPs of the motoneuron (Figures 1.13 and 1.15) rise smoothly from zero to achieve their maxima and minima. The chief components of the current pulses in Figure 3.6 can be approximated using various mathematical expressions. One approximation is a triangular pulse, which is studied in the next section on repetitive stimulation. However, a commonly employed approximation is a function that has been called an *alpha function* by Jack et al. (1985) and is proportional to a gamma density. For this approximation to a synaptic input current, we put

$$I(t) = kte^{-\alpha t}, \qquad \alpha > 0, \tag{3.17}$$

which corresponds to delivering a total charge of k/α^2 to the cell, as the reader may verify.

Assuming $V(0) = 0$, we find that the response of the Lapicque model neuron to such a current pulse is, from (3.6),

$$V(t) = \frac{k}{C} e^{-t/\tau} \int_0^t t' e^{(1/\tau - \alpha)t'}\, dt'. \tag{3.18}$$

Now put

$$\beta = 1/\tau - \alpha. \tag{3.19}$$

Using the rule for *integration by parts*

$$\int_0^t u'v\, dt' = [uv]_0^t - \int_0^t uv'\, dt', \tag{3.20}$$

we finally obtain

$$V(t) = \frac{ke^{-t/\tau}}{\beta C}\left[te^{\beta t} - \frac{(e^{\beta t} - 1)}{\beta}\right], \qquad \beta \neq 0. \tag{3.21}$$

If $\alpha = 1/\tau$, then $\beta = 0$ and the following result is obtained:

$$V(t) = \frac{kt^2}{2C} e^{-t/\tau}, \qquad \beta = 0. \tag{3.22}$$

Notice that if $\alpha \to \infty$ and $k = \alpha^2$, the pulse given by (3.17) approaches a delta function. Thus the rise time becomes smaller as $\alpha \to \infty$. EPSPs in response to inputs of the form of (3.17) for various α are drawn in Figure 3.7.

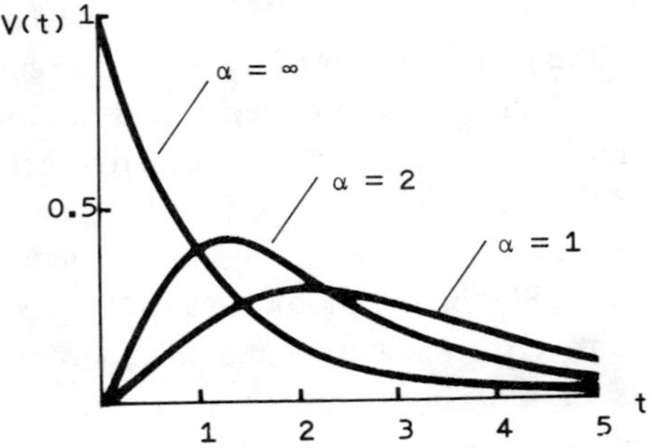

Figure 3.7. EPSP's in the Lapicque model neuron for input currents of the form of an alpha function (3.17) with $k = C$, $\tau = 1$.

3.4 Subthreshold repetitive excitation

Delta-function input currents

We start with a simple situation – a train of impulse currents at regular intervals. If these occur a time interval T apart, we have

$$I(t) = kC \sum_{n=1}^{\infty} \delta(t - nT), \tag{3.23}$$

where k is a constant and C is the membrane capacitance. After the first input event, $V(t)$ will jump from zero to k, then decay exponentially to $k\exp[-T/\tau]$ and the second input event takes $V(t)$ to $k\exp[-T/\tau] + k$, then $V(t)$ decays to $(k\exp[-T/\tau] + k)\exp[-T/\tau]$, and so forth (see Figure 3.8).

We see that the values of $V(t)$, just before and after the nth impulse current is delivered, are

$$V(nT^-) = ke^{-T/\tau}[1 + e^{-T/\tau} + \cdots + e^{-(n-2)T/\tau}], \qquad n \geq 2, \tag{3.24}$$

and

$$V(nT^+) = V(nT^-) + k. \tag{3.25}$$

These geometric series may be summed but it is immediate that if a steady state prevails and we denote the *maxima and minima in the steady state* by V_{max} and V_{min}, then we must have

$$V_{max} e^{-T/\tau} = V_{min}, \tag{3.26}$$

and

$$V_{max} = V_{min} + k. \tag{3.27}$$

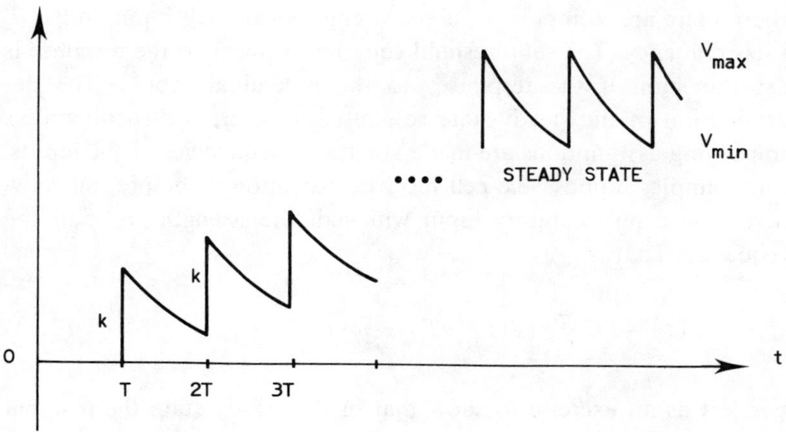

Figure 3.8. Response of the Lapicque model to a repetitive train of impulse currents each of which causes $V(t)$ to jump by k. It is assumed that the steady-state maxima and minima are below threshold for action-potential generation.

Solving these, we get

$$V_{\text{max}} = \frac{k}{1 - e^{-T/\tau}}, \tag{3.28A}$$

$$V_{\text{min}} = \frac{ke^{-T/\tau}}{1 - e^{-T/\tau}}. \tag{3.28B}$$

If the maxima in the steady state do not rise above a threshold level θ (assumed constant), then the neuron will never fire. Hence a *necessary condition for firing* is

$$\frac{k}{1 - e^{-T/\tau}} \geq \theta. \tag{3.29}$$

If we rearrange this expression, we get a value (the reciprocal of the T value) for the *critical frequency* of inputs that is necessary to make the cell fire

$$f_{\text{crit}} = \frac{1}{\tau \ln\left(\dfrac{1}{1 - k/\theta}\right)}. \tag{3.30}$$

Suppose the neuron receives *multiple inputs* each of which is periodic and with constant magnitude. Then the input current has the form

$$I(t) = C \sum_{j=1}^{m} k_j \sum_{n=1}^{\infty} \delta(t - nT_j), \tag{3.31}$$

where there are m inputs. k_j is the strength of the jth input and $1/T_j$ is its frequency. The subthreshold equation is linear so the response is just the sum of the responses to the individual inputs. The determination of the steady-state response, however, is difficult unless simplifying assumptions are made about the frequencies of the inputs. For example, suppose the cell receives excitation as before, but now there is also an inhibitory input with half the strength and half the frequency. That is,

$$I(t) = kC \sum_{n=1}^{\infty} \left(\delta(t - nT) - \tfrac{1}{2}\delta(t - 2nT)\right). \tag{3.32}$$

It is left as an exercise to show that in the steady state the maxima and minima are

$$V_{\max} = \frac{k\left[1 + \frac{1}{2}e^{-T/\tau}\right]}{\left[1 - e^{-2T/\tau}\right]}, \tag{3.33}$$

$$V_{\min} = V_{\max} - k. \tag{3.34}$$

Triangular current pulses

The results obtained by Curtis and Eccles (1959), which are shown in Figure 3.6 for the current that flows during postsynaptic potentials, suggest that a good approximating function for the current would be a triangular one. A particular form for such a current is, for a waveform with period T,

$$I(t) = \begin{cases} Jt/a, & 0 \le t \le a, \\ J - J(t-a)/2a, & a < t \le 3a, \\ 0, & 3a < t \le T, \end{cases} \tag{3.35}$$

with an EPSP occurring if J is positive and an IPSP if J is negative. The train of input pulses and the response $V(t)$ to a single input together with the steady-state response are sketched in Figure 3.9. The calculation of the response is quite involved and provides a good test of manipulative skill with Laplace transforms. The steady-state maxima and minima are

$$V_{\max} = J\tau\left(3 - \frac{t_2}{a}\right)/2C, \tag{3.36}$$

$$V_{\min} = J\tau\frac{t_1}{aC}, \tag{3.37}$$

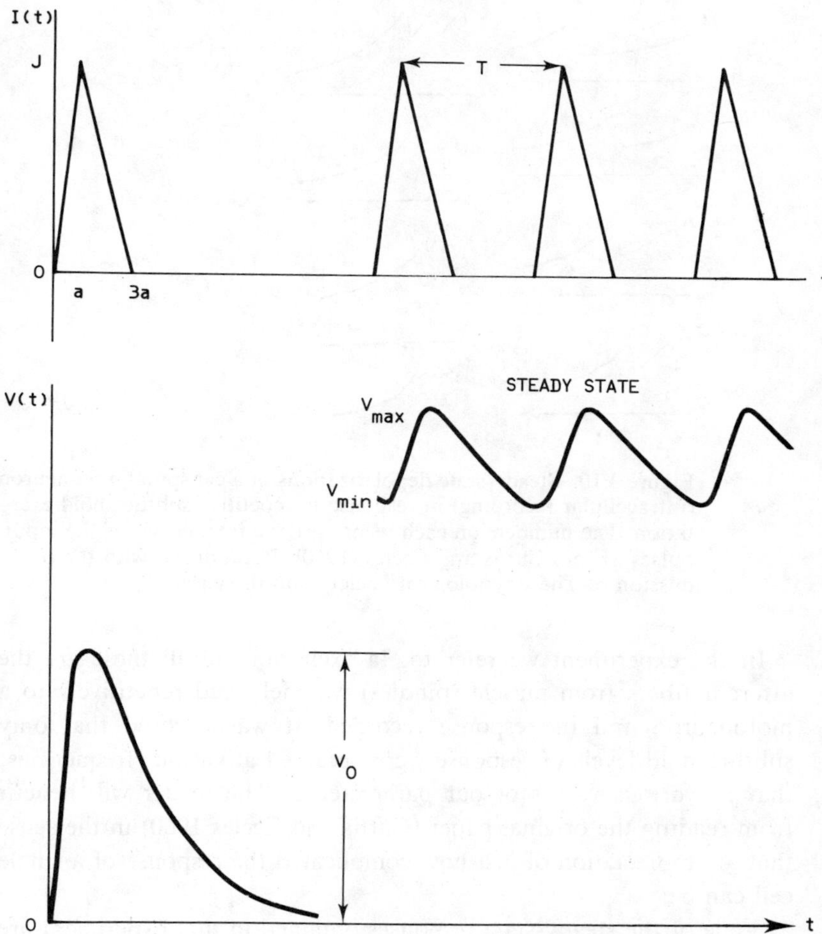

Figure 3.9. The upper graph is the train of triangular current pulses representing repetitive synaptic excitation of the model neuron. The response to the first current pulse has amplitude V_0, and eventually a steady state prevails with maxima $V_{\max}$ and minima $V_{\min}$.

where

$$t_1 = \tau \ln\left[\frac{2e^{T/\tau} - 3e^{a/\tau} + e^{3a/\tau}}{2(e^{T/\tau} - 1)}\right], \tag{3.38}$$

$$t_2 = \tau \ln\left[\frac{2 - 3e^{a/\tau} + e^{-T/\tau}e^{3a/\tau}}{e^{-T/\tau} - 1}\right]. \tag{3.39}$$

These results turn out to be very useful and provide us with a means of comparing some experimental results on cat spinal motoneurons with those predicted by our simple neuron model.

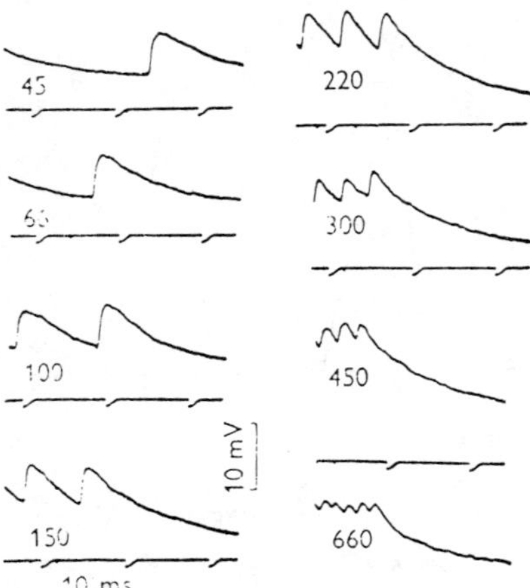

Figure 3.10. Steady-state depolarizations of a cat spinal motoneuron (intracellular recording) in response to repetitive subthreshold excitation. The numbers on each record are the frequencies of the input pulses. [From Curtis and Eccles (1960). Reproduced with the permission of The Physiological Society and the authors.]

In the experiment we refer to, Ia excitation (recall these are the afferent fibers from muscle spindles) was delivered repetitively to a motoneuron and its response recorded. It was ensured that only subthreshold levels of response were obtained at various frequencies; that is, various values of our parameter T. The reader will benefit from reading the original paper (Curtis and Eccles 1960) in the sense that an appreciation of just how complicated the response of a single cell can be.

Some of the steady-state responses obtained in that experiment are shown in Figure 3.10. Curtis and Eccles (1960) plotted the ratio $(V_{\max} - V_{\min})/V_0$ against T, and this is shown in Figure 3.11. To compare these results with the Lapicque model predictions, the quantities $(V_{\max} - V_{\min})/V_0$ at various T are plotted against T using appropriate parameter values ($a = 0.5$ ms, $\tau = 5$ ms, $J = 3.3 \times 10^{-8}$ A, $C = 5 \times 10^{-9}$ F; but note that the results do not depend on the values chosen for J or C) from the formulas above.

It can be seen that the predictions of the Lapicque model agree rather well with the experimental data in this situation. In Figure 3.11 a dashed line occurs at small values of T. This was drawn in by Curtis and Eccles (1960) who seemed to think that the limiting behavior was linear and represented a stage at which "maximal transmitter output rate had been attained at input frequencies of 300/s." This transmitter output rate refers to the release of the chemical, called a

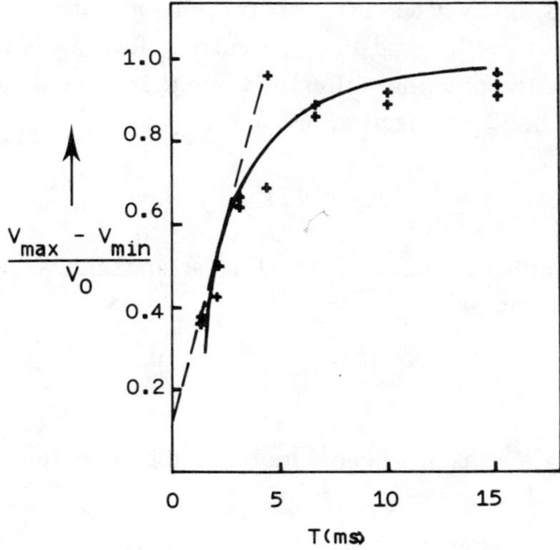

Figure 3.11. Values of $(V_{max} - V_{min})/V_0$ at various input frequencies. The crosses are experimental points (Curtis and Eccles 1960). The smooth curve is the set of results for the Lapicque model with triangular current pulses [adapted from Walsh and Tuckwell (1978)]. The dashed line was drawn in by Curtis and Eccles (1960) as an extrapolation of results to small T-values.

neurotransmitter, from the presynaptic terminal, which attaches at receptors on the postsynaptic cell and results in the occurrence of the postsynaptic potentials (EPSP or IPSP). It seems from the calculated and experimental results of Figure 3.11 that nothing can be inferred about the transmitter release rate in this situation since good agreement between theory and experiment is obtained with a constant size input current pulse.

3.5 Strength–duration curve for constant applied current

If $V(t)$ is given by (3.10), $V(t) = IR(1 - \exp(-t/\tau))$ when a constant current is applied, there will be a critical value of I below which a constant threshold θ is never reached. Constant currents greater than the critical value will cause the cell to emit a train of action potentials at regularly spaced time intervals. The term employed for the critical current is the *rheobase current* I_{Rh}. When currents above I_{Rh} are plotted against the time for which they must be applied in order to elicit an action potential, the curve obtained is called a *strength–duration curve*.

We will find the strength-duration curve for the Lapicque model. We will not, however, assume that the membrane resistance is an ohmic one (that is, one whose current is proportional to the potential difference). Instead, we assume that the current flowing is a general function $I_m(V)$ of the potential difference. We then have the differential equation for the depolarization

$$C\frac{dV}{dt} + I_m(V) = I, \qquad V < \theta,\ V(0) = 0, \tag{3.40}$$

where I is the applied current. Since I is a constant we may still separate variables and get

$$dt = \frac{C\,dV}{I - I_m(V)}. \tag{3.41}$$

If $T_\theta(I)$ is the time taken to generate an action potential when the applied current is I, then we must have

$$T_\theta(I) = C\int_0^\theta \frac{dV}{I - I_m(V)}. \tag{3.42}$$

From this equation we can obtain the strength–duration curve for various current–voltage relations, but first we simplify the problem.

3.5.1 Dimensionless variables and phase portrait

It is often desirable to reduce the number of parameters or constants by introducing *dimensionless variables*. As a simple example consider the Lapicque model with an ohmic resistance so that the depolarization satisfies

$$\frac{dV}{dt} + \frac{V}{\tau} = I(t), \qquad V < \theta, \tag{3.43}$$

where $\tau = RC$ is the time constant. We let $t' = t/\tau$, so that t' is dimensionless, representing time measured in units of τ. Then the equation becomes simply, with $V(t) = v(t')$, $I(t) = i(t')$,

$$\frac{dv}{dt'} + v = i(t'), \tag{3.44}$$

so that τ no longer appears and does not have to be carried along in any calculations that we do. At the end of the calculation we can always go back to the original variables if necessary.

We introduce dimensionless voltage, current, and time in the general equation (3.40). First, we let

$$\overline{V} = V/\theta, \qquad I_m(V) = \bar{I}_m(\overline{V}), \tag{3.45A}$$

so that the threshold is $\overline{V} = 1$. The unit of current is chosen as the rheobase current, which is given by

$$I_{\mathrm{Rh}} = I_m(\theta). \tag{3.45B}$$

Then (3.40) becomes

$$\frac{C\theta}{I_{\mathrm{Rh}}}\frac{d\bar{V}}{dt} + \frac{\bar{I}_m(\bar{V})}{I_{\mathrm{Rh}}} = \frac{I}{I_{\mathrm{Rh}}}, \qquad \bar{V} < 1, \bar{V}(0) = 0. \tag{3.46}$$

If we now measure time in units of the "time constant,"

$$\tau = C\theta / I_{\mathrm{Rh}}, \tag{3.47}$$

so that

$$t' = t/\tau, \tag{3.48A}$$

and we put

$$v(t') = \bar{V}(t), \tag{3.48B}$$

$$f(v) = \bar{I}_m(\bar{V})/I_{\mathrm{Rh}}, \tag{3.48C}$$

$$i = I/I_{\mathrm{Rh}}, \tag{3.48D}$$

then the differential equation becomes

$$dv/dt' = -f(v) + i, \qquad v < 1, v(0) = 0. \tag{3.49}$$

In the dimensionless system, the rheobase current has value unity so $f(v)$ has a maximum value of unity for $0 \le v < 1$. For the ohmic resistance case $f(v) = v$ and, in general, $f(v)$ will have the shape shown in Figure 3.12.

Stability analysis

To obtain insight into the general problem of the conditions for nerve-cell firing, we perform a phase analysis of Equation (3.49). In Figure 3.13 is shown dv/dt plotted against v for various i. When

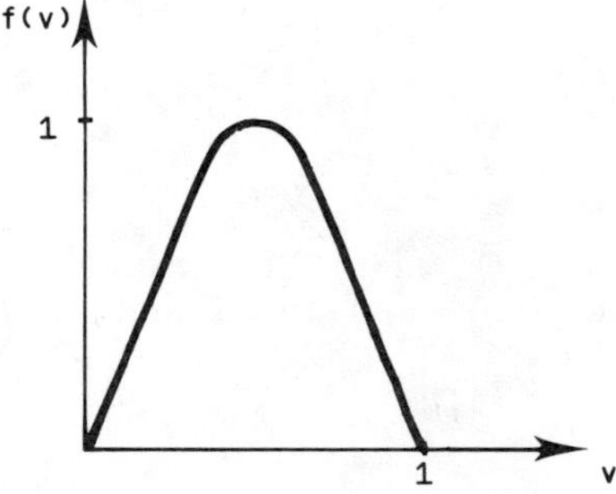

Figure 3.12. A typical current–voltage relation in dimensionless variables.

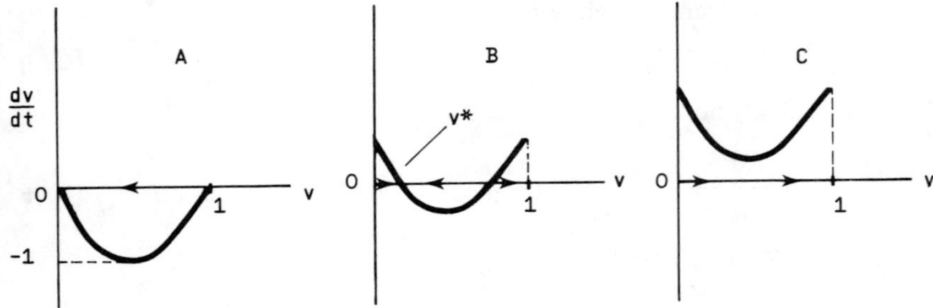

Figure 3.13. Stability analysis for the Lapicque model with a nonlinear current–voltage relation for three values of the input current. A – No input current. B – Input current between zero and rheobase. C – Input current just greater than rheobase.

$i = 0$ as in Figure 3.13A, $v(t')$ will return to zero (resting level) for *all* initial voltage values between 0 and 1. This is indicated by the arrow pointing to the left (direction of decreasing v) on the v-axis. If i is larger so $0 < i < 1$, as in Figure 3.13B, there is a *critical point* (one at which $dv/dt' = 0$) at $v = v^*$. This point is (locally) *asymptotically stable* since it is approached as $t' \to \infty$ if $v(0)$ is just below or just above v^*. If $v(0) = 0$, the threshold for firing cannot be reached. If perchance $v(0)$ were greater than the value at the second critical point, then threshold would be reached. In Figure 3.13C, the applied current i is just greater than the rheobase value (unity) and the dv/dt'-versus-v curve is above the v-axis. Hence threshold will be reached from $v(0) = 0$ for any current greater than rheobase.

3.5.2 Determining the strength–duration curve

We now return to the problem of finding the relation between the strength of the applied current and the time it takes for V to get from zero to θ. In the new units, the firing time is

$$t_\theta = T_\theta/\tau, \tag{3.50}$$

and t_θ is found from

$$t_\theta(i) = \int_0^1 \frac{dv}{i - f(v)}. \tag{3.51}$$

For the ohmic resistance case, $f(v) = v$, $0 \le v < 1$, and

$$t_\theta(i) = \int_0^1 \frac{dv}{i - v}$$

$$= -\ln\left[1 - \frac{1}{i}\right], \qquad i > 1. \tag{3.52}$$

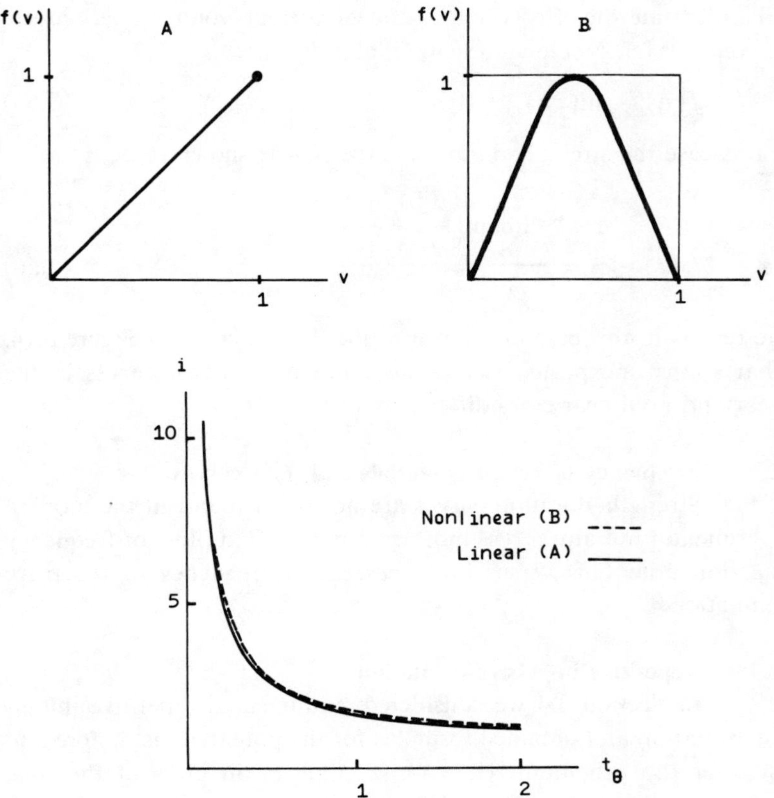

Figure 3.14. The strength–duration curves, from Noble and Stein (1966), for a linear (A) and nonlinear (B) current–voltage relation. The functions $f(v)$ are shown above. The nonlinear example is Equation (3.55).

Hence we get an explicit strength–duration relation

$$i = 1/(1 - \exp[-t_\theta]). \tag{3.53}$$

The result is plotted in Figure 3.14 as the solid curve. If we return to the original variables, we may write

$$I = I_{\mathrm{Rh}}/(1 - \exp[-T_\theta/\tau]), \tag{3.54}$$

which was the relation derived by Lapicque (1907). This simple relation was found to provide a good approximation to the experimental strength–duration curves for many nerve cells. However, it was found that the parameters of the model that gave the best fit to data depended on the manner in which the stimulus was applied. (For example whether current was injected at a point or uniformly, say, in the case of a cylindrical squid axon.)

To illustrate the effects of a nonlinear current–voltage relationship, we consider [cf. Noble and Stein (1966)] the case

$$f(v) = \sin(\pi v), \qquad 0 \le v < 1. \tag{3.55}$$

In this case the strength–duration curve can be shown to be

$$t_\theta(i) = \frac{1 + \dfrac{2}{\pi}\arctan\left(\dfrac{1}{\sqrt{i^2-1}}\right)}{\sqrt{i^2-1}}. \tag{3.56}$$

The results in this case are shown as the dashed curve in Figure 3.14. What seems unexpected is that the strength–duration curves in the linear and nonlinear cases differ very little.

3.6 Frequency of action potentials and f/I curves

Strength–duration curves are not often found in the modern experimental literature. It is more common to find plots of frequency of action potentials versus input current or frequency of repetitive stimulation.

3.6.1 Repetitive impulsive excitation

In Section 3.4 we considered subthreshold repetitive impulsive excitation and obtained formulas for the potential just before and just after the nth input. Here we let time be in units of the time constant τ. Then the value of the depolarization just after the nth input is

$$V(nT^+) = k\sum_{m=0}^{n-1} e^{-mT}. \tag{3.57}$$

If the threshold is $\theta =$ constant and the input frequency is greater than the critical value given in (3.30), then the time taken to generate an action potential from rest is

$$T_\theta = T\inf\{n|V(nT^+) \ge \theta\}, \tag{3.58}$$

where $\inf\{n|A\}$ here means the smallest value of n which makes A true. Summing the geometric series in (3.57), we have

$$T_\theta = T\inf\{n|k(1-e^{-nT})/(1-e^{-T}) \ge \theta\}. \tag{3.59}$$

By rearranging the inequality,

$$T_\theta = T\inf\left\{n|n \ge \frac{1}{T}\ln\left[\frac{1}{1-\rho(1-e^{-T})}\right]\right\}, \tag{3.60}$$

where

$$\rho = \theta / k \tag{3.61}$$

is the ratio of threshold to EPSP amplitude.

Letting

$$x = \frac{1}{T}\ln\left[\frac{1}{1-\rho(1-e^{-T})}\right], \tag{3.62}$$

we may write

$$T_\theta = ([x]+1)T, \tag{3.63}$$

where $[x]$ denotes the largest integer in $[0, x)$. As the input frequency becomes large, $T \to 0$, and we have the *asymptotic relation*

$$x \underset{T\to 0}{\sim} \frac{\rho(1-e^{-T})}{T}. \tag{3.64}$$

Generally, we have, if we incorporate a refractory period T_R (in units of τ), that the *frequency of firing* f is in units of $1/\tau$,

$$f = \frac{1}{([x]+1)T + T_R}. \tag{3.65}$$

For $\rho = 1$, 1.5, 2, 3, 4, and 5 the results are plotted in Figure 3.15 with f versus input frequency. Note that there is a sudden increase to a

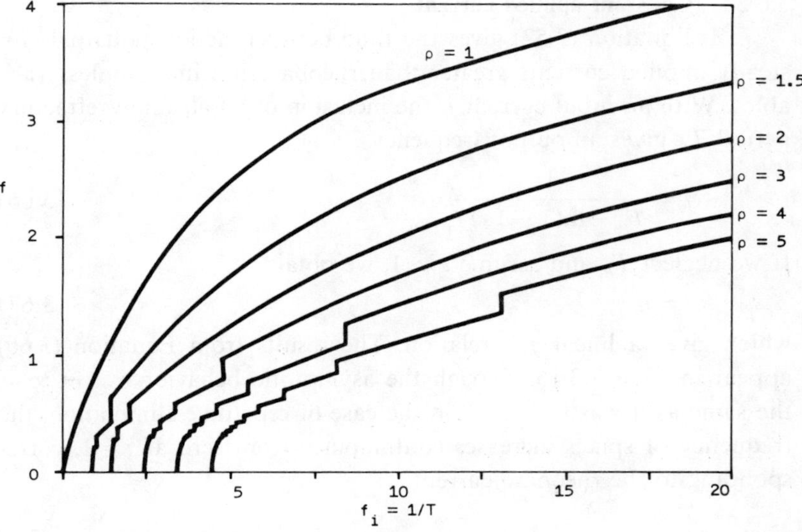

Figure 3.15. Variation in firing rate f with increasing input frequency $1/T$ of repetitive inputs with various values of ρ, the ratio of threshold depolarization to amplitude of the EPSP.

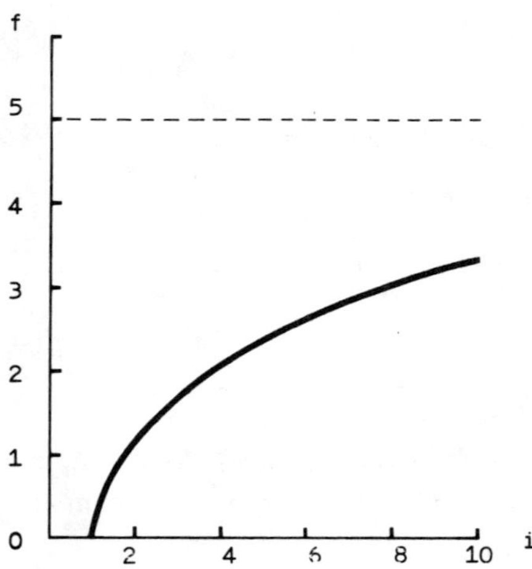

Figure 3.16. Frequency of action potentials versus strength of applied current for the Lapicque model. Both f and i are dimensionless and a maximum firing rate of 5 is due to an assumed refractory period of $T_R = 0.2$.

positive firing rate at a particular input frequency and that there are sequences of discontinuities as various numbers of input events are necessary to cause a threshold crossing. As $T \to 0$, the firing rate must approach $1/\{([\rho]+1)T + T_R\}$.

3.6.2 Constant applied current

Equation (3.52) gives the time between action potentials for steady applied currents greater than rheobase in dimensionless variables. With an input current i, the inclusion of an absolute refractory period T_R gives an output frequency

$$f = \frac{1}{T_R - \ln(1 - 1/i)}. \tag{3.66}$$

If we neglect T_R and assume $i \gg 1$, we obtain

$$f \simeq i, \tag{3.67}$$

which gives a linear f/i relation. The results from Equation (3.66) appear in Figure 3.16. Though the asymptotic behavior as $i \to \infty$ is the same as it was for $T \to 0$ in the case of repetitive stimulation, the frequency of spikes increases continuously from zero at $i = 1$, corresponding to the rheobase current.

3.7 Graphical determination of the spike train

When a constant current is applied, the time T_θ between action potentials may be found, sometimes exactly, as in the previous two sections. Determining T_θ is more difficult when the applied current is not constant. Scharstein (1979) has shown how the output train of spikes from a Lapicque model neuron can be obtained by graphical or numerical methods for an *arbitrary* input current.

A linear current–voltage relation is assumed and in dimensionless variables we have

$$dv/dt + v = i(t), \qquad v < 1, \tag{3.68}$$

where the threshold is unity, the current is in units of the rheobase current, and t is in time constants. Let the times of occurrence of action potentials be $\{t_k,\ k = 1, 2, \ldots\}$. *The problem is to determine this sequence.*

Since it is assumed that v is reset to zero after each spike, the potential (depolarization) at times between the kth and $(k+1)$th spikes is given by

$$v(t) = \int_{t_k}^{t} \exp[-(t - t')]\, i(t')\, dt', \qquad t_k \le t < t_{k+1}. \tag{3.69}$$

This is rewritten as the difference of two integrals

$$v(t) = \int_0^t \exp[-(t - t')]\, i(t')\, dt' - \int_0^{t_k} \exp[-(t - t')]\, i(t')\, dt'. \tag{3.70}$$

Multiplying the second integral by $1 = \exp(t_k - t_k)$ gives

$$v(t) = \int_0^t \exp[-(t - t')]\, i(t')\, dt'$$
$$- \exp[-(t - t_k)] \int_0^{t_k} \exp[-(t_k - t')]\, i(t')\, dt'. \tag{3.71}$$

Now define

$$u(t) = \int_0^t \exp[-(t - t')]\, i(t')\, dt', \tag{3.72}$$

which is the value the depolarization would have at time t if there was no threshold—sometimes called the *low pass filtered input*. Then (3.70) becomes

$$v(t) = u(t) - u(t_k)\exp[-(t - t_k)], \qquad t_k < t < t_{k+1}. \tag{3.73}$$

The $(k+1)$th action potential is generated when $v(t)$ reaches the threshold of unity for the first time after t_k. Hence there exists the following implicit relation between the times of occurrence of the kth and $(k+1)$th spikes, obtained by putting $t = t_{k+1}$ in (3.73),

$$u(t_{k+1}) - 1 = u(t_k)\exp[-(t_{k+1} - t_k)]. \tag{3.74}$$

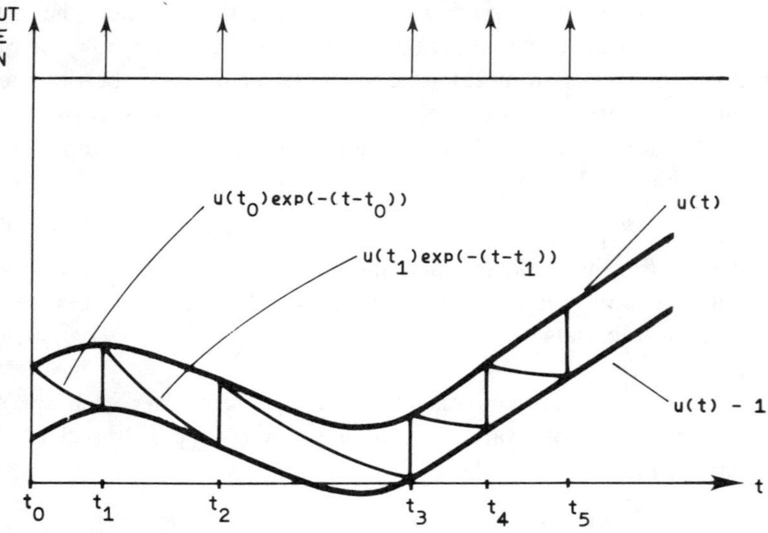

Figure 3.17. Method of graphically obtaining the sequence of spike times from the response in the absence of spike generation. [Adapted from Scharstein (1979).]

In most cases graphical or numerical methods must be employed to solve this equation for the t_k's. The graphical method is illustrated in Figure 3.17. First, we find the function $u(t)$, which depends on the input current up to time t – this is the upper curve in the figure. We also find $u(t)-1$, which is the lower curve.

Suppose the kth spike has been emitted at $t = t_k$. Then between t_k and t_{k+1} the voltage is given by (3.73). When the depolarization $u(t)-u(t_k)\exp(-(t-t_k))$ reaches 1, the next spike is generated. That is, we wait until

$$u(t_k)\exp[-(t-t_k)] = u(t)-1. \tag{3.75}$$

In Figure 3.17, a spike occurs at t_0. We look at the exponentially decaying function $u(t_0)\exp(-(t-t_0))$ and see where it hits $u(t)-1$. When it does, at $t=t_1$, the next spike is generated. Then v is reset to zero and we wait until $u(t_1)\exp(-(t-t_1))$ hits $u(t)-1$ to generate the next spike at t_2, and so forth.

Constant input current

We will verify the previously obtained relation (3.52) between t_k and t_{k+1} in the case of a constant current of magnitude $i>1$. In this case, from (3.72),

$$\begin{aligned} u(t) &= i\exp(-t)\int_0^t \exp(t')\,dt' \\ &= i(1-\exp(-t)). \end{aligned} \tag{3.76}$$

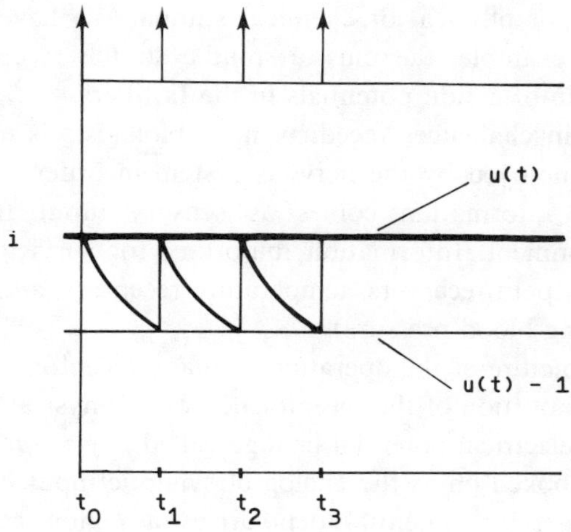

Figure 3.18. Graphical method employed for the steady state under constant current injection.

Substituting in (3.74), we obtain

$$i[1-\exp(-t_{k+1})]-1=i[1-\exp(-t_k)]\exp[-(t_{k+1}-t_k)]. \tag{3.77}$$

Rearranging this gives

$$t_{k+1}-t_k=-\ln(1-1/i), \tag{3.78}$$

which shows that the time interval between spikes is independent of k and agrees with (3.52).

We now make an important observation exemplified in this simple case, which will be useful in the next section when we consider cyclic inputs. Suppose we let $t \to \infty$ in the absence of threshold when the input current is a constant i. Then $u(t) \to i$. If we now use the graphical construction we get the situation in Figure 3.18. If the zeroth spike occurred at t_0, then the next one will occur when $i\exp[-(t-t_0)]=i-1$. This gives $t_1-t_0=-\ln(1-1/i)$, which leads to $t_{k+1}-t_k=-\ln(1-1/i)$, $i>1$. We will see shortly how this generalizes so that we can use the steady-state component of the low pass filtered input to determine the sequence of spikes with particular reference to cyclic inputs. This is a surprising result.

3.8 Cyclic inputs

We have investigated the response of the Lapicque model to steady applied currents and to repetitive sequences of current pulses. It is also important to study the effects of cyclic inputs mainly in connection with *receptor* neurons. These are nerve cells that respond as a consequence of physical or chemical stimuli. We have already encountered one example, the muscle spindle stretch receptor that gives rise to a train of action potentials in the Ia fibers.

One of the main challenges faced by neurobiologists is to understand the *code* employed by the nervous system in order to convey *information*. The information comes as sensory input from the organism's environment. Information important for survival comes, for example, from pain receptors, temperature receptors, and baroreceptors (monitoring blood pressure).

The paradigm picture of the operation of many receptor systems is as follows. At one portion of the receptor neuron, a physical stimulus produces a local electrical-potential change called a *generator potential*. This can be looked on as the analog of synaptic input in central neurons. The generator potential depolarizes a region called the *trigger zone* at which action potentials are usually instigated when a threshold condition for firing is reached. The existence of trigger zones is well established for some cells. The classic demonstration of the existence of this preferred site of action-potential generation was made for the stretch receptor of the lobster. This is shown in Figure 3.19, where the extracellular recording of the action potential is given at various parts of the cell. It can be seen that in response to stretch, the action potential is first recorded at the point B, about 500 μm from the cell body, which is implicated as being at the trigger zone of the receptor.

The physical stimulus (stretch in this case) is therefore converted or mapped into a train of action potentials that form the representation to the organism of the properties of the stimulus. Thus the train of action potentials is the internal representation of that part of the "external world" monitored by the receptor. The representation process is called *encoding* and the spike-generating region of the cell is often called the *encoder*. In the case of muscle stretch the encoding is relatively well understood physiologically and its mathematical description fairly straightforward. For example, Figure 3.20 shows that the firing rate of the slowly adapting stretch receptor of the lobster is, over a considerable range, a *linear* function of the length of the muscle. Similar results are obtained for tonically discharging muscle spindles in the cat soleus muscle (Granit 1958).

Lapicque model with sinusoidal input current

There have been several experimental studies on the responses of receptors to sinusoidally varying stimuli (Knight 1972a, b; Poppele and Chen 1972; Ascoli et al. 1978).

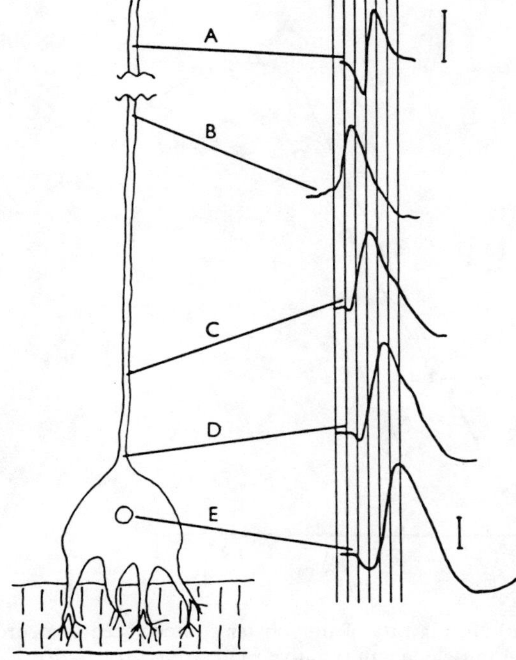

Figure 3.19. Demonstration of the existence of a trigger zone for action-potential generation in the lobster stretch receptor neuron. In response to stretch an action potential forms first at B and spreads along the axon and back to the soma and dendrites. Time markers are 0.1 ms apart; voltage markers, 0.5 mV. [From Edwards and Ottoson (1958). Reproduced with the permission of The Physiological Society and the authors.]

The Lapicque model with a *sinusoidally varying input current* can be used to obtain some insight into the experimentally observed phenomena.

The equation satisfied by the depolarization $V(t)$ is

$$C\frac{dV}{dt} + \frac{V}{R} = I_0 + I_1\cos(\omega t + \phi), \qquad V(0) = 0,\ I_0,\ I_1 > 0, \tag{3.79}$$

for $V < \theta$. To rid the calculation of some constants, we again work with dimensionless variables

$$\begin{aligned} &t' = t/\tau, \qquad \tau = RC, \qquad \omega' = \omega\tau, \\ &V' = V/\theta, \\ &I_0' = I_0/R_{\mathrm{Rh}}, \qquad I_1' = I_1/I_{\mathrm{Rh}}, \end{aligned} \tag{3.80}$$

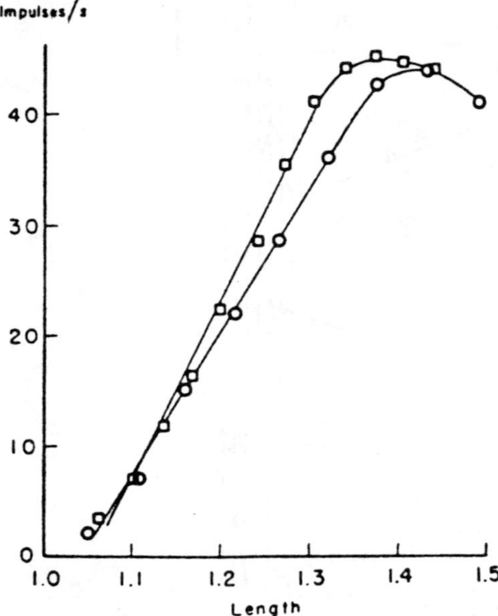

Figure 3.20. Firing rate of the lobster stretch receptor neuron as a function of muscle length (relative to unstretched length). The circles and squares are for two different receptors. [From Terzuolo and Washizu (1962). Reproduced with the permission of The American Physiological Society and the authors.]

where $I_{\mathrm{Rh}} = \theta/R$ is the rheobase current. The threshold is now unity and, hopefully without confusion, we drop the primes on the new variables so that

$$\frac{dV}{dt} + V = I_0 + I_1 \cos(\omega t + \phi), \qquad V(0) = 0,\ V < 1. \tag{3.81}$$

We call ω the *angular frequency*, I_1 the *amplitude* of the oscillatory component, ϕ its *phase*, and I_0 the *constant component* of the input current.

The solution of (3.81) can be found by direct integration or by using Laplace transforms. Use of the latter is left as an exercise. We proceed directly with the help of the standard integrals

$$\int e^{ax} \sin(bx)\, dx = \frac{e^{ax}}{a^2 + b^2}[a \sin(bx) - b \cos(bx)], \tag{3.82}$$

$$\int e^{ax} \cos(bx)\, dx = \frac{e^{ax}}{a^2 + b^2}[a \cos(bx) + b \sin(bx)]. \tag{3.83}$$

The required solution is

$$V(t) = e^{-t}\int_0^t e^{t'}[I_0 + I_1\{\cos(\omega t')\cos\phi - \sin(\omega t')\sin\phi\}]\,dt'. \qquad (3.84)$$

Evaluating these integrals and rearranging gives

$$V(t) = I_0(1-e^{-t}) + \frac{I_1}{1+\omega^2}[\cos(\omega t+\phi) + \omega\sin(\omega t+\phi) - (\cos\phi + \omega\sin\phi)e^{-t}]. \qquad (3.85)$$

As $t \to \infty$, the exponentially decaying terms go to zero and $V(t)$ settles down to

$$\tilde{V}(t) = I_0 + \frac{I_1}{1+\omega^2}[\cos(\omega t+\phi) + \omega\sin(\omega t+\phi)], \qquad (3.86)$$

which has period $2\pi/\omega$.

3.8.1 Condition for firing

The threshold for action potentials is unity so that the condition for firing must be that the maximum of the values of $\tilde{V}(t)$ over one period $2\pi/\omega$ as given by (3.86) exceeds unity. Note that in the absence of the oscillatory component the condition is just $I_0 \geq 1$, because we have scaled currents by the rheobase current. We rewrite the asymptotic expression in amplitude and phase form

$$\tilde{V}(t) = I_0 + \frac{I_1 A}{1+\omega^2}\cos(\omega t + \phi - \psi), \qquad (3.87)$$

which requires that

$$A\cos\psi = 1,$$

$$A\sin\psi = \omega,$$

whereupon

$$A = \sqrt{1+\omega^2},$$

and

$$\psi = \arctan(\omega).$$

Since $\cos(\omega t + \phi - \psi)$ has a maximum value of 1, we see that the maximum value of $V(t)$ over one period of the input is

$$\max_{0 \le t < 2\pi/\omega} \tilde{V}(t) = I_0 + \frac{I_1}{\sqrt{1+\omega^2}}. \tag{3.88}$$

Hence the *necessary and sufficient condition for firing* is

$$I_0 + \frac{I_1}{\sqrt{1+\omega^2}} \ge 1. \tag{3.89}$$

In the rest of this section we assume that this condition is met.

3.8.2 Determination of the sequence of spike times

As we mentioned in the last section, the computed spike train is the same whether we use the voltage $V(t)$ given by (3.85) in the absence of a threshold, or by the voltage of (3.86) when the transient parts of the response have died away. We will now show that this is true for cyclic inputs.

Recall the relation between the time of occurrence t_{i+1} of the $(i+1)$th spike and t_i,

$$u(t_{i+1}) - 1 = u(t_i)\exp[-(t_{i+1} - t_i)],$$

where $u(t)$ is the voltage in the absence of a threshold. Thus $u(t)$ is the same as $V(t)$ given by (3.85). Using this latter equation, we find the following implicit relation between t_{i+1} and t_i,

$$\begin{aligned} I_0(1 - e^{-t_{i+1}}) + \gamma\big[\cos(\omega t_{i+1} + \phi) + \omega \sin(\omega t_{i+1} + \phi) \\ - (\cos\phi + \omega\sin\phi)e^{-t_{i+1}}\big] - 1 \\ = \big\{ I_0(1 - e^{-t_i}) + \gamma\big[\cos(\omega t_i + \phi) + \omega\sin(\omega t_i + \phi) \\ - (\cos\phi + \omega\sin\phi)e^{-t_i}\big]\big\} e^{-(t_{i+1} - t_i)}, \end{aligned} \tag{3.90}$$

where $\gamma = I_1/(1+\omega^2)$. The terms involving the transient components cancel as we note that

$$\begin{aligned} -I_0 e^{-t_{i+1}} - \gamma(\cos\phi + \omega\sin\phi)e^{-t_{i+1}} \\ = \big[-I_0 e^{-t_i} - \gamma(\cos\phi + \omega\sin\phi)e^{-t_i}\big]e^{-(t_{i+1}-t_i)}. \end{aligned} \tag{3.91}$$

Hence we obtain the same sequence of spike times whether we use the complete solution (3.85) or the steady-state solution (3.86). Since the latter is less complicated, it is better to use it rather than the complete solution.

We employ the version of the steady-state solution given by (3.87) to obtain

$$I_0 + \frac{I_1}{\sqrt{1+\omega^2}}\cos(\omega t_{i+1} + \phi - \psi) - 1$$
$$= \left[I_0 + \frac{I_1}{\sqrt{1+\omega^2}}\cos(\omega t_i + \phi - \psi)\right]\exp[-(t_{i+1} - t_i)], \tag{3.92}$$

as an implicit relation between t_{i+1} and t_i. We can now consider the spiking response of the Lapicque model neuron as a function of the parameters I_0, I_1, ω, and ϕ, which describe the properties of the cyclic input current. *Also, we can specify relationships between t_{i+1} and t_i and see which values of the input parameters could lead to such a relation.*

Example 1: *a graphical investigation*

To illustrate the usefulness of Scharstein's method, we find the output sequence of spikes graphically for two sets of input-parameter values. In Figure 3.21A we do this for $I_0 = 2$, $I_1 = \sqrt{2}$, $\omega = 1$, and, for convenience, we set $\phi - \psi = 3\pi/2$ so that the input current is two units at $t = 0$. The curves $u(t) = 2 + \cos(t + 3\pi/2)$ and $u(t) - 1$ are all we need. We place a spike at the origin and observe the decaying exponential until it meets $u(t) - 1$ whereupon the next spike at $t = t_1$ is emitted. At t_1 we "go back up" to the upper curve and again follow the exponential decay until it meets $u(t) - 1$ at time t_2, the time of the next spike, and so forth. The output train of spikes is shown by arrows below the graphical construction and on the right a histogram of interspike intervals has been collected.

We notice a *burst* of about four spikes with comparatively short *interspike intervals* followed by a silence, which terminates with another burst of activity. The histogram of interspike intervals shows a bimodal distribution with a primary peak at short interspike times (the bursts) and a secondary peak at larger T values (the silences). The main peak occurs at approximately the position of the arrow on the histogram, which corresponds to the interspike time that would arise if the input current were constant at $I_0 = 2$, as calculated from Equation (3.78).

In Figure 3.21B we see the effect of doubling the mean value of the input current I_0, whereas all the other parameter values remain the same. Many more spikes occur on each rising phase of the response in the absence of threshold. The time between bursts is shorter even

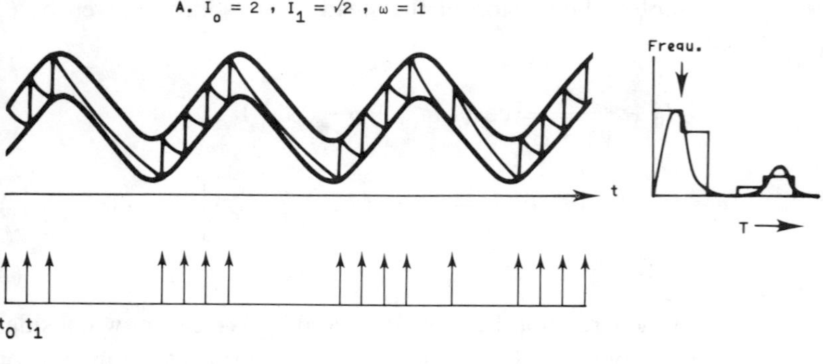

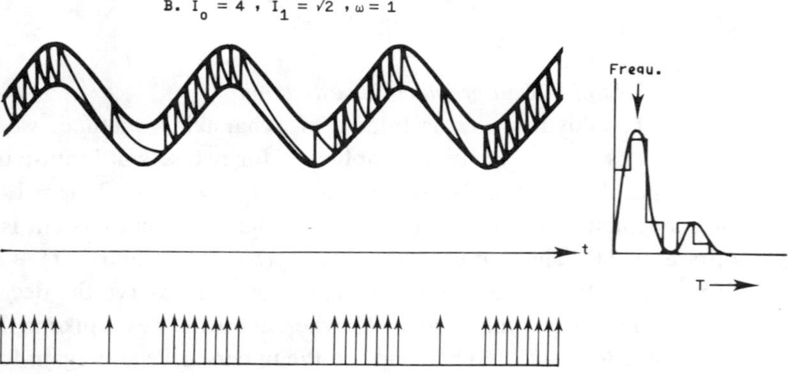

Figure 3.21. Graphical investigation of spike trains as a function of input parameters for cyclic inputs. Corresponding histograms of interspike intervals are shown on the right.

though the frequency of oscillations is unaltered. The histogram of interspike intervals is still *bimodal*, both peaks occurring at smaller values.

Example 2: *an "exact" calculation*

In the last example only a rough calculation of spike times was made by visually inspecting the curves $u(t)$ and $u(t)-1$. We can calculate the interspike times more accurately. Choosing the parameter values, $I_0 = 2$, $I_1 = \sqrt{2}$, $\omega = 1$, and $\phi + \psi = 3\pi/2$, and setting $t_0 = 0$ in formula (3.92), we get an equation for t_1,

$$1 + \cos(t_1) = 3e^{-t_1}.$$

Suitable root-finding methods can be employed to find accurate solutions especially with the aid of a computer.

Example 3: *phase locking*

Roughly speaking, the term phase locking means that the output spikes bear a fixed phase relationship with the input. We will consider a simple example, namely that in which an output spike occurs at each period of the input. To see how this may happen, consider Figure 3.22. The stated input–output relationship is inserted graphically. In the upper figure an apparently haphazard sequence of spike times occurs. In the lower figure we see a spike at $t = 0$, $t = 2\pi/\omega$, $t = 4\pi/\omega$, and so forth, and we say that the phase of the *output is locked* (*or locked in*) *to that of the input*.

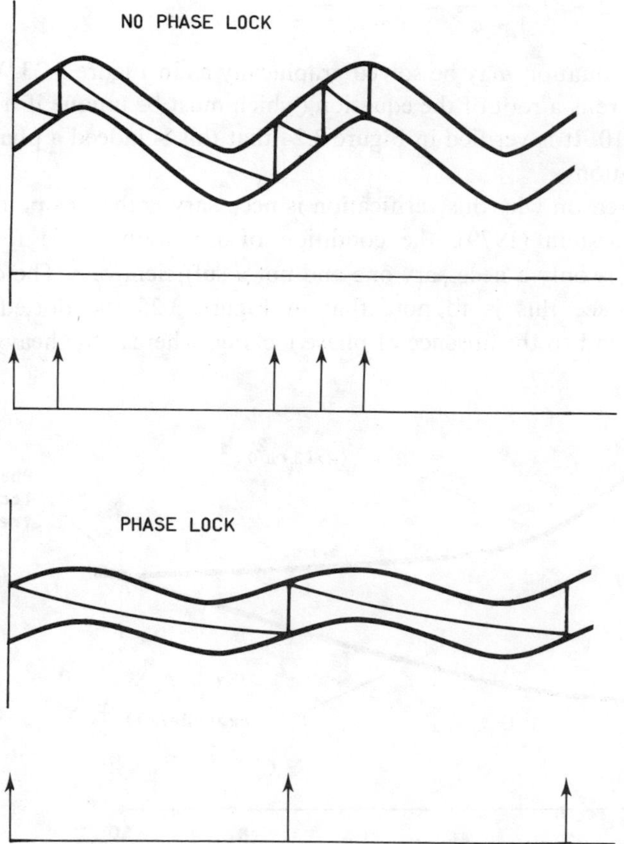

Figure 3.22. Conditions for simple phase locking using the graphical method.

The mathematical condition under which this occurs can be obtained as follows. If we imagine that there is a spike at $t_0 = 0$, then this is t_i in formula (3.92) and $t_1 = 2\pi/\omega = t_{i+1}$. Without loss of generality we set $\phi - \psi = 0$ and obtain the requirement, after rearrangement, that

$$I_0 + \frac{I_1}{\sqrt{1+\omega^2}} = \frac{1}{1-e^{-2\pi/\omega}}. \tag{3.93}$$

We cannot expect this condition to be fulfilled unless I_0, I_1, and ω bear special relationships to each other, as can be judged by looking at the two examples of Figure 3.22. Suppose we specify the values $I_0 = 2$ and $I_1 = \sqrt{2}$. Then Equation (3.93) becomes a requirement for ω, the angular frequency of the input,

$$2 + \sqrt{\frac{2}{1+\omega^2}} = \frac{1}{1-e^{-2\pi/\omega}}.$$

This equation may be solved graphically as in Figure 3.23. We find that there is a root of the equation (which must be unique if it occurs) at $\omega = 10$. It is verified in Figure 3.24 that this is indeed a phase-locking relation.

The reason why this verification is necessary is that, as pointed out by Scharstein (1979), the condition of Equation (3.93) for phase locking is only a necessary one and not a sufficient one. The quickest way to see this is to note that in Figure 3.25 the dotted curves correspond to the absence of phase locking, whereas the heavy curves

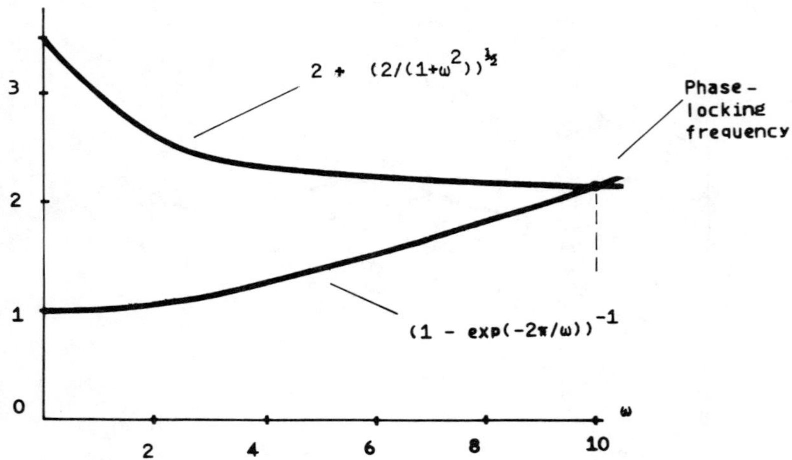

Figure 3.23. Graphical method of determination of the angular frequency at which phase locking occurs as in the text example 3.

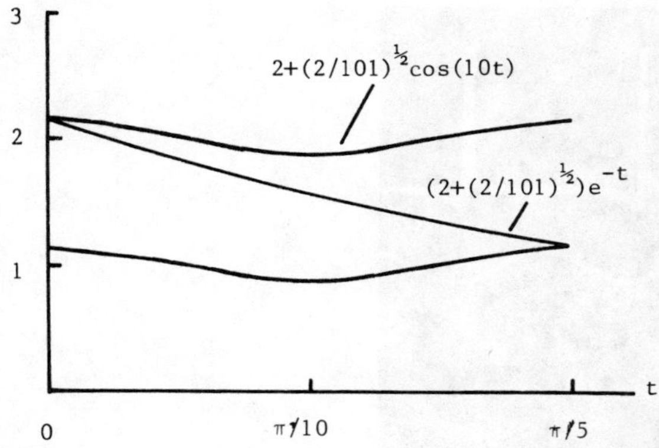

Figure 3.24. Demonstrating that the condition obtained for phase locking really does lead to phase locking.

correspond to phase locking; yet in both cases the condition (3.93) is fulfilled. Hence, in addition to that condition, one must ascertain that there is no value of t smaller than $2\pi/\omega$, where $u(t)$ intersects $u(t)-1$.

There have been many experimental and theoretical studies of phase locking. The general case of phase locking has been taken as the occurrence of m spikes for every n input cycles, and the condition for

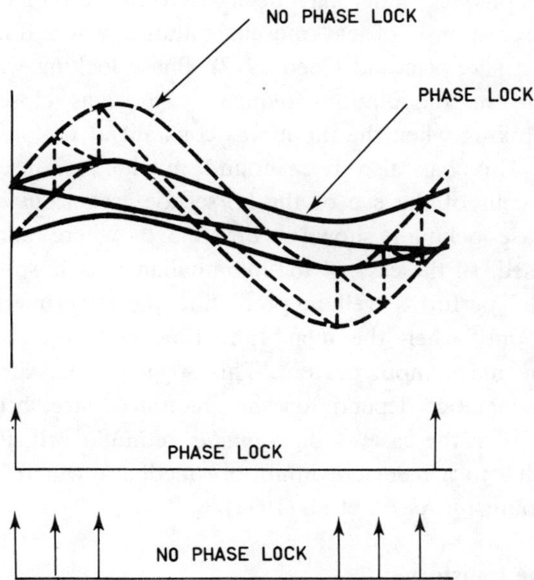

Figure 3.25. How the conditions for phase locking are necessary but not sufficient.

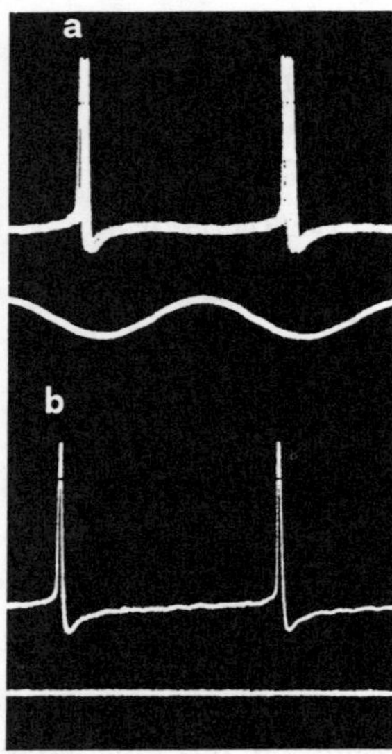

Figure 3.26. (a) – Phase locking in the *Limulus* photoreceptor [from Ascoli et al. (1977)]. Many traces were superimposed. (b) – Light level is same as the average in (a).

this has been found by Rescigno et al. (1970). A complete analysis was subsequently performed by Keener, Hoppensteadt, and Rinzel (1981). The Lapicque model has been used to investigate phase locking in the response of cat muscle spindles when undergoing sinusoidal input (Poppele and Chen 1972). Phase locking was found to occur when the modulation frequency $2\pi\omega$ was close to the frequency of spiking when the input was constant at the value I_0 in our notation. This has also been found in the response of the photoreceptor cells of the eye of the horseshoe crab (*Limulus*). An example of phase locking is shown in Figure 3.26, where many traces are superimposed. In the case of the mammalian muscle spindle, the Lapicque model performs well in predicting the experimental phenomenon but only when the membrane time constant is made a function of the mean input current. This suggests that the passive membrane conductance depends on the amount of stretch (Poppele and Chen 1972). In the case of the *Limulus* retinular cell, there is a complication due to a recurrent inhibitory feedback which has been taken into account by Ascoli et al. (1977).

3.9 Laplace transforms

The method of Laplace transforms is useful in applied mathematics for solving differential or similar equations. The usefulness is as follows. From the equation we derive an expression for the

Laplace transform of the solution. In some cases having the Laplace transform enables us to immediately find the solution because, when the method works, we will be able to identify the function from a table of the corresponding transforms. A few simple examples will make this clear. Laplace transforms are used frequently in later chapters.

Suppose $f(t)$ is a function defined for $0 \leq t < \infty$.

Definition

The Laplace transform of $f(t)$, denoted by $f_L(s)$, is defined as

$$f_L(s) = \int_0^\infty e^{-st} f(t)\, dt. \tag{3.94}$$

To this definition is attached the proviso that the integral exists. Functions that become infinite too fast for large t are not able to be Laplace transformed. Sometimes we put $\mathscr{L}\{f(t)\}$ for the right-hand side of (3.94).

The variable s is called the *transform variable*. In actuality it is allowed to take any value in the complex plane but this has no importance for the user of tables of Laplace transforms.

The Laplace transform of the derivative

We use integration by parts to find

$$f'_L(s) = \int_0^\infty f'(t) e^{-st}\, dt = \left[f(t)e^{-st}\right]_0^\infty + s\int_0^\infty f(t)e^{-st}\, dt$$
$$= -f(0) + sf_L(s). \tag{3.95}$$

Hence we see that the Laplace transform of the derivative of a function can be expressed in terms of the Laplace transform of the function and the initial value $f(0)$ of the function. When one performs a similar calculation on the Laplace transform of the second derivative $f''(t)$, one finds that its transform can be expressed in terms of the Laplace transform of $f(t)$ and the values of $f(0)$ and $f'(0)$. Thus, as can be verified by integration,

$$f''_L(s) = s^2 f_L(s) - sf(0) - f'(0). \tag{3.96}$$

Formulas such as this and the previous one for $f'_L(s)$ are useful when we solve differential equations.

Example: solution of a differential equation

We will show how Laplace transforms can be used to solve the initial value problem

$$\frac{d^2f}{dt^2} + 2\frac{df}{dt} + 2f = \delta(t - \pi), \qquad t > 0, \tag{3.97}$$

with initial data $f(0) = f'(0) = 0$.

Table 3.2. *Commonly encountered Laplace transforms*[a]

Function	Transform
$f(t)$	$f_L(s)$
1	$\frac{1}{s}$, $s > 0$
e^{at}	$\frac{1}{s-a}$, $s > a$
$\sin at$	$\frac{a}{s^2+a^2}$, $s > 0$
t^n, $n =$ positive integer	$\frac{n!}{s^{n+1}}$, $s > 0$
t^p, $p > -1$	$\frac{\Gamma(p+1)}{s^{p+1}}$, $s > 0$
$\cos at$	$\frac{s}{s^2+a^2}$, $s > 0$
$\sinh at$	$\frac{a}{s^2-a^2}$, $s > \|a\|$
$\cosh at$	$\frac{s}{s^2-a^2}$, $s > \|a\|$
$e^{at}\sin bt$	$\frac{b}{(s-a)^2+b^2}$, $s > a$
$e^{at}\cos bt$	$\frac{s-a}{(s-a)^2+b^2}$, $s > a$
$t^n e^{at}$, $n =$ positive integer	$\frac{n!}{(s-a)^{n+1}}$, $s > a$
$H(t-c)$	$\frac{e^{-cs}}{s}$, $s > 0$
$H(t-c)f(t-c)$	$e^{-cs}f_L(s)$
$e^{ct}f(t)$	$f_L(s-c)$
$f(ct)$	$\frac{1}{c}f_L\left(\frac{s}{c}\right)$
$\int_0^t f(t-\tau)g(\tau)\,d\tau$	$f_L(s)g_L(s)$
$\delta(t-c)$	e^{-cs}
$f^{(n)}(t)$	$s^n f_L(s) - s^{n-1}f(0) \cdots - f^{(n-1)}(0)$
$(-t)^n f(t)$	$f_L^{(n)}(s)$

[a]Adapted from Boyce and Di Prima (1977).

Using the table of transforms (Table 3.2), we operate on the left- and right-hand sides of the differential equation with the Laplace transform operator. Importantly, we note that *the initial data are automatically incorporated in the solution*. We have

$$s^2 f_L + 2s f_L + 2 f_L = e^{-\pi s}, \tag{3.98}$$

because $f(0) = f'(0) = 0$. Hence

$$f_L(s) = \frac{e^{-\pi s}}{s^2 + 2s + 2} = \frac{e^{-\pi s}}{(s+1)^2 + 1}. \tag{3.99}$$

We see by inspection of the table that the inverse transform of $1/(1 + (s+1)^2)$ is $\mathrm{e}^{-t}\sin(t)$. Since this transform is multiplied by $e^{-\pi s}$, we see that the solution is shifted to the right by π. Hence the inverse transform of (3.99), which we may denote by $\mathscr{L}^{-1}\{f_L(s)\}$, is

$$f(t) = H(t - \pi) e^{-(t-\pi)} \sin(t - \pi). \tag{3.100}$$

4

Linear cable theory for nerve cylinders and dendritic trees: steady-state solutions

4.1 Introduction

In Chapter 3 the Lapicque model, in which a lumped circuit consisting of a resistance and capacitance in parallel, was employed to represent the entire nerve cell. Though that simple model is not without usefulness, when we cast our minds back to the anatomical facts we reviewed concerning motoneurons in Chapter 1, we realize that we must extend the model if we wish to incorporate the realities of neuronal structure.

For example, we might ask what is the relative effectiveness of synapses close to the soma, compared to those of the same strength on distal parts of the dendritic tree. Or we might wish to inquire how branching affects the integration of various inputs. The model developed in this chapter will help us answer these kinds of questions within a mathematical framework, which is not conceptually difficult, though it has a cumbersome nature when the geometry of the cell is complicated. The various portions of the dendritic tree and the axon are now regarded as *passive nerve cylinders* and the equations satisfied by the electric potential are the partial differential equations of linear cable theory. (The term *passive* here means that the membrane conductance is fixed.)

We will here bridge the gap between the simple lumped-circuit model of Chapter 3 and the more complicated Hodgkin–Huxley model considered later. It must be emphasized that in this chapter and the next we deal only with *subthreshold responses*: that is, levels of excitation less than those required to generate action potentials. In fact, the model is only expected to be valid for responses that are, roughly speaking, of about half the strength required for action-potential generation. The reason for this is that for stronger responses the membrane conductance starts to change too much for the nerve cylinders to be considered passive. To handle suprathreshold responses in a satisfactory manner requires more complicated systems of equations. We may summarize the details of the kinds of model considered in this book as follows.

(i) *Lapicque type.* Lumped circuit for whole neuron. Linear or nonlinear first-order differential equation satisfied by the membrane potential. Suitable only for subthreshold responses. Threshold may be imposed.

(ii) *Passive cable type.* Spatially distributed circuit. Linear partial differential equation for the membrane potential. Suitable for (weak) subthreshold responses. Threshold may be imposed.

(iii) *Hodgkin–Huxley type.* Spatially distributed circuit. Membrane potential and ionic conductances satisfy a coupled system of nonlinear partial differential equations. Suitable for subthreshold and suprathreshold responses. The threshold property is a natural property of the equations.

In this chapter we will derive the linear cable equation and discuss the various boundary conditions that may be imposed on it in the context of nerve cylinders. We will obtain various steady-state solutions that enable membrane properties to be obtained from experimental results. We will also investigate solutions for dendritic trees. The more difficult time-dependent problems are treated in Chapters 5 and 6 where we include the method of reducing the tree to an "equivalent" cylinder. For a review of the history of cable theory, see Rall (1977).

4.2 Derivation of the cable equation for nerve cylinders

We will derive the cable equation satisfied by the *depolarization* of a nerve cylinder, often appropriate for unmyelinated axons and portions of dendrites. *Axial symmetry* is assumed and we only consider one space variable, the distance x along the cylinder. [A treatment in cylindrical coordinates is given in Rall (1969b).]

Figure 4.1 shows a segment of nerve cylinder. Here the inner radius is a and the membrane thickness is δ. Inside the membrane is a conducting fluid (the *axoplasm* in the case of an axon, or simply intracellular fluid), and the external medium is taken to be a uniform bathing medium, also with conducting properties. The membrane itself is assumed to have an electrical resistance and capacitance.

At a distance x along the cylinder (with $x = 0$ some convenient reference point), the electrical potential is assumed to be the same at all interior points on a circular cross section. Thus the state of the system at all such interior points is characterized by $V_i(x, t)$, the

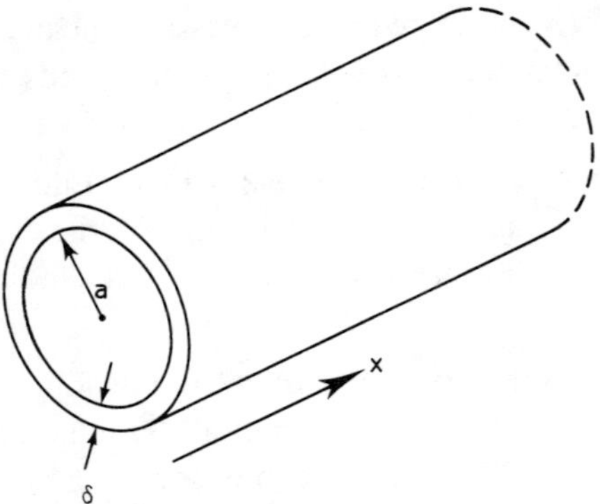

Figure 4.1. Geometrical details of a segment of a nerve cylinder.

electrical potential at x (relative to a point at infinity) at time t. Similarly, the potential outside is $V_0(x, t)$. The *transmembrane potential*, or just *membrane potential*, is

$$V_m(x, t) = V_i(x, t) - V_0(x, t). \tag{4.1}$$

If the nerve cylinder is *uniformly at rest*, then, with the convention that the external potential is zero, we have

$$V_0(x, t) = 0, \tag{4.2A}$$

$$V_m(x, t) = V_{i, R} = V_{m, R}, \tag{4.2B}$$

where $V_{m, R}$ is the *resting membrane potential*, assumed constant. The *depolarization*, relative to the resting level, is

$$V(x, t) = V_m(x, t) - V_{m, R}, \tag{4.3}$$

and negative values of $V(x, t)$ are referred to as *hyperpolarization*.

We saw in Chapter 2 how the classical theory of membrane potentials predicts that there may be ionic fluxes through the membrane in the resting state. These are counteracted by various transport mechanisms. Such considerations make for complications that may be sidestepped in the following way.

Let V_i^* and V_0^* be the differences between the interior and exterior potentials and their resting values,

$$\begin{aligned} V_i^*(x, t) &= V_i(x, t) - V_{i, R}, \\ V_0^*(x, t) &= V_0(x, t) - V_{0, R}. \end{aligned} \tag{4.4}$$

Then

$$\begin{aligned} V_i^*(x, t) - V_0^*(x, t) &= (V_i(x, t) - V_{i, R}) - (V_0(x, t) - V_{0, R}) \\ &= V_i(x, t) - V_0(x, t) - (V_{i, R} - V_{0, R}) \\ &= V_m(x, t) - V_{m, R} \\ &= V(x, t). \end{aligned} \tag{4.5}$$

We will work with the quantities $V_i^*(x, t)$ and $V_0^*(x, t)$ and assume that when they have their resting values, in the absence of external current sources, there is no net current through the membrane's representative circuit. However, we will drop the asterisks on V_i^* and V_0^*.

We assume that the nerve cylinder is ensheathed in a cylindrically shaped extracellular space. It is further assumed that the following three kinds of current may occur. One is through the membrane, the *transmembrane current*; the other two are longitudinal currents inside and outside the cell. We adopt the conventions that the longitudinal currents are positive in the direction of increasing x and that the transmembrane currents are positive in the *inward* direction.

Consider a small segment of nerve cylinder and surrounding medium with an electrical-circuit diagram as shown in Figure 4.2. Here, for the segment between x and $x + \Delta x$, the internal and external resistances to longitudinal current are $R_i(\Delta x)$ and $R_0(\Delta x)$, where the dependence on Δx is indicated explicitly. Similarly, the resistance to transmembrane current is $R_m(\Delta x)$ and the membrane capacitance is $C_m(\Delta x)$. Two kinds of applied current are shown. These are $I_{i,A}(\Delta x)$ and $I_{0,A}(\Delta x)$ applied to the intracellular and extracellular compartments. At x at time t, the longitudinal internal and external currents are $I_i(x, t)$ and $I_0(x, t)$, respectively.

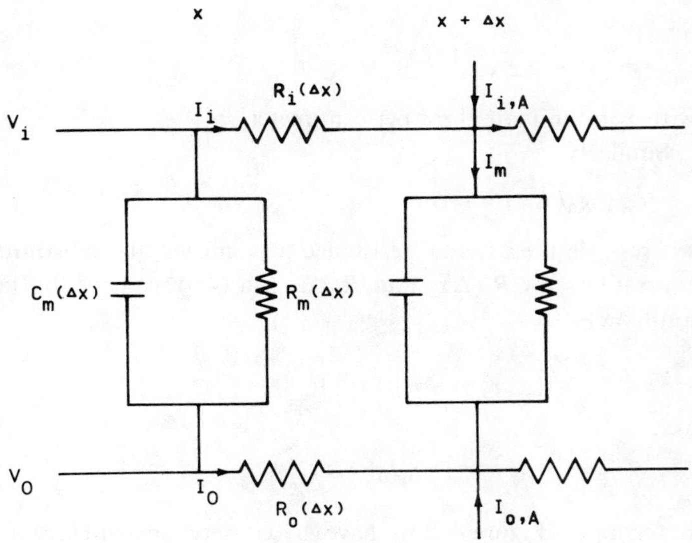

Figure 4.2. Circuit diagram representing a small segment of a nerve cylinder with membrane, intracellular and extracellular compartments, and applied currents.

We assume that the resistances in the circuit are ohmic (i.e., they obey Ohm's law), so

$$\begin{aligned} V_i(x+\Delta x, t) &= V_i(x, t) - R_i(\Delta x) I_i(x, t), \\ V_0(x+\Delta x, t) &= V_0(x, t) - R_0(\Delta x) I_0(x, t). \end{aligned} \tag{4.6}$$

Recall that the electrical resistance R of a piece of material is proportional to its length l (in the direction of the current) and inversely proportional to the area A through which the current flows. Thus, in general,

$$R = \rho l / A, \tag{4.7}$$

where the proportionality constant is called the *resistivity* or *specific resistance* of the material. It is measured in ohms centimeters if lengths are in centimeters and resistances are in ohms.

Between x and $x + \Delta x$, the resistance of the internal medium to longitudinal current is that of a cylinder of length Δx and area πa^2. Hence

$$R_i(\Delta x) = \rho_i \Delta x / \pi a^2, \tag{4.8}$$

where ρ_i is the resistivity of the internal medium. In terms of the internal resistance per unit length, which is

$$r_i = \rho_i / \pi a^2, \tag{4.9}$$

we have

$$R_i = r_i \Delta x, \tag{4.10}$$

with r_i in units of ohms per centimeter.

Similarly,

$$R_0(\Delta x) = r_0 \Delta x, \tag{4.11}$$

where r_0 is the external resistance per unit length. Substituting these expressions for $R_i(\Delta x)$ and $R_0(\Delta x)$ in (4.6) we have, on taking the limit $\Delta x \to 0$,

$$\begin{aligned} \frac{\partial V_i}{\partial x}(x, t) &= -r_i I_i(x, t), \\ \frac{\partial V_0}{\partial x}(x, t) &= -r_0 I_0(x, t). \end{aligned} \tag{4.12}$$

Referring to Figure 4.2, we have, by conservation of current,

$$\begin{aligned} I_i(x+\Delta x, t) &= I_i(x, t) + I_m(x+\Delta x, t) + I_{i,A}(x+\Delta x, t), \\ I_0(x+\Delta x, t) &= I_0(x, t) - I_m(x+\Delta x, t) + I_{0,A}(x+\Delta x, t). \end{aligned} \tag{4.13}$$

The transmembrane current is

$$I_m = \frac{V_0 - V_i}{R_m(\Delta x)} + C_m(\Delta x)\frac{\partial}{\partial t}[V_0 - V_i]. \tag{4.14}$$

Assuming the membrane thickness δ is small compared to the inner radius of the nerve cylinder, the transmembrane resistance between x and $x + \Delta x$ is that of a slab of material of length δ and area $2\pi a\,\Delta x$. Hence

$$R_m(\Delta x) = \rho_m \delta / 2\pi a\,\Delta x, \tag{4.15}$$

where ρ_m is the resistivity of the membrane material. This is written as

$$R_m(\Delta x) = r_m / \Delta x, \tag{4.16}$$

where

$$r_m = \rho_m \delta / 2\pi a \tag{4.17}$$

is the membrane resistance of unit length times unit length (in ohms centimeters). If c_m is the membrane capacitance per unit length (in farads per centimeter), then

$$C_m(\Delta x) = c_m\,\Delta x. \tag{4.18}$$

Thus

$$I_m = \frac{(V_0 - V_i)\,\Delta x}{r_m} + c_m \frac{\partial}{\partial t}[V_0 - V_i]\,\Delta x. \tag{4.19}$$

Substituting in (4.13), rearranging, and letting $\Delta x \to 0$, we get

$$\frac{\partial I_i}{\partial x} = I_{i,A}(x,t) + [V_0(x,t) - V_i(x,t)]/r_m + c_m \frac{\partial}{\partial t}[V_0(x,t) - V_i(x,t)], \tag{4.20}$$

and

$$\frac{\partial I_0}{\partial x} = I_{0,A}(x,t) - [V_0(x,t) - V_i(x,t)]/r_m - c_m \frac{\partial}{\partial t}[V_0(x,t) - V_i(x,t)]. \tag{4.21}$$

Differentiating (4.12) with respect to x, we have

$$\frac{\partial^2 V_i}{\partial x^2} = -r_i \frac{\partial I_i}{\partial x}, \qquad \frac{\partial^2 V_0}{\partial x^2} = -r_0 \frac{\partial I_0}{\partial x}. \tag{4.22}$$

Substituting (4.20) and (4.21) in (4.22), we get

$$\frac{\partial^2 V_i}{\partial x^2} = -r_i\left[I_{i,A} + \frac{V_0 - V_i}{r_m} + c_m\frac{\partial}{\partial t}(V_0 - V_i)\right],$$
$$\frac{\partial^2 V_0}{\partial x^2} = -r_0\left[I_{0,A} - \frac{(V_0 - V_i)}{r_m} - c_m\frac{\partial}{\partial t}(V_0 - V_i)\right]. \tag{4.23}$$

Subtracting the second of these equations from the first and rearranging gives the *cable equation for the depolarization*,

$$r_m c_m \frac{\partial V}{\partial t} = \frac{r_m}{r_i + r_0}\frac{\partial^2 V}{\partial x^2} - V + \frac{r_m}{r_i + r_0}[r_i I_{i,A} - r_0 I_{0,A}], \tag{4.24}$$

where $I_{i,A}(x,t)$ and $I_{0,A}(x,t)$ are the applied current densities (per unit length) in the intracellular and extracellular compartments. Often it is assumed that the *extracellular medium is perfectly conducting* so that $r_0 = 0$ and $V_0(x,t) = 0$ for all x and all t. In this case the cable equation becomes

$$r_m c_m \frac{\partial V}{\partial t} = \frac{r_m}{r_i}\frac{\partial^2 V}{\partial x^2} - V + r_m I_A, \tag{4.25}$$

where $I_A(x,t)$ is the applied current density (per unit length). For earlier derivations of the cable equation in a neurophysiological setting, see Hodgkin and Rushton (1946), Jack et al. (1985), and Rall (1977).

4.3 Boundary and initial conditions

We suppose for now that we are concerned with a nerve cylinder of length l cm, so that for the cable equation $0 \le x \le l$. Equations (4.24) and (4.25) are linear partial differential equations, which are first order in t and second order in x. They are classified as of parabolic type and have a unique solution if we specify suitable *initial data* for $t = 0$ and certain *boundary conditions* at $x = 0$ and $x = l$.

The initial data describes the depolarization present at the beginning of the experiment for all relevant values of x. Thus we will have

$$V(x,0) = v(x), \qquad 0 \le x \le l. \tag{4.26}$$

In the often-considered case of a cylinder initially in a *uniform resting state*, the initial data is

$$v(x) = 0, \qquad 0 \le x \le l. \tag{4.27}$$

Boundary conditions

There are several boundary conditions that may be imposed upon the cable equation at $x = 0$ and $x = l$. In all the cases we look at, we will consider the simpler, second cable equation (4.25).

Notation. The following notation will often be employed. Partial differentiation with respect to x or t is indicated by subscripting those variables. Thus we write

$$V_t(x,t) = \frac{\partial V(x,t)}{\partial t}, \qquad V_x(x,t) = \frac{\partial V(x,t)}{\partial x}, \qquad V_{xx} = \frac{\partial^2 V(x,t)}{\partial x^2}. \tag{4.28}$$

(i) *Voltage clamp.* A *voltage clamp* is the holding of the electrical potential at some fixed value. If the voltage at $x = 0$ is clamped such that the depolarization there is V_c, then the boundary condition is

$$V(0,t) = V_c, \qquad t > 0, \tag{4.29}$$

or if applied at $x = l$,

$$V(l,t) = V_c, \qquad t > 0. \tag{4.30}$$

(ii) *Sealed end.* The term *sealed end* means that there is no longitudinal current at the end. The longitudinal current is, from (4.12),

$$I_i(x,t) = -V_x(x,t)/r_i. \tag{4.31}$$

Hence if $I_i = 0$ we must have zero space derivative. If the end at $x = 0$ is sealed,

$$V_x(0,t) = 0, \qquad t > 0, \tag{4.32}$$

or if the end at $x = l$ is sealed,

$$V_x(l,t) = 0, \qquad t > 0. \tag{4.33}$$

A sealed end is sometimes referred to as an *open-circuit* termination.

(iii) *Killed end.* A *killed end* refers to the sudden ending of the nerve-cylinder membrane without a terminal covering membrane. The intracellular fluid therefore ends abruptly and abuts the extracellular fluid. If the end at $x = 0$ is killed, then for $x < 0$ the depolarization is zero, as is the resting potential. Hence the appropriate boundary condition is

$$V(0,t) = 0, \qquad t > 0, \tag{4.34}$$

or if the end at $x = l$ is killed,

$$V(l,t) = 0, \qquad t > 0. \tag{4.35}$$

A killed end is sometimes referred to as a *short-circuit* termination.

(iv) *Current injection at an end.* If there is a current of magnitude $I(t)$ injected *into the end* at $x=0$ (in the positive x-direction), then from (4.12) the boundary condition is

$$V_x(0,t) = -r_i I(t), \qquad t>0. \tag{4.36}$$

If the current $I(t)$ is injected into the end at $x=l$, we must make allowance for our convention that longitudinal currents are positive in the positive x-direction. Hence

$$V_x(l,t) = r_i I(t), \qquad t>0. \tag{4.37}$$

(v) *Natural termination.* The term natural termination is defined as the sealing of the end of a nerve cylinder with the nerve membrane itself. To obviate considering the actual geometrical details, it is convenient to lump the terminal membrane. Its resistance R_e will be that of a slab of length δ, area πa^2, and resistivity ρ_m. Hence

$$R_e = \rho_m \delta / \pi a^2. \tag{4.38}$$

Similarly, its capacitance is

$$C_e = \pi a^2 C_m, \tag{4.39}$$

where C_m is the membrane capacitance per unit area.

The natural termination has the circuit diagram shown in Figure 4.3. By conservation of current at $x=0$, the current through the terminal membrane at $x=0^-$, must be equal to the longitudinal current at $x=0^+$. Thus

$$-\frac{1}{r_i}V_x(0,t) = -\left[C_e V_t(0,t) + \frac{V(0,t)}{R_e}\right], \qquad t>0, \tag{4.40}$$

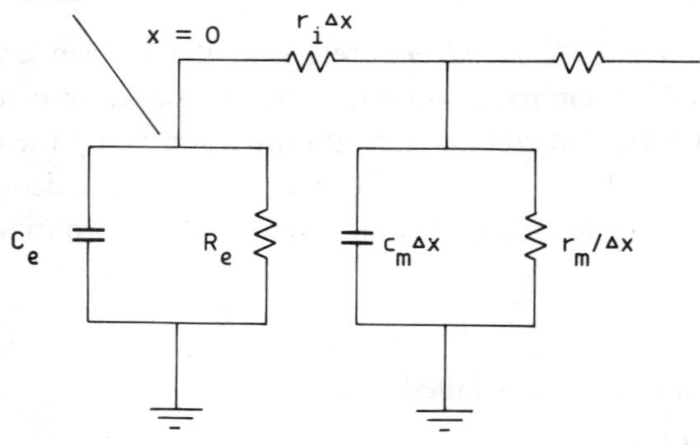

Figure 4.3. Electrical circuit representing a natural termination of a nerve cylinder at $x=0$.

or

$$\frac{V(0,t)}{R_e} + C_e V_t(0,t) - \frac{1}{r_i} V_x(0,t) = 0, \qquad t > 0. \tag{4.41}$$

If the terminal membrane is at $x = l$, then allowance for signs must be made. The corresponding boundary condition is

$$\frac{V(l,t)}{R_e} + C_e V_t(l,t) + \frac{1}{r_i} V_x(l,t) = 0, \qquad t > 0. \tag{4.42}$$

(vi) *Lumped-soma termination.* A lumped soma refers to the treating of the soma as an equipotential surface and regarding it as a single resistance R_s and capacitance C_s attached to a nerve cylinder as in the case of a natural termination. Thus the boundary condition for a lumped soma at $x = 0$ is

$$\frac{V(0,t)}{R_s} + C_s V_t(0,t) - \frac{1}{r_i} V_x(0,t) = 0, \qquad t > 0. \tag{4.43}$$

If a current $I_s(t)$ is applied at $x = 0^-$, as indicated in Figure 4.4, then we must have

$$\frac{V(0,t)}{R_s} + C_s V_t(0,t) - \frac{1}{r_i} V_x(0,t) = I_s(t), \qquad t > 0. \tag{4.44}$$

It will be seen that for a semiinfinite cylinder attached to a lumped soma this equation may be written as (Rall 1960)

$$\frac{V(0,t)}{R_s} + C_s V_t(0,t) - \frac{\lambda\rho}{R_s} V_x(0,t) = I_s(t), \tag{4.45}$$

where λ is the characteristic length or space constant (see the next

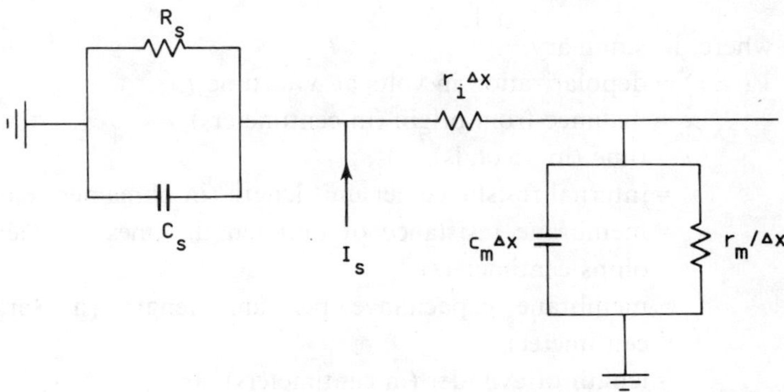

Figure 4.4. Electrical circuit showing application of current at a lumped soma at $x = 0^-$ leading to Equation (4.44).

section) and ρ is the ratio of the cylinder (dendritic) input conductance to the somatic conductance. However, the general use of the boundary condition in this form leads to inconsistencies (see Section 6.1). The mathematical solutions will, apart from numerical scaling factors, be the same whether (4.44) or (4.45) is employed.

(vii) *Infinite cylinders.* Some nerve cylinders, in particular certain axons, are sufficiently long to be considered *semiinfinite* $(0 \le x < \infty)$ or infinite $(-\infty < x < \infty)$. In the semiinfinite case the boundary condition at $x = 0$ may be any of those discussed so far. The boundary conditions at infinitely large x (positive or negative) are usually that V *remains bounded*; that is,

$$\lim_{|x| \to \infty} |V(x, t)| < \infty, \qquad t > 0. \tag{4.46}$$

The advantage of using the semiinfinite or infinite cylinders as approximations is that the mathematical formulas for the solutions are usually in closed form (as opposed to series) and are hence more easily evaluated.

4.4 Dimensionless variables and typical constants

The cable equations (4.24) and (4.25) contain physical constants that may be reduced in number or eliminated entirely by changing to dimensionless variables. We will consider the case when the external medium is perfectly conducting ($r_0 = 0$) and write (4.25) as

$$r_m c_m \frac{\partial V}{\partial t} = \frac{r_m}{r_i} \frac{\partial^2 V}{\partial x^2} - V + r_m I, \qquad 0 < x < l,\ t > 0, \tag{4.25$'$}$$

where, in summary,

$V(x, t)$ = depolarization in volts at x at time t,
x = distance from origin (in centimeters),
t = time (in seconds),
r_i = internal resistance per unit length (in ohms per centimeter),
r_m = membrane resistance of unit length times unit length (in ohms centimeters),
c_m = membrane capacitance per unit length (in farads per centimeter),
l = length of cylinder (in centimeters), and
$I(x, t)$ = applied current density in (amperes per centimeter).

It is convenient to note that $I(x, t)\, dx\, dt$ is the charge passing through

$[x, x+dx)$ in $[t, t+dt)$ so that

$$I(x,t) = \frac{\partial^2 Q}{\partial x\,\partial t}(x,t), \tag{4.47}$$

where $Q(x,t)$ is the charge passing through $[0, x)$ in $[0, t)$. We now let

$$X = x/\alpha, \tag{4.48A}$$

$$T = t/\beta, \tag{4.48B}$$

where α is a distance and β is a time so that X and T are dimensionless space and time variables. We also let

$$\overline{V}(X,T) = V(x,t), \tag{4.49A}$$

$$\overline{Q}(X,T) = Q(x,t). \tag{4.49B}$$

Then

$$\frac{\partial V}{\partial t} = \frac{\partial \overline{V}}{\partial T}\frac{dT}{dt} = \frac{1}{\beta}\frac{\partial \overline{V}}{\partial T}, \tag{4.50A}$$

$$\frac{\partial V}{\partial x} = \frac{\partial \overline{V}}{\partial X}\frac{dX}{dx} = \frac{1}{\alpha}\frac{\partial \overline{V}}{\partial X}, \tag{4.50B}$$

$$\frac{\partial^2 V}{\partial x^2} = \frac{1}{\alpha^2}\frac{\partial^2 \overline{V}}{\partial X^2}, \tag{4.50C}$$

$$\frac{\partial^2 Q}{\partial x\,\partial t} = \frac{1}{\alpha\beta}\frac{\partial^2 \overline{Q}}{\partial X\,\partial T}. \tag{4.50D}$$

Substituting these expressions in (4.25′) gives

$$\frac{r_m c_m}{\beta}\frac{\partial \overline{V}}{\partial T} = \frac{r_m}{r_i}\frac{1}{\alpha^2}\frac{\partial^2 \overline{V}}{\partial X^2} - \overline{V} + \frac{r_m}{\alpha\beta}\frac{\partial^2 \overline{Q}}{\partial X\,\partial T}. \tag{4.51}$$

We now let

$$\beta = r_m c_m, \tag{4.52A}$$

$$\alpha^2 = r_m/r_i, \tag{4.52B}$$

which makes the coefficients of $\partial\overline{V}/\partial T$ and $\partial^2\overline{V}/\partial X^2$ both unity. In more standard notation β is denoted

$$\tau = r_m c_m, \tag{4.53A}$$

the *membrane time constant*, and α is denoted

$$\lambda = \sqrt{r_m/r_i}, \tag{4.53B}$$

the membrane *characteristic length or space constant*. Thus Equation

(4.51) becomes

$$\frac{\partial \bar{V}}{\partial T} = \frac{\partial^2 \bar{V}}{\partial X^2} - \bar{V} + \frac{1}{\lambda c_m} \frac{\partial^2 \bar{Q}}{\partial X \partial T}. \tag{4.54}$$

If we let

$$\bar{c}_m = \lambda c_m, \tag{4.55A}$$

which is the capacitance of a characteristic length of membrane, and let

$$\bar{I}(X,T) = \frac{\partial^2 \bar{Q}}{\partial X \partial T} = \lambda \tau I(x,t), \tag{4.55B}$$

we get

$$\frac{\partial \bar{V}}{\partial T} = \frac{\partial^2 \bar{V}}{\partial X^2} - \bar{V} + \frac{\bar{I}}{\bar{c}_m}. \tag{4.56}$$

By using subscripts to denote partial differentiation, this may be written as

$$\bar{V}_T = \bar{V}_{XX} - \bar{V} + \bar{I}/\bar{c}_m, \tag{4.56A}$$

where $\bar{V}$ is in volts, $\bar{c}_m$ is in farads, and $\bar{I}$ is in coulombs.

It will sometimes prove useful to have this equation in a form that shows the explicit dependence of the forcing term on the diameter of the cylinder:

$$\bar{V}_T = \bar{V}_{XX} - \bar{V} + k\bar{I}/d^{3/2}, \tag{4.57}$$

where d is the *diameter measured in centimeters*, k is a constant of the membrane defined by

$$k = \frac{2}{\pi C_m}\left(\frac{\rho_i}{\delta \rho_m}\right)^{1/2}, \tag{4.57A}$$

and $\bar{I}$ is in coulombs per unit dimensionless time per unit dimensionless distance. k is measured in farads^{-1} centimeters$^{3/2}$.

If, however, diameter is also expressed in units of the space constant,

$$\bar{d} = d/\lambda, \tag{4.57B}$$

then the cable equation in dimensionless space and time variables becomes

$$\bar{V}_T = \bar{V}_{XX} - \bar{V} + k'\bar{I}/\bar{d}^3, \tag{4.57C}$$

where k' is a further constant of the membrane,

$$k' = \frac{16}{\pi C_m}\left(\frac{\rho_i}{\delta\rho_m}\right)^2$$

with units F^{-1}.

A further simplification can be made by changing to a dimensionless voltage variable $U(X, T)$. Let $\bar{V}^*$ be the steady-state voltage for the space-clamped cell ($\bar{V}_{XX} = 0$) in response to a current density of unity ($\bar{I} = 1$ C), so $\bar{V}^* = 1/\bar{c}_m$. Then set

$$U = \bar{V}/V^*. \tag{4.58}$$

The dimensionless voltage U satisfies the equation,

$$U_T = U_{XX} - U + J, \tag{4.59}$$

where J is numerically equal to $\bar{I}$ but is dimensionless. Note that if $0 \le x \le l$ and $0 \le t \le s$, then $0 \le X \le L$ and $0 \le T \le S$, where $L = l/\lambda$ and $S = s/\tau$.

Space and time constants in terms of fundamental constants

It is useful to show the dependence of the space and time constants on diameter d. We first introduce the membrane capacitance per unit area C_m, which is related to c_m by

$$C_m = c_m/\pi d.$$

The time constant is

$$\tau = \rho_m \delta C_m,$$

which shows that τ *does not depend on diameter*. For the space constant we find

$$\lambda = \frac{1}{2}\left(\frac{\rho_m \delta}{\rho_i}\right)^{1/2} \sqrt{d},$$

so that λ *is proportional to the square root of the diameter*.

In terms of the resistance of a unit area of membrane, $R_m = \rho_m \delta$, these expressions become

$$\tau = R_m C_m,$$

$$\lambda = \frac{1}{2}\left(\frac{R_m}{\rho_i}\right)^{1/2} \sqrt{d}.$$

In Table 4.1 some representative values of some constants are shown. These are based in part on Rall (1977), which contains much more information. See also Eccles (1964) and Chapter 6 of the present volume.

Table 4.1. *Typical constants*

	Squid giant axon	Cat spinal motoneuron (soma and dendrites)
R_m (Ω cm^2)	1000	2500
C_m (F cm^2)	1×10^{-6}	2×10^{-6}
ρ_i (Ω cm)	30	70
d (μm)	500	10 (primary dendrite)
r_i (Ω cm^{-1})	1.5×10^4	8.9×10^7
τ (ms)	1	5
λ (μm)	6500	1000

4.5 Steady-state cable equation

If a current is applied to a nerve cylinder for a sufficiently long time, a state of equilibrium may be attained in which the electrical potential, for all space points x, does not vary in time. Such an equilibrium is called a *steady state*.

Consider the cable equation (4.59), which we now write as

$$V_t = V_{xx} - V + I, \qquad 0 < x < L,\ t > 0, \tag{4.60}$$

where x and t are the dimensionless space and time variables from the last section and I is dimensionless current density. Suppose that the applied current density and the depolarization approach steady-state values so that

$$I(x,t) \underset{t\to\infty}{\longrightarrow} \tilde{I}(x), \tag{4.61}$$

$$V(x,t) \underset{t\to\infty}{\longrightarrow} \tilde{V}(x). \tag{4.62}$$

Then $V_t(x,t) \to 0$ as $t \to \infty$, and the steady-state solution satisfies the second-order linear ordinary differential equation

$$-\frac{d^2\tilde{V}}{dx^2} + \tilde{V} = \tilde{I}, \qquad 0 < x < L. \tag{4.63}$$

The boundary conditions for this last equation will be obtained as the limiting form (as $t \to \infty$) of those imposed on the partial differential equation (4.60).

There are two types of problems we wish to consider. In the first situation, current is injected only at the ends (one or both) of the cylinder, so that the applied current appears only in the boundary conditions. In the second situation, current may also be applied along the cylinder itself so that the applied current appears as a term on the right-hand side of (4.63). Before proceeding to solve these problems we take a brief look at second-order linear ordinary differential equations.

4.6 Second-order linear ordinary differential equations

Equation (4.63), which we need to solve for the steady-state response, is a second-order linear ordinary differential equation whose general standard form is

$$\frac{d^2y}{dx^2} + p(x)\frac{dy}{dx} + q(x)y = g(x), \qquad a < x < b. \tag{4.64}$$

If $g(x) = 0$ for all $x \in (a, b)$, then the equation is called *homogeneous*; otherwise it is *nonhomogeneous*.

Two main kinds of problems arise in connection with such equations. The first kind is called an *initial-value problem*, where the value of y and its first derivative y' are specified at the *same* point x_0, where $a < x_0 < b$. The second kind of problem is called a *boundary-value problem*, where y and/or its derivatives satisfy given conditions at two *different* points x_1 and x_2. Usually x_1 and x_2 are the endpoints of the interval. *In this chapter we are concerned with boundary-value problems for (4.63).*

The general solution $y_h(x)$ of the *homogeneous equation* can be expressed as a linear combination of two *linearly independent* solutions $y_1(x)$ and $y_2(x)$. Thus

$$y_h(x) = c_1 y_1(x) + c_2 y_2(x), \tag{4.65}$$

where c_1 and c_2 are constants. Whether y_1 and y_2 are linearly independent [in which case (4.65) is, in fact, the general solution of the homogeneous equation] can be ascertained by computing their *Wronskian*, $W_{y_1,y_2}(x)$, defined as

$$W_{y_1,y_2}(x) = \begin{vmatrix} y_1 & y_2 \\ y_1' & y_2' \end{vmatrix} = y_1 y_2' - y_1' y_2. \tag{4.66}$$

If the Wronskian is nonzero on (a, b), then it can be concluded that y_1 and y_2 are linearly independent.

There is one more fact that we need. Suppose $y_{nh}(x)$ is *any* solution of the nonhomogeneous equation. Then the general solution of the nonhomogeneous equation is

$$\begin{aligned} y(x) &= y_h(x) + y_{nh}(x) \\ &= c_1 y_1(x) + c_2 y_2(x) + y_{nh}(x). \end{aligned} \tag{4.67}$$

General solution of $-y'' + y = 0$

The homogeneous equation corresponding to (4.63) is

$$-d^2y/dx^2 + y = 0. \tag{4.68}$$

If we try a solution $y = e^{rx}$, we obtain $y' = re^{rx}$ and $y'' = r^2 e^{rx}$.

Substituting in (4.68) gives

$$(r^2-1)e^{rx}=0. \tag{4.69}$$

To make this an identity we require $r=+1$ or $r=-1$. Thus e^{-x} and e^x are solutions of (4.68). In particular, if we set

$$y_1(x)=e^{-x}, \tag{4.70}$$

$$y_2(x)=e^{x}, \tag{4.71}$$

then a computation of the Wronskian gives

$$W_{y_1,y_2}(x)=2,$$

so the Wronskian of these solutions is never zero. Hence y_1 and y_2 are linearly independent and the general solution of (4.68) is

$$y_h(x)=c_1e^{-x}+c_2e^{x}. \tag{4.72}$$

General solution of the nonhomogeneous equation

By letting a solution of the nonhomogeneous equation be of the form of the solution (4.65) of the homogeneous equation, except that c_1 and c_2 may depend on x, a solution of the nomhomogeneous equation (4.64) is given by [Boyce and DiPrima (1977), page 123],

$$y_{nh}(x)=\int^x\frac{[y_1(t)y_2(x)-y_1(x)y_2(t)]}{W_{y_1,y_2}(t)}g(t)\,dt, \tag{4.73}$$

where y_1 and y_2 are a pair of linearly independent solutions of the homogeneous equation.

We may use formula (4.73) to find an expression for a particular solution (not the general solution) of the nonhomogeneous equation,

$$-y''+y=g(x), \tag{4.74}$$

which has the same form as (4.63). We have, using $y_1=e^{-x}$, $y_2=e^{x}$, $W_{y_1,y_2}(x)=2$,

$$y_{nh}(x)=\tfrac{1}{2}\int^x(e^{t-x}-e^{x-t})g(t)\,dt. \tag{4.75}$$

Let us summarize the above in terms of $\tilde{V}(x)$.

Summary

The general solution of the steady-state equation

$$-d^2\tilde{V}/dx^2+\tilde{V}=\tilde{I}(x), \qquad 0<x<L, \tag{4.76}$$

is

$$V(x)=c_1e^{-x}+c_2e^{x}+V_{nh}(x), \tag{4.77}$$

where

$$V_{nh}(x) = \tfrac{1}{2}\left[e^{-x}\int^x e^t\tilde{I}(t)\,dt - e^x\int^x e^{-t}\tilde{I}(t)\,dt\right]. \qquad (4.78)$$

The values of c_1 and c_2 are found from the boundary conditions.

4.7 Current injection at an end: steady-state solutions

When no current is applied except at a terminal, Equation (4.63) is

$$d^2\tilde{V}/dx^2 - \tilde{V} = 0, \qquad 0 < x < L. \qquad (4.79)$$

We suppose the current is injected at $x = 0$ and has attained its steady-state constant value I_0. The boundary condition at $x = 0$ is, from (4.36),

$$\left.\frac{d\tilde{V}}{dx}\right|_{x=0} = -\bar{r}_i I_0, \qquad (4.80)$$

where $\bar{r}_i$ is the internal resistance per characteristic length. We consider three cases.

(i) *Semiinfinite cylinder.* As we have seen, the general solution of (4.79) is

$$\tilde{V}(x) = c_1 e^{-x} + c_2 e^x, \qquad (4.81)$$

where c_1 and c_2 are constants to be determined. The condition (4.46) requires $c_2 = 0$. Then

$$\frac{d\tilde{V}}{dx}\Big|_{x=0} = -c_1 e^{-x}\Big|_{x=0} = -c_1 = -\bar{r}_i I_0.$$

Thus the solution is

$$\tilde{V}(x) = \bar{r}_i I_0 e^{-x}, \qquad 0 \le x < \infty. \qquad (4.82)$$

Since

$$\tilde{V}(0) = \bar{r}_i I_0, \qquad (4.83)$$

this may be written as

$$\tilde{V}(x) = \tilde{V}(0)e^{-x}, \qquad 0 \le x < \infty. \qquad (4.84)$$

The steady-state solution is shown in Figure 4.5. The depolarization decays exponentially in such a way that at a physical distance of λ cm, it is e^{-1} times its value at $x = 0$. Hence the term "space constant" for λ. Note that if λ is small, the decay is rapid, whereas if λ is large, the attenuation is slow. That is, *responses propagate further when the space constant is large*.

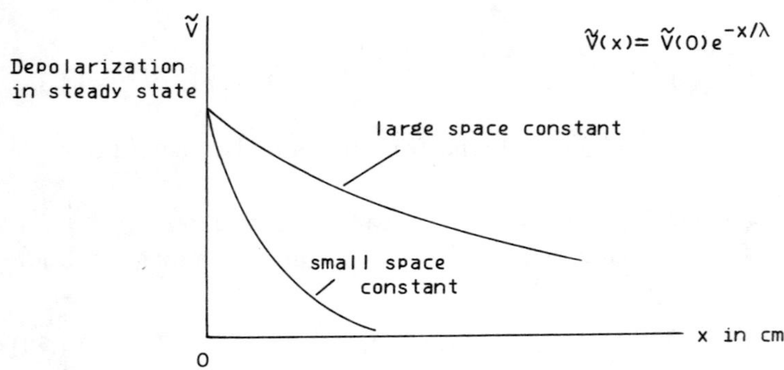

Figure 4.5. Steady-state depolarization for a semiinfinite cylinder with current injection at $x = 0$. Distances are measured in centimeters.

The *input resistance* at $x = 0$ for the semiinfinite cable is defined as the ratio of the steady-state voltage at $x = 0$ to the constant applied current. Thus

$$R_\infty = \tilde{V}(0)/I_0. \tag{4.85}$$

Hence

$$R_\infty = \bar{r}_i, \tag{4.86}$$

where $\bar{r}_i$ is the internal resistance per characteristic length,

$$\bar{r}_i = \lambda r_i. \tag{4.87}$$

This gives the expression, when $r_0 = 0$,

$$R_\infty = (r_i r_m)^{1/2}. \tag{4.88}$$

For the squid axon $\lambda \simeq 0.65$ cm and $r_i \simeq 15{,}000\ \Omega\ \mathrm{cm}^{-1}$ so $R_\infty \simeq 9750$ Ω. The *input* conductance G_∞ is the reciprocal of R_∞.

The injection of a constant current at a terminal and measurement of the steady-state response is sufficient to enable the constants r_i and r_m to be found. First, the value of λ can be found from the decay of potential – it is equal to the distance at which the depolarization reaches $1/e$ of its value at the point $x = 0$, presumed to be the point of current injection. Since $\tilde{V}(0) = \bar{r}_i I_0$ and I_0 is known, knowledge of $\tilde{V}(0)$ enables $\bar{r}_i$ to be found (and hence r_i). Finally, since $\lambda = \sqrt{r_m/r_i}$, the value of r_m can now be found.

(ii) *Finite cylinder, sealed end at $x = L$.* We now consider the differential equation (4.79) with the same boundary condition (4.80) at

$x = 0$, but now with the constraint

$$\left.\frac{d\tilde{V}}{dx}\right|_{x=L} = 0, \tag{4.89}$$

where $L = l/\lambda$.

The general solution (4.81) is still valid but it is convenient to turn to another pair of linearly independent solutions. The hyperbolic cosine of x, $\cosh x$, is defined as

$$\cosh x = \frac{e^x + e^{-x}}{2}, \tag{4.90}$$

and the hyperbolic sine, $\sinh x$, is

$$\sinh x = \frac{e^x + e^{-x}}{2}. \tag{4.91}$$

It can be verified by computing their Wronskian that these provide a pair of linearly independent solutions of (4.79). Thus we now write the solution as

$$\tilde{V}(x) = c_1 \cosh x + c_2 \sinh x. \tag{4.92}$$

We have

$$\frac{d\tilde{V}}{dx} = c_1 \sinh x + c_2 \cosh x. \tag{4.93}$$

At $x = 0$, $\sinh x = 0$ and $\cosh x = 1$, so that application of the boundary condition (4.80) gives

$$\left.\frac{d\tilde{V}}{dx}\right|_{x=0} = c_2 = -\bar{r}_i I_0. \tag{4.94}$$

At $x = L$,

$$\left.\frac{d\tilde{V}}{dx}\right|_{x=L} = c_1 \sinh L + c_2 \cosh L = 0, \tag{4.95}$$

which gives

$$c_1 = -c_2 \coth L, \tag{4.96}$$

where $\coth x = \cosh x / \sinh x$. Thus the required solution is

$$\tilde{V}(x) = \bar{r}_i I_0 [\coth L \cosh x - \sinh x]. \tag{4.97}$$

Since

$$\tilde{V}(0) = \bar{r}_i I_0 \coth L, \tag{4.98}$$

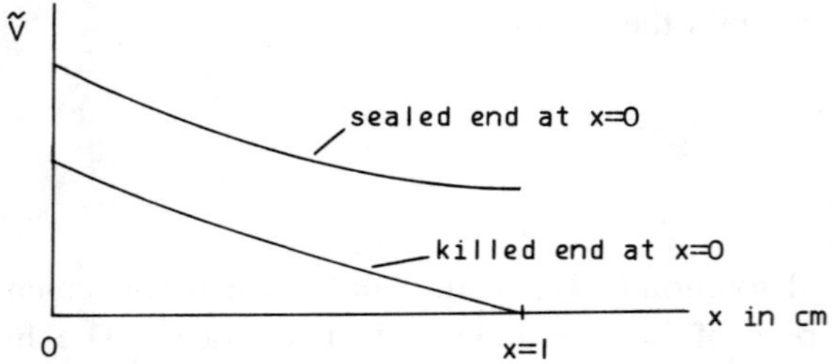

Figure 4.6. The steady-state depolarization in a finite cylinder with current injected at $x = 0$ in the cases of a sealed end at $x = l$ and a killed end at $x = l$.

we have, on using the identity,

$$\cosh(A - B) = \cosh A \cosh B - \sinh A \sinh B, \tag{4.99}$$

that our solution can be written as

$$\tilde{V}(x) = \tilde{V}(0)\frac{\cosh(L - x)}{\cosh L}, \qquad 0 \le x \le L. \tag{4.100}$$

Here $\tilde{V}(x)$ decreases as x increases to approach, with eventually zero slope, the value $\tilde{V}(0)/\cosh L$ at $x = L$, as indicated in Figure 4.6. The input resistance $\tilde{V}(0)/I_0$ is

$$R_{\text{sealed}} = \bar{r}_i \coth L. \tag{4.101}$$

Since $\coth L > 1$ for all $L < \infty$, the input resistance for a cylinder with a sealed end at $x = L$ is larger than that of the semiinfinite cylinder.

To obtain the physical constants r_i and r_m, all that are needed are the values of $\tilde{V}(x)$ at $x = 0$ and $x = L$. The ratio of these two quantities gives $\tilde{V}(0)/\tilde{V}(L) = \cosh L$, so that L and hence λ can be found from the physical length l of the cylinder. Once L is known, the value of $\tilde{V}(0) = \bar{r}_i I_0 \coth L$ can be used to find $\bar{r}_i$ and r_i. Then r_m can be found from the value of λ and r_i.

(iii) *Finite cylinder, killed end at* $x = L$. We now solve (4.79) with the boundary condition at $x = L$,

$$\tilde{V}(L) = 0, \tag{4.102}$$

and with the same condition as before at $x = 0$. Equations (4.92)–(4.94) apply again but we now have, using (4.102),

$$\tilde{V}(L) = c_1 \cosh L + c_2 \sinh L = 0, \tag{4.103}$$

so that

$$c_1 = -c_2 \tanh L. \tag{4.104}$$

The solution is

$$\tilde{V}(x) = \bar{r}_i I_0 [\tanh L \cosh x - \sinh x], \tag{4.105}$$

and

$$\tilde{V}(0) = \bar{r}_i I_0 \tanh L. \tag{4.106}$$

On using the identity,

$$\sinh(A - B) = \sinh A \cosh B - \cosh A \sinh B, \tag{4.107}$$

this rearranges to give

$$\tilde{V}(x) = \tilde{V}(0) \frac{\sinh(L-x)}{\sinh L}, \qquad 0 \le x \le L. \tag{4.108}$$

Here $V(x)$ decreases from the value $\tilde{V}(0)$ at $x = 0$ to zero at $x = L$, as indicated in Figure 4.6. The input resistance is

$$R_{\text{killed}} = \bar{r}_i \tanh L, \tag{4.109}$$

and, since $\tanh L < 1$ for all $L < \infty$, the input resistance with a killed end at $x = L$ is smaller than that of the semiinfinite cylinder. Thus we have the inequality

$$R_{\text{killed}} < R_{\infty} < R_{\text{sealed}}. \tag{4.110}$$

Note that the values of the corresponding depolarizations (if $I_0 > 0$) at $x = 0$ also satisfy this inequality.

To obtain the constants r_i and r_m from the measurement of the steady-state response in the case of a killed end it is sufficient to have the values of the depolarization at $x = 0$ and $x = L/2$. Utilizing the relation $\sinh(L) = 2\sinh(L/2)\cosh(L/2)$, we find $\tilde{V}(0)/2\tilde{V}(L/2) = \cosh(L/2)$. Thus L and hence λ can be found. Thereupon r_i is found from $\tilde{V}(0) = \bar{r}_i I_0 \tanh(L)$ and r_m can be obtained from the values of λ and r_i.

Table 4.2 gives a summary of the steady-state results obtained thus far and is styled after Weidmann (1952).

Table 4.2. *Summary of steady-state results for current I_0 injected at $x = 0$*

	Semiinfinite cable	Finite cable, sealed at $x = l$	Finite cable, killed at $x = l$
Voltage at $x = 0$, $= \tilde{V}(0)$	$I_0 r_i \lambda$	$I_0 r_i \lambda \coth(l/\lambda)$	$I_0 r_i \lambda \tanh(l/\lambda)$
Input resistance $= \tilde{V}(0)/I_0$	$r_i \lambda$	$r_i \lambda \coth(l/\lambda)$	$r_i \lambda \tanh(l/\lambda)$
Voltage as function of x, $\simeq \tilde{V}(x)$	$\tilde{V}(0) e^{-x/\lambda}$	$\dfrac{\tilde{V}(0)\cosh[(l-x)/\lambda]}{\cosh(l/\lambda)}$	$\dfrac{\tilde{V}(0)\sinh[(l-x)/\lambda]}{\sinh(l/\lambda)}$

r_i is the resistance per unit length of internal medium (in ohms per centimeter); $\lambda = \sqrt{r_m/r_i}$ is the space constant (in centimeters); and r_m is the membrane resistance of unit length times unit length (in ohms centimeters). I_0 is measured in amperes, $V(x)$ in volts, and l is the total length (in centimeters).

4.8 Current injection along the cylinder: steady-state solutions using the Green's function

We now suppose that instead of having current injection at a terminal of the nerve cylinder, the current is applied along the cylinder itself. In the steady state the depolarization satisfies

$$-d^2\tilde{V}/dx^2 + \tilde{V} = \tilde{I}, \qquad 0 < x < L, \tag{4.111}$$

with appropriate boundary conditions at $x = 0$ and $x = L$.

Since we will be interested in several forms for the input-current function $\tilde{I}(x)$, it is very useful to have a general method for solving (4.111) with given boundary conditions. One such method, the method of Green's functions, we already mentioned in connection with first-order equations in Chapter 3. Once we have the Green's function, it is a matter of inserting the particular $\tilde{I}(x)$ of interest in a formula that involves a single integration.

Definition of the Green's function

The Green's function $G(x; y)$ for Equation (4.111), with given boundary conditions, is the solution of the equation

$$-d^2G/dx^2 + G = \delta(x - y), \qquad 0 < x < L, 0 < y < L, \tag{4.112}$$

which satisfies the given boundary conditions. Thus the Green's function is the steady-state response to a constant current applied at a single space point y. Here y is regarded as fixed. Physically, this corresponds to the case of injecting current through an electrode of zero (or infinitesimal) width. The Green's function is very useful as the following result shows.

Solution to (4.111) *using the Green's function*
The solution of

$$-\frac{d^2\tilde{V}}{dx^2}+\tilde{V}=\tilde{I}, \qquad 0<x<L, \tag{4.113}$$

with the *unmixed homogeneous boundary conditions*,

$$\begin{aligned} a_1\tilde{V}(0)+a_2\tilde{V}'(0)&=0,\\ b_1\tilde{V}(L)+b_2\tilde{V}'(L)&=0, \end{aligned} \tag{4.114}$$

is

$$\tilde{V}(x)=\int_0^L G(x;\,y)\tilde{I}(y)\,dy, \tag{4.115}$$

where $G(x;\,y)$ is the solution of (4.112), which also satisfies the boundary conditions (4.114).

Note that the boundary conditions are called *unmixed* because the conditions at $x=0$ and $x=L$ are separate, and they are called homogeneous because the right-hand sides of (4.114) are zero.

Proof. We establish formula (4.115) as follows. We have, on differentiating $V(x)$,

$$\begin{aligned} \tilde{V}'(x)&=\frac{d}{dx}\int_0^L G(x;\,y)\tilde{I}(y)\,dy\\ &=\int_0^L\frac{dG}{dx}(x;\,y)\tilde{I}(y)\,dy, \end{aligned} \tag{4.116}$$

and differentiating again,

$$\tilde{V}''(x)=\int_0^L\frac{d^2G}{dx^2}(x;\,y)\tilde{I}(y)\,dy. \tag{4.117}$$

Then, using (4.116) and (4.117), we get

$$\begin{aligned} -\tilde{V}''(x) + \tilde{V}(x) &= \int_0^L [-G'' + G]\tilde{I}(y)\,dy \\ &= \int_0^L \delta(x-y)\tilde{I}(y)\,dy \\ &= \tilde{I}(x), \end{aligned} \tag{4.118}$$

on utilizing the last property of the delta function in Section 2.16.1. Hence $\tilde{V}(x)$ satisfies (4.113). It also satisfies the boundary conditions, for example at $x = 0$, because

$$\begin{aligned} a_1\tilde{V}(0) + a_2\tilde{V}'(0) &= \int_0^L [a_1G(0;\,y) + a_2G'(0;\,y)]\,\tilde{I}(y)\,dy \\ &= 0, \end{aligned} \tag{4.119}$$

since

$$a_1G(0;\,y) + a_2G'(0;\,y) = 0. \tag{4.120}$$

Determination of the Green's function: an example

To illustrate how to find the Green's function for the steady-state cable equations, we will consider the case of a semiinfinite cylinder with a sealed end at $x = 0$. Thus, in this case, the Green's function $G(x;\,y)$ is the solution of

$$-G'' + G = \delta(x-y), \qquad 0 < x < \infty,\, 0 < y < \infty, \tag{4.121}$$

satisfying the boundary conditions

$$\begin{aligned} &G'(0;\,y) = 0, \\ &|G(\infty;\,y)| < \infty. \end{aligned} \tag{4.122}$$

From Equations (4.77) and (4.78) we see that

$$\begin{aligned} G(x;\,y) = c_1e^{-x} + c_2e^{x} \\ -\tfrac{1}{2}\left[e^{x}\int^{x} e^{-t}\,\delta(t-y)\,dt - e^{-x}\int^{x} e^{t}\,\delta(t-y)\,dt\right]. \end{aligned} \tag{4.123}$$

If $x < y$ the integrals do not include $\delta(t-y)$ so the contributions from the terms involving the integrals are zero. Hence

$$G(x;\,y) = c_1e^{-x} + c_2e^{x}, \qquad 0 < x < y. \tag{4.124}$$

When $x > y$, the range of integration takes in the point of concentration of $\delta(t-y)$. The integrals are

$$\int^{x} e^{-t}\,\delta(t-y)\,dt = e^{-y}, \qquad x > y, \tag{4.125}$$

and

$$\int^{x} e^{t}\,\delta(t-y)\,dt = e^{y}, \qquad x > y. \tag{4.126}$$

Thus

$$G(x;\, y) = c_1 e^{-x} + c_2 e^{x} - \tfrac{1}{2}\left[e^{x-y} - e^{-(x-y)L}\right], \qquad x > y. \tag{4.127}$$

Now, as $x \to \infty$ with $x > y$,

$$G(x;\, y) \to e^{x}\left[c_2 - \tfrac{1}{2}e^{-y}\right]. \tag{4.128}$$

Thus if $G(x;\, y)$ is to be bounded at $x = \infty$, we require

$$c_2 = \tfrac{1}{2}e^{-y}. \tag{4.129}$$

Since near $x = 0$ formula (4.124) applies, we have

$$G'(0;\, y) = \left(-c_1 e^{-x} + c_2 e^{x}\right)\Big|_{x=0} = -c_1 + c_2 = 0. \tag{4.130}$$

Thus $c_1 = c_2$ and the required Green's function is

$$G(x;\, y) = \begin{cases} \tfrac{1}{2}e^{-y}\left[e^{-x} + e^{x}\right], & 0 < x < y, \\ \tfrac{1}{2}e^{-x}\left[e^{-y} + e^{y}\right], & y < x < \infty. \end{cases} \tag{4.131}$$

More neatly,

$$G(x;\, y) = \begin{cases} e^{-y}\cosh x, & 0 < x < y, \\ e^{-x}\cosh y, & y < x < \infty. \end{cases} \tag{4.132}$$

The Green's function, for fixed y, is sketched in Figure 4.7, where in the upper part of the figure a schematic representation of the delta function is shown.

A general formula for the Green's function

The Green's function for Equation (4.111) can, for given boundary conditions, be calculated in the manner illustrated above. However, the following general result applies (Stakgold, 1967, page 66).

The Green's function $G(x;\, y)$ satisfying

$$-G'' + G = \delta(x-y), \qquad 0 < x < L,\, 0 < y < L, \tag{4.133}$$

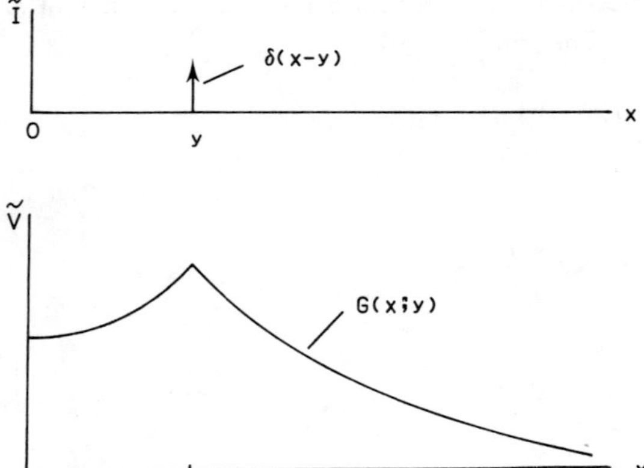

Figure 4.7. The Green's function for the steady-state cable equation in the case of a semiinfinite cylinder with a sealed end at $x = 0$. The upper part of the figure is a schematic representation of the steady-state current density, which is a delta function concentrated at $x = y$.

with boundary conditions of the type (4.114), is given by

$$G(x;\, y) = \begin{cases} -\dfrac{u_1(x)u_2(y)}{W_{u_1,\,u_2}(y)}, & 0 \le x < y, \\[2ex] -\dfrac{u_1(y)u_2(x)}{W_{u_1,\,u_2}(y)}, & y < x \le L, \end{cases} \tag{4.134}$$

where u_1 and u_2 are linearly independent solutions of the homogeneous equation, $-u'' + u = 0$, with u_1 satisfying the boundary condition at $x = 0$ and u_2 satisfying the boundary condition at $x = L$. The Wronskian $W_{u_1,\,u_2}(y)$ is given by formula (4.66).

Summary of the Green's function for the steady-state cable equation

By using (4.134) it is possible to find expressions for the Green's function in all the cases of interest. There are seven of them all together and the results are summarized in Table 4.3.

Example

An axon, which may be considered semiinfinite, has a sealed end at $x = 0$. The steady-state current density is given by

$$\tilde{I}(x) = \begin{cases} I_0, & x_0 - \varepsilon/2 < x < x_0 + \varepsilon/2, \\ 0, & \text{otherwise}, \end{cases}$$

where $0 < x_0 - \varepsilon/2$ and distances are in characteristic lengths. Find the steady-state depolarization.

Table 4.3. *Green's functions for the steady-state cable equation, $-\tilde{V}'' + \tilde{V} = \tilde{I}$, for various cable lengths and boundary conditions.*

Case	Cable on	Left b.c.	Right b.c.	$G(x;\,y),\ x<y$	$G(x;\,y),\ x>y$
1	$(-\infty,\infty)$	$\tilde{V}=0$	$\tilde{V}=0$	$\frac{1}{2}e^{x-y}$	$\frac{1}{2}e^{y-x}$
2	$(0,\infty)$	$\tilde{V}'=0$	$\tilde{V}=0$	$\cosh xe^{-y}$	$\cosh ye^{-x}$
3	$(0,\infty)$	$\tilde{V}=0$	$\tilde{V}=0$	$\sinh xe^{-y}$	$\sinh ye^{-x}$
4	$(0,L)$	$\tilde{V}'=0$	$\tilde{V}'=0$	$\dfrac{\cosh x\cosh(L-y)}{\sinh L}$	$\dfrac{\cosh y\cosh(L-x)}{\sinh L}$
5	$(0,L)$	$\tilde{V}=0$	$\tilde{V}'=0$	$\dfrac{\sinh x\cosh(L-y)}{\cosh L}$	$\dfrac{\sinh y\cosh(L-x)}{\cosh L}$
6	$(0,L)$	$\tilde{V}'=0$	$\tilde{V}=0$	$\dfrac{\cosh x\sinh(L-y)}{\cosh L}$	$\dfrac{\cosh y\sinh(L-x)}{\cosh L}$
7	$(0,L)$	$\tilde{V}=0$	$\tilde{V}=0$	$\dfrac{\sinh x\sinh(L-y)}{\sinh L}$	$\dfrac{\sinh y\sinh(L-x)}{\sinh L}$

In the first case the cable is infinite, in the next two cases it is semiinfinite, and in the remaining cases finite. Here b.c. = boundary condition.

Solution

To solve this problem, we use formula (4.115),

$$\tilde{V}(x) = \int_0^\infty G(x;\,y)\tilde{I}(y)\,dy,$$

where $G(x;\,y)$ is the Green's function for *Case 2* in Table 4.3. Since $\tilde{I}(y)$ vanishes outside the interval $(x_0-\varepsilon/2,\ x_0+\varepsilon/2)$, we have

$$\tilde{V}(x) = I_0\int_{x_0-\varepsilon/2}^{x_0+\varepsilon/2} G(x;\,y)\,dy.$$

There are three solution regions.

(i) $x < x_0 - \varepsilon/2$. Here y is always greater than x. Hence we use $G(x;\,y)$, as in the second to last column in Table 4.3,

$$\begin{aligned}\tilde{V}(x) &= I_0\int_{x_0-\varepsilon/2}^{x_0+\varepsilon/2}\cosh xe^{-y}\,dy\\ &= I_0\cosh x\left[-e^{-y}\right]_{x_0-\varepsilon/2}^{x_0+\varepsilon/2}\\ &= -I_0\cosh x\left[e^{-(x_0+\varepsilon/2)} - e^{-(x_0-\varepsilon/2)}\right]\\ &= 2I_0e^{-x_0}\cosh x\sinh(\varepsilon/2).\end{aligned}$$

(ii) $x_0 - \varepsilon/2 < x < x_0 + \varepsilon/2$. Here the region of integration splits into two parts. In the first $x_0 - \varepsilon/2 < y < x$ and in the second $x < y < x_0 + \varepsilon/2$. Hence, using the appropriate form for the Green's

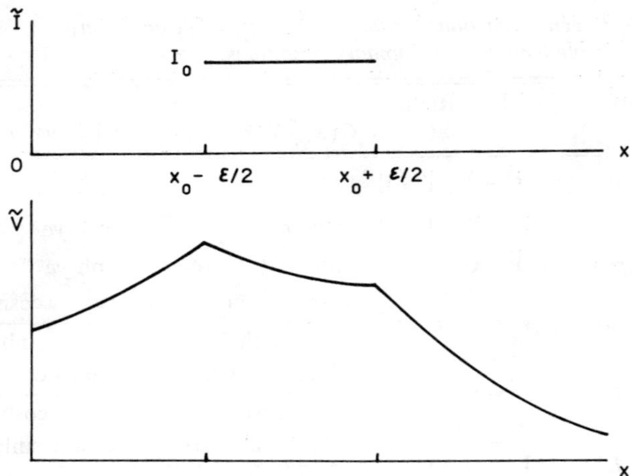

Figure 4.8. Steady-state current density (upper part) and depolarization for the worked example of the text.

function in each region,

$$\begin{aligned}\tilde{V}(x) &= I_0\left[e^{-x}\int_{x_0-\varepsilon/2}^{x}\cosh y\,dy + \cosh x\int_{x}^{x_0+\varepsilon/2} e^{-y}\,dy\right]\\ &= I_0\big[e^{-x}(\sinh x - \sinh(x_0-\varepsilon/2))\\ &\qquad -\cosh x\big(e^{-(x_0+\varepsilon/2)} - e^{-x}\big)\big].\end{aligned}$$

(iii) $x > x_0 + \varepsilon/2$. Here y is always less than x. Hence

$$\begin{aligned}\tilde{V}(x) &= I_0 e^{-x}\int_{x_0-\varepsilon/2}^{x_0+\varepsilon/2}\cosh y\,dy\\ &= I_0 e^{-x}\left[\sinh(x_0+\varepsilon/2) - \sinh(x_0-\varepsilon/2)\right].\end{aligned}$$

The solution is sketched in Figure 4.8.

The use of input conductances

The following analysis is useful in understanding certain boundary conditions that are sometimes applied to the cable equation. Consider a cable, for example one with finite length and sealed ends at $x = 0$ and $x = L$. We have seen that the Green's function for the steady state with input at y is

$$G(x;y) = \begin{cases}\cosh x\cosh(L-y)/\sinh L, & x < y,\\ \cosh y\cosh(L-x)/\sinh L, & x \geq y.\end{cases}$$

The derivative of G is thus

$$G'(x;y) = \begin{cases}\sinh x\cosh(L-y)/\sinh L, & x < y,\\ -\cosh y\sinh(L-x)/\sinh L, & x > y.\end{cases}$$

Thus the left-hand derivative at y (or just to the left of y at y^-) is

$$G'(y^-;\, y) = \sinh y \cosh(L-y)/\sinh L,$$

and the right-hand derivative (at y^+) is

$$G'(y^+;\, y) = -\cosh y \sinh(L-y)/\sinh L.$$

Recalling that the axial current is, in the steady state, $(-dV/dx)/\bar{r}_i$, we see that the longitudinal current into the cylinder of length $L-y$ to the right of the point y is then

$$I_i^+ = \cosh y \sinh(L-y)/\bar{r}_i \sinh L,$$

whereas the axial current into the cylinder of length y to the left of the point y is (with allowance for the fact that this is in the direction of decreasing x)

$$I_i^- = \sinh y \cosh(L-y)/\bar{r}_i \sinh L.$$

The ratio of these currents is

$$I_i^+/I_i^- = \coth y/\coth(L-y).$$

From (4.101) the input resistance of a cylinder of length L with a sealed end is $\bar{r}_i \coth L$. Therefore if we denote by R^- and R^+ the input resistances of the segments to the left and right of y and call their reciprocals, the input conductances G^- and G^+, we obtain $I_i^+/I_i^- = G^+/G^-$.

Thus the current injected at the point y divides into two portions, to the right and left, proportional to the input conductances to the right and left of y. Furthermore, using the identity (4.107), we get

$$I_i^+ + I_i^- = 1/\bar{r}_i.$$

Now the total applied current I_A is the integral of the current density $I(x)$ (in amperes per centimeter)

$$I_A = \int_0^l I(x)\, dx$$

which, using capitals for electrotonic distances, becomes on considering $-V'' + V = \delta(X-Y) \equiv r_m I(x)$,

$$I_A = \frac{\lambda}{r_m}\int_0^L \delta(X-Y)\, dX = \frac{\lambda}{r_m} = \frac{1}{\bar{r}_i} = I_i^+ + I_i^-.$$

Thus the total applied current divides into left and right components proportional to the respective input conductances. This is the basis of Equation (4.45).

Furthermore, suppose current to the left is blocked off by virtue of its being a terminus. Then $I_i^+ = 1/\bar{r}_i$. Thus a delta function applied to the membrane induces an axial current of $1/\bar{r}_i$. Therefore it is useful to observe that an axial current that is a delta function at a terminus will give $\bar{r}_i$ times the response to a delta function applied through the membrane.

4.9 The steady-state potential over dendritic trees

Thus far the theory has concerned single nerve cylinders. To mathematically describe the electrophysiology of most nerve cells, it is necessary to take into account their often complicated anatomical structure. As seen in Chapter 1, for example, the spinal motoneuron has about 13 primary dendrites originating at its soma and there are several branchings from each primary dendrite.

We assume here that a cable equation can be applied to each segment of each dendritic tree between branch points. For simplicity, we also assume that the cylinder diameter does not change along a segment, though tapering can be taken into account by formulating the cable equation with a radius that depends on distance along the cylinder. (See Section 6.6).

In most dendritic trees the number of daughter cylinders after a branch point is two. Such structures are called *binary trees* and we will restrict our attention to these. Figure 4.9 indicates the nomenclature we shall employ for such trees. One ending, usually associated with the soma, is called the *origin* and the cylinder emanating from the origin is called the *primary cylinder or trunk*. Points of bifurcation are called *branch points or nodes*. The endings of the tree distal to the origin are called *terminals*. The cylinders prior to the terminals are called *preterminal cylinders*. The *order of branching n* is the maximum

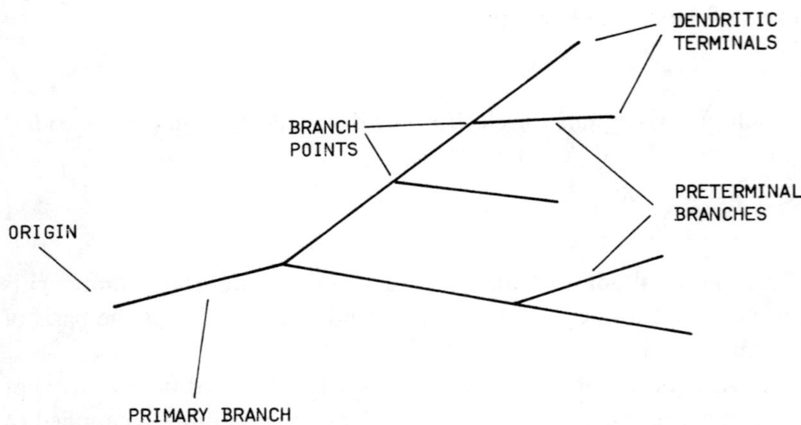

Figure 4.9. Terminology employed in reference to dendritic trees. Here the order of branching is 3.

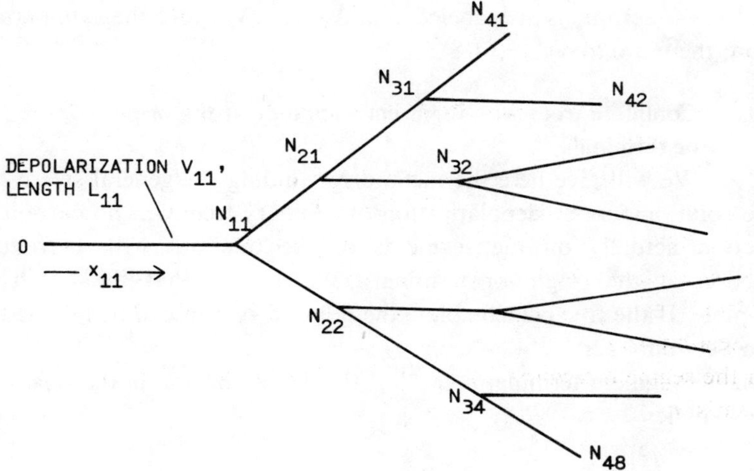

Figure 4.10. Notation to be employed for a complete binary dendritic tree. Here the order of branching is 3.

of the numbers of branch points encountered as one proceeds from origin to terminals.

To the left of the node N_{11} the space coordinate is x_{11} and the total length of the corresponding cylinder is L_{11}. All distances are in terms of the characteristic lengths of the individual cylinders. The depolarization of the primary branch is V_{11} with steady-state value $\tilde{V}_{11}(x_{11})$, but for the rest of this chapter we will drop the tilde on the steady-state values.

In general, we have

The cylindrical segment terminating on the right at N_{jk} has length L_{jk} and on it the distance from the left-hand end is $x_{jk} \in [0, L_{jk}]$, with all distances in characteristic lengths λ_{jk}. On this segment the depolarization has the steady-state value $V_{jk}(x_{jk})$. If the order of branching is n, then $j = 1, 2, \ldots, n+1$ and the corresponding values of k are $k = 1, 2, \ldots, 2^{j-1}$.

A *complete binary tree* is one in which the number of branch points is the same from the origin to each terminal. If this is not the case, the tree is called *incomplete*. In order to introduce additional notation, a complete tree of order 3 is shown in Figure 4.10. The origin is marked 0. The first branch point is N_{11} and the cylinder to the left of N_{11} is called cylinder (1,1) – the primary branch. At the second-order branch points the nodes are N_{21} and N_{22} and the cylinders preceding them are (2,1) and (2,2). The third-order branch points are N_{31}, N_{32}, N_{33}, N_{34} with cylinders (3,1), (3,2), (3,3), and (3,4) preceding them.

Finally, the terminals are labeled $N_{41}, N_{42}, \ldots, N_{48}$, and the cylinders before them are $(4,1), \ldots, (4,8)$.

4.10 Complete trees with input currents only at the origin or terminals

We will give here the method for finding the general steady-state solution for the depolarization of a binary tree with no current injection actually on the segments of the tree but with current injection at the origin or terminals. We assume that the tree is complete. If the tree is incomplete the reduced system can be handled as in Section 4.12.

On the segment terminating at N_{jk}, the depolarization in the steady state satisfies

$$-\frac{d^2V_{jk}}{dx_{jk}^2} + V_{jk} = 0, \qquad 0 < x_{jk} < L_{jk}. \tag{4.135}$$

As seen above, the general solution of these equations may be written as

$$V_{jk}(x_{jk}) = c_{jk}\cosh x_{jk} + d_{jk}\sinh x_{jk}. \tag{4.136}$$

The problem is to find all the constants c_{jk} and d_{jk}. If the order of branching is n, then the number of constants to be determined is $N = 2[1 + 2 + \cdots + 2^n]$, or

$$N = 2[2^{n+1} - 1], \qquad n = 0, 1, 2, \ldots. \tag{4.137}$$

Boundary conditions

In order to determine the N constants, we will need N linearly independent equations. These are obtained as follows.

Condition at the origin. The boundary condition at the origin may be written in the general form

$$\alpha_{11}V_{11}(0) + \beta_{11}V_{11}'(0) = \gamma_{11}, \tag{4.138}$$

where α_{11}, β_{11}, and γ_{11} are constants that depend on the choice of condition at the origin. This form covers the possibilities of current injection, voltage clamp (including killed end), sealed end, a mixed condition, or a lumped-soma termination with input current that has reached a steady-state value. From (4.136),

$$V_{jk}'(x_{jk}) = c_{jk}\sinh x_{jk} + d_{jk}\cosh x_{jk}. \tag{4.139}$$

Since $V_{11}(0) = c_{11}$ and $V_{11}'(0) = d_{11}$, Equation (4.138) can be written as

$$\alpha_{11}c_{11} + \beta_{11}d_{11} = \gamma_{11}. \tag{4.140}$$

Conditions at the branch points. At each branch point there are three conditions that arise from the requirements of continuity of the electric potential and conservation of current. The first two conditions arise because discontinuous potentials can only arise from infinite currents, which are excluded on physical grounds. The other condition states that at a branch point, the axial current from the previous cylinder all goes into the two cylinders beyond the branch point.

At the node N_{jk}, the *condition of continuity* of potential requires that the potential at $x_{jk} = L_{jk}$ in the (j, k)-cylinder equals the potential at $x_{j+1,2k-1} = 0$ and $x_{j+1,2k} = 0$ in the daughter cylinders $(j+1, 2k-1)$ and $(j+1, 2k)$. Thus we have

$$\begin{aligned} V_{jk}(L_{jk}) &= V_{j+1,2k-1}(0), \\ V_{jk}(L_{jk}) &= V_{j+1,2k}(0). \end{aligned} \tag{4.141}$$

Here $j = 1, 2, \dots, n$ and the corresponding k-values are $k = 1, 2, \dots, 2^{j-1}$. There are $2[1 + 2 + \cdots + 2^{n-1}] = 2[2^n - 1]$ such equations. Using (4.136) and (4.139) these become

$$\begin{aligned} c_{jk}\cosh L_{jk} + d_{jk}\sinh L_{jk} - c_{j+1,2k-1} &= 0, \\ c_{jk}\cosh L_{jk} + d_{jk}\sinh L_{jk} - c_{j+1,2k} &= 0, \end{aligned} \tag{4.142}$$

where $j = 1, 2, \dots, n$ and $k = 1, 2, \dots, 2^{j-1}$.

The axial current at the point x_{jk} on cylinder (j, k) is, in the steady state,

$$I_{i,jk}(x_{jk}) = -\frac{1}{\bar{r}_{i,jk}} \frac{dV_{jk}(x_{jk})}{dx_{jk}} \tag{4.143}$$

where $\bar{r}_{i,jk}$ is the internal resistance of a characteristic length (λ_{jk}) of cylinder (j, k). For *conservation of current* we require

$$I_{i,jk}(L_{jk}) = I_{i;\,j+1,2k-1}(0) + I_{i;\,j+1,2k}(0), \tag{4.144}$$

where $j = 1, 2, \dots, n$ and $k = 1, 2, \dots, 2^{j-1}$. Using (4.139) for the derivative, these equations become

$$\begin{aligned} &\bar{r}_{i,jk}^{-1}\left[c_{jk}\sinh L_{jk} + d_{jk}\cosh L_{jk}\right] \\ &\quad = \bar{r}_{i;\,j+1,2k-1}^{-1} d_{j+1,2k-1} + \bar{r}_{i;\,j+1,2k}^{-1} d_{j+1,2k}. \end{aligned} \tag{4.145}$$

We set, for $j = 1, 2, \dots, n$ and $k = 1, 2, \dots, 2^{j-1}$,

$$\kappa_{jk} = \sinh(L_{jk})/\bar{r}_{i,jk}, \tag{4.146}$$

$$\kappa_{jk}^{*} = \cosh(L_{jk})/\bar{r}_{i,jk}, \tag{4.147}$$

$$\sigma_{j+1,2k-1} = 1/\bar{r}_{i;\,j+1,2k-1}, \tag{4.148}$$

$$\sigma_{j+1,2k} = 1/\bar{r}_{i;\,j+1,2k}, \tag{4.149}$$

so that Equations (4.145) become

$$\kappa_{jk}c_{jk} + \kappa_{jk}^{*}d_{jk} - \sigma_{j+1,2k-1}d_{j+1,2k-1} - \sigma_{j+1,2k}d_{j+1,2k} = 0, \tag{4.150}$$

where $j = 1, 2, \dots, n$ and $k = 1, 2, \dots, 2^{j-1}$. There are $2^n - 1$ such equations.

Table 4.4. *The number of equations and constants (N) for the complete dendritic tree, steady-state equations*

n	N
0	2
1	6
2	14
3	30
4	62
5	126
6	254

Conditions at the terminals. We allow for the possibility of a general boundary at each terminal. Thus there are 2^n equations of the form:

$$\alpha_{n+1,k}V_{n+1,k}(L_{n+1,k}) + \beta_{n+1,k}V'_{n+1,k}(L_{n+1,k}) = \gamma_{n+1,k}, \tag{4.151}$$

where the α's, β's, and γ's are constants. Using (4.136) and (4.139), these may be written as

$$\alpha^*_{n+1,k}c_{n+1,k} + \beta^*_{n+1,k}d_{n+1,k} = \gamma_{n+1,k}, \qquad k = 1,2,\ldots,2^n, \tag{4.152}$$

where

$$\alpha^*_{n+1,k} = \alpha_{n+1,k}\cosh(L_{n+1,k}) + \beta_{n+1,k}\sinh(L_{n+1,k}), \tag{4.153}$$

$$\beta^*_{n+1,k} = \alpha_{n+1,k}\sinh(L_{n+1,k}) + \beta_{n+1,k}\cosh(L_{n+1,k}). \tag{4.154}$$

Altogether we have for Equations (4.140), (4.142), (4.150), and (4.152), a total of $1 + 2[2^n - 1] + 2^n - 1 + 2^n = 4 \cdot 2^n - 2 = 2[2^{n+1} - 1] = N$ equations for the N unknown constants. The values of N for various n are given in Table 4.4. [However, one-third of the nodal equations may be dropped, as from (4.142), $c_{j+1,2k-1} = c_{j+1,2k}$, $j = 1,2,\ldots,n$ and $k = 1,2,\ldots,2^{j-1}$.] Note that in the most extensively arborized dendritic trees of Purkinje cells of the cerebellum, orders of branching rarely exceed $n = 5$.

Form of the equations

The N equations for the N unknown constants may be written in the standard form for a linear system,

$$\mathbf{Ax} = \mathbf{b}, \tag{4.155}$$

where $\mathbf{A}$ is an $N \times N$ matrix, $\mathbf{x}$ is an N-vector whose entries are the unknown constants, and $\mathbf{b}$ is an N-vector. If we set

$$\mathbf{x} = [c_{11}c_{21} \cdots c_{n+1,2^n}d_{11}d_{21} \cdots d_{n+1,2^n}]^{\mathrm{T}}, \tag{4.156}$$

where T denotes transpose, then the structure of the matrix **A** is

$$\mathbf{A} \sim \left[\begin{array}{c} \text{boundary conditions at the origin} \\ \hdashline 3[2^n-1] \text{ nodal boundary conditions} \\ \hdashline 2^n \text{ terminal boundary conditions} \\ \hdashline \end{array}\right]. \tag{4.157}$$

Examples and methods of solution

(i) *Explicit solution when $n=1$.* When there is one order of branching there are six unknown constants and the linear system (4.155) is, setting $C_{ij} = \cosh L_{ij}$ and $S_{ij} = \sinh L_{ij}$,

$$\begin{bmatrix} \alpha_{11} & 0 & 0 & \beta_{11} & 0 & 0 \\ C_{11} & -1 & 0 & S_{11} & 0 & 0 \\ C_{11} & 0 & -1 & S_{11} & 0 & 0 \\ \kappa_{11} & 0 & 0 & \kappa_{11}^* & -\sigma_{21} & -\sigma_{22} \\ 0 & \alpha_{21}^* & 0 & 0 & \beta_{21}^* & 0 \\ 0 & 0 & \alpha_{22}^* & 0 & 0 & \beta_{22}^* \end{bmatrix} \begin{bmatrix} x_1 \\ x_2 \\ x_3 \\ x_4 \\ x_5 \\ x_6 \end{bmatrix} = \begin{bmatrix} \gamma_{11} \\ 0 \\ 0 \\ 0 \\ \gamma_{21} \\ \gamma_{22} \end{bmatrix}. \tag{4.158}$$

To be specific, let us consider the case of current injection of magnitude I_0 at the origin with killed ends at the terminals as in Figure 4.11. At the origin we have $V_{11}'(0) = -\bar{r}_{11} I_0 = \gamma_{11}$, so $\alpha_{11} = 0$ and $\beta_{11} = 1$. Since we require zero depolarization at the terminals, the boundary conditions there are $V_{21}(L_{21}) = V_{22}(L_{22}) = 0$. Hence from (4.151), $\alpha_{21} = \alpha_{22} = 1$, $\beta_{21} = \beta_{22} = 0$, and $\gamma_{21} = \gamma_{22} = 0$. Hence

$$\begin{aligned} \alpha_{21}^* &= \cosh L_{21} \doteq C_{21}, \\ \alpha_{22}^* &= \cosh L_{22} \doteq C_{22}, \\ \beta_{21}^* &= \sinh L_{21} \doteq S_{21}, \\ \beta_{22}^* &= \sinh L_{22} \doteq S_{22}. \end{aligned}$$

Thus the system to be solved is

$$\begin{bmatrix} 0 & 0 & 0 & 1 & 0 & 0 \\ C_{11} & -1 & 0 & S_{11} & 0 & 0 \\ C_{11} & 0 & -1 & S_{11} & 0 & 0 \\ \kappa_{11} & 0 & 0 & \kappa_{11}^* & -\sigma_{21} & -\sigma_{22} \\ 0 & C_{21} & 0 & 0 & S_{21} & 0 \\ 0 & 0 & C_{22} & 0 & 0 & S_{22} \end{bmatrix} \begin{bmatrix} x_1 \\ x_2 \\ x_3 \\ x_4 \\ x_5 \\ x_6 \end{bmatrix} = \begin{bmatrix} \gamma_{11} \\ 0 \\ 0 \\ 0 \\ 0 \\ 0 \end{bmatrix}. \tag{4.159}$$

From the first equations,

$$x_4 = \gamma_{11}.$$

We note that $x_2 = x_3$ and the second and fourth equations give

$$C_{11} x_1 - x_2 + S_{11}\gamma_{11} = 0,$$

$$\kappa_{11} x_1 + \kappa_{11}^* \gamma_{11} - \sigma_{21} x_5 - \sigma_{22} x_6 = 0.$$

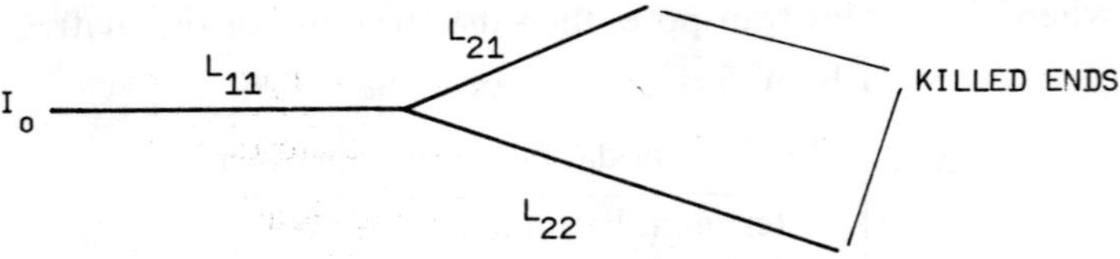

Figure 4.11.

Since, from the last two equations,

$$x_5 = -C_{21}x_2/S_{21},$$
$$x_6 = -C_{22}x_2/S_{22},$$

we have two equations in two unknowns

$$x_2 = C_{11}x_1 + S_{11}\gamma_{11},$$
$$\kappa_{11}x_1 + \kappa_{11}^*\gamma_{11} + \sigma_{21}C_{21}x_2/S_{21} + \sigma_{22}C_{22}x_2/S_{22} = 0.$$

These are solved for x_1 and x_2 whereupon all the constants may be found. Then, on cylinder (j, k),

$$V_{jk}(x_{jk}) = c_{jk}\cosh(x_{jk}) + d_{jk}\sinh(x_{jk}),$$

where

$$c_{11} = \bar{r}_{11}I_0 \frac{\left[\dfrac{1}{\bar{r}_{11}} + \tanh(L_{11})\left\{\dfrac{\coth(L_{21})}{\bar{r}_{21}} + \dfrac{\coth(L_{22})}{\bar{r}_{22}}\right\}\right]}{\left[\dfrac{\tanh(L_{11})}{\bar{r}_{11}} + \left\{\dfrac{\coth(L_{21})}{\bar{r}_{21}} + \dfrac{\coth(L_{22})}{\bar{r}_{22}}\right\}\right]},$$

$$d_{11} = -\bar{r}_{11}I_0, \tag{4.160}$$
$$c_{21} = -\bar{r}_{11}I_0\sinh(L_{11}) + \cosh(L_{11})c_{11},$$
$$d_{21} = -\coth(L_{21})c_{21},$$
$$c_{22} = c_{21},$$
$$d_{22} = -\coth(L_{22})c_{21}.$$

The voltage at the origin is c_{11}, so the "input resistance" of the tree (as seen from the origin) is

$$R_{\text{killed},\, n=1} = \frac{\bar{r}_{11}\left[\dfrac{1}{\bar{r}_{11}} + \tanh(L_{11})\left\{\dfrac{\coth(L_{21})}{\bar{r}_{21}} + \dfrac{\coth(L_{22})}{\bar{r}_{22}}\right\}\right]}{\left[\dfrac{\tanh(L_{11})}{\bar{r}_{11}} + \left\{\dfrac{\coth(L_{21})}{\bar{r}_{21}} + \dfrac{\coth(L_{22})}{\bar{r}_{22}}\right\}\right]}. \tag{4.161}$$

In the limit as the lengths of the daughter cylinders (2,1) and (2,2) approach zero, we have

$$R_{\text{killed},\, n=1} \to R_{\text{killed, n=0}} = \bar{r}_{11}\tanh(L_{11}), \tag{4.162}$$

as it must. Also, as $L_{11} \to \infty$,

$$R_{\text{killed},\, n=1} \to R_{\text{semiinfinite}} = \bar{r}_{11}. \tag{4.163}$$

Note that by suitable renaming of variables, the above solution covers current injection at either terminal with a killed origin and a killed other terminal.

(ii) *Form of the system when* $n = 2$. When $n = 2$ (two orders of branching) the linear system (4.157) has $N = 14$ equations. The system has the form

$$\begin{bmatrix}
\alpha_{11} & 0 & 0 & 0 & 0 & 0 & 0 & \beta_{11} & 0 & 0 & 0 & 0 & 0 & 0 \\
C_{11} & -1 & 0 & 0 & 0 & 0 & 0 & S_{11} & 0 & 0 & 0 & 0 & 0 & 0 \\
C_{11} & 0 & -1 & 0 & 0 & 0 & 0 & S_{11} & 0 & 0 & 0 & 0 & 0 & 0 \\
\kappa_{11} & 0 & 0 & 0 & 0 & 0 & 0 & \kappa_{11}^* & -\sigma_{21} & -\sigma_{22} & 0 & 0 & 0 & 0 \\
0 & C_{21} & 0 & -1 & 0 & 0 & 0 & 0 & S_{21} & 0 & 0 & 0 & 0 & 0 \\
0 & C_{21} & 0 & 0 & -1 & 0 & 0 & 0 & S_{21} & 0 & 0 & 0 & 0 & 0 \\
0 & \kappa_{21} & 0 & 0 & 0 & 0 & 0 & 0 & \kappa_{21}^* & 0 & -\sigma_{31} & -\sigma_{32} & 0 & 0 \\
0 & 0 & C_{22} & 0 & 0 & -1 & 0 & 0 & 0 & S_{22} & 0 & 0 & 0 & 0 \\
0 & 0 & C_{22} & 0 & 0 & 0 & -1 & 0 & 0 & S_{22} & 0 & 0 & 0 & 0 \\
0 & 0 & \kappa_{22} & 0 & 0 & 0 & 0 & 0 & 0 & \kappa_{22}^* & 0 & 0 & -\sigma_{33} & -\sigma_{34} \\
0 & 0 & 0 & \alpha_{31}^* & 0 & 0 & 0 & 0 & 0 & 0 & \beta_{31}^* & 0 & 0 & 0 \\
0 & 0 & 0 & 0 & \alpha_{32}^* & 0 & 0 & 0 & 0 & 0 & 0 & \beta_{32}^* & 0 & 0 \\
0 & 0 & 0 & 0 & 0 & \alpha_{33}^* & 0 & 0 & 0 & 0 & 0 & 0 & \beta_{33}^* & 0 \\
0 & 0 & 0 & 0 & 0 & 0 & \alpha_{34}^* & 0 & 0 & 0 & 0 & 0 & 0 & \beta_{34}^*
\end{bmatrix}$$

$$\times \begin{bmatrix} x_1 \\ x_2 \\ x_3 \\ x_4 \\ x_5 \\ x_6 \\ x_7 \\ x_8 \\ x_9 \\ x_{10} \\ x_{11} \\ x_{12} \\ x_{13} \\ x_{14} \end{bmatrix} = \begin{bmatrix} \gamma_{11} \\ 0 \\ 0 \\ 0 \\ 0 \\ 0 \\ 0 \\ 0 \\ 0 \\ 0 \\ \gamma_{31} \\ \gamma_{32} \\ \gamma_{33} \\ \gamma_{34} \end{bmatrix}, \tag{4.164}$$

where again we have used the abbreviation $C_{ij} = \cosh L_{ij}$ and $S_{ij} = \sinh L_{ij}$. From the form of the system for $n = 2$, the form for larger values of n can be written down by following the pattern of the nonzero entries in the coefficient matrix.

Methods of solution

(1) *Elimination methods.* The system (4.164) can be solved by Gaussian elimination. In this sytematic procedure, the first equation is used to eliminate x_1 from the second, third,... . Then the second equation is used to eliminate x_2 from the third, fourth, and so on, etc. equations. This procedure is continued until the matrix of coefficients is in upper triangular form (i.e., all elements below the principal diagonal are zeros).

The Gaussian elimination method may seem tedious by hand calculation. However, the 14×14 matrix of (4.164) can be quickly reduced to a 7×7 coefficient matrix, by utilizing the following equations:

$$x_2 = x_3,$$

$$x_4 = x_5,$$

$$x_6 = x_7,$$

$$x_{10+j} = [\gamma_{3j} - \alpha_{3j} x_{3+j}]/\beta_{3j}, \qquad j = 1,\dots,4.$$

We then get

$$\begin{bmatrix} \alpha_{11} & 0 & 0 & 0 & \beta_{11} & 0 & 0 \\ C_{11} & -1 & 0 & 0 & S_{11} & 0 & 0 \\ 0 & C_{21} & -1 & 0 & 0 & S_{21} & 0 \\ 0 & C_{22} & 0 & -1 & 0 & 0 & S_{22} \\ \kappa_{11} & 0 & 0 & 0 & \kappa_{11}^* & -\sigma_{21} & -\sigma_{22} \\ 0 & \kappa_{21} & \tau_1 & 0 & 0 & \kappa_{21}^* & 0 \\ 0 & \kappa_{22} & 0 & \tau_2 & 0 & 0 & \kappa_{22}^* \end{bmatrix} \begin{bmatrix} x_1 \\ x_2 \\ x_4 \\ x_6 \\ x_8 \\ x_9 \\ x_{10} \end{bmatrix} = \begin{bmatrix} \gamma_{11} \\ 0 \\ 0 \\ 0 \\ 0 \\ \sigma_1 \\ \sigma_2 \end{bmatrix},$$

where

$$\tau_1 = -(\sigma_{31}\beta_1 + \sigma_{32}\beta_2),$$

$$\tau_2 = -(\sigma_{33}\beta_3 + \sigma_{34}\beta_4),$$

$$\sigma_1 = \sigma_{31}\alpha_1 + \sigma_{32}\alpha_2,$$

$$\sigma_2 = \sigma_{33}\alpha_3 + \sigma_{34}\alpha_4,$$

$$\alpha_j = \gamma_{3j}/\beta_{3j}, \qquad j = 1,\dots,4,$$

$$\beta_j = -\alpha_{3j}/\beta_{3j}, \qquad j = 1,\dots,4.$$

The depolarization at the soma is

$$x_1 = (\gamma_{11} - \beta_{11} x_8)/\alpha_{11},$$

where

$$x_8 = \frac{\gamma_2 + \sigma_{21}x_9 + \sigma_{22}x_{10}}{\kappa_{11}^{**}},$$

$$x_9 = \frac{\nu_2 - \rho_1\sigma_{22}x_{10}}{\nu_1},$$

$$x_{10} = \frac{\sigma_2^* + \tau_2 C_{22}\gamma_1 - \rho_2\gamma_2 - \dfrac{\rho_2\sigma_{21}\nu_2}{\nu_1}}{\kappa_{22}^* + \tau_2 S_{22} + \rho_2\sigma_{22} - \dfrac{\rho_2\sigma_{21}\rho_1\sigma_{22}}{\nu_1}},$$

$$\gamma_1 = -\frac{C_{11}\gamma_{11}}{\alpha_{11}},$$

$$\gamma_2 = -\frac{\kappa_{11}\gamma_{11}}{\alpha_{11}},$$

$$\kappa_{11}^{**} = \kappa_{11}^* - \frac{\kappa_{11}\beta_{11}}{\alpha_{11}},$$

$$\sigma_1^* = \sigma_1 + \kappa_{21}\gamma_1,$$
$$\sigma_2^* = \sigma_2 + \kappa_{22}\gamma_1,$$

$$S_{11}^* = S_{11} - \frac{C_{11}\beta_{11}}{\alpha_{11}},$$

$$\rho_1 = (\kappa_{21} + \tau_1 C_{21})\frac{S_{11}^*}{\kappa_{11}^{**}},$$

$$\rho_2 = (\kappa_{22} + \tau_2 C_{22})\frac{S_{11}^*}{\kappa_{11}^{**}},$$

$$\nu_1 = \kappa_{21}^* + \tau_1 S_{21} + \sigma_{21}\rho_1,$$
$$\nu_2 = \sigma_1^* + \tau_1 C_{21}\gamma_1 - \rho_1\gamma_2.$$

When there are more than two orders of branching, it will be impracticable to perform Gaussian elimination by hand. However, most computer centers will have in their software libraries, routines that will perform Gaussian elimination, incorporating such refinements as pivoting in order to obtain the most accurate solutions.

The results of two worked examples are shown in Figures 4.12A and 4.12B. Figure 4.12A shows the steady-state potential over a dendritic tree with one order of branching when there is current injection at the soma, in the cases of both sealed and killed terminals. The physical lengths of the cylinders are $l_{11} = 200$, $l_{21} = 100$, $l_{22} = 200$ μm, with corresponding diameters $d_{11} = 10$, $d_{21} = d_{22} = 6$ μm. The electrical constants are $\rho_i = 70\ \Omega$ cm, $R_m = 2500\ \Omega\ \text{cm}^2$, and $I_0 = 5 \times 10^{-4}$ A. Note the drastic effect of the killed terminals on the voltage.

In Figure 4.12B is shown the steady-state voltage over a dendritic tree with three orders of branching when there is current injection at the soma and the dendritic terminals are sealed. The anatomical details of the tree are shown in the lower part of the figure and are based in part on the spinal motoneuron data of Figure 1.6.

(2) *Iterative methods.* It can be shown [see, for example, Steinberg (1974)] that when the coefficient matrix of a linear system contains a large proportion of zero elements (in which case the matrix is called *sparse*), the iterative methods may be more efficient than Gaussian elimination in the sense that the number of arithmetic operations to achieve a solution of desired accuracy is less. One such iterative method is called *Jacobi iteration*. To solve the system (4.164) by this procedure, we first rewrite the system such that each unknown is the subject of its own equation. For example, the first four equations are now

$$\begin{aligned} x_1 &= (\gamma_{11} - \beta_{11}x_8)/\alpha_{11}, \\ x_2 &= C_{11}x_1 + S_{11}x_8, \\ x_3 &= C_{11}x_1 + S_{11}x_8, \\ x_4 &= C_{21}x_2 + S_{21}x_9. \end{aligned} \tag{4.165}$$

Figure 4.12A. Steady-state voltage distribution for a binary tree with one order of branching and current injection at the soma. The top set of results is for sealed terminals, and the bottom set for killed terminals. For parameter values see the text.

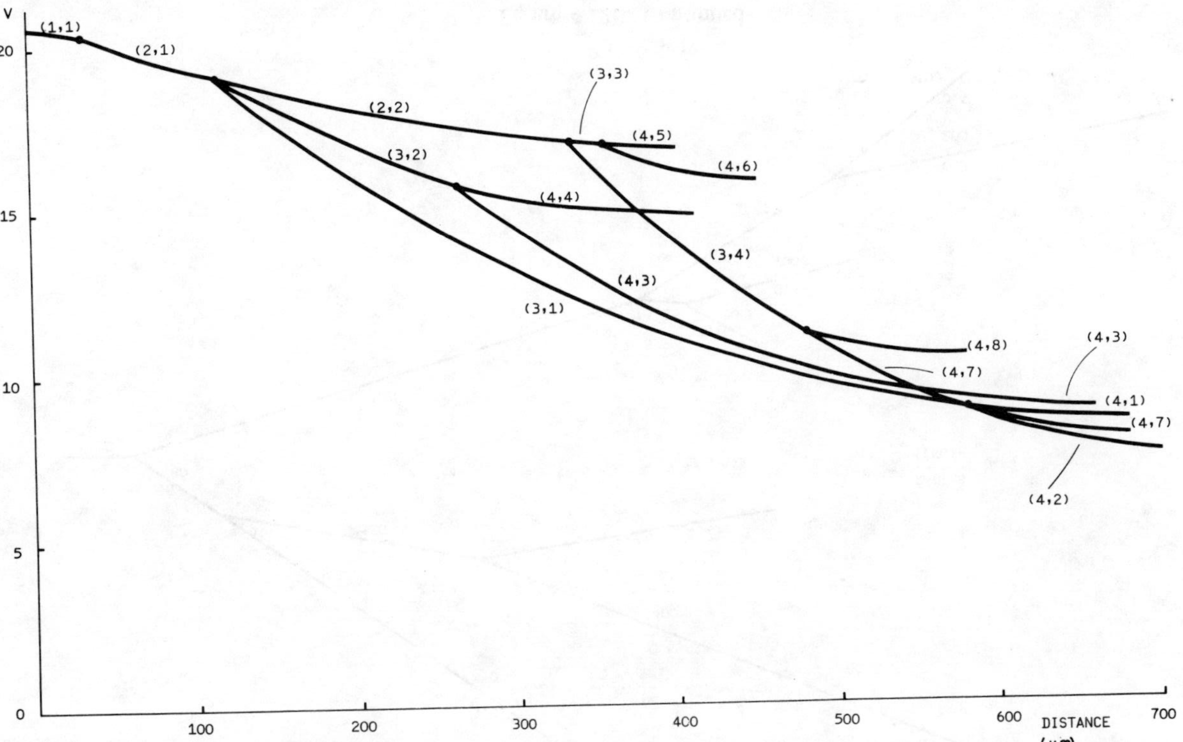

Figure 4.12B. Steady-state voltage over a dendritic tree with three orders of branching. The tree is sketched in the lower part of the figure. The pairs of numbers are the lengths and radii of the dendritic segments. Solution obtained by Gaussian elimination via a computer library routine.

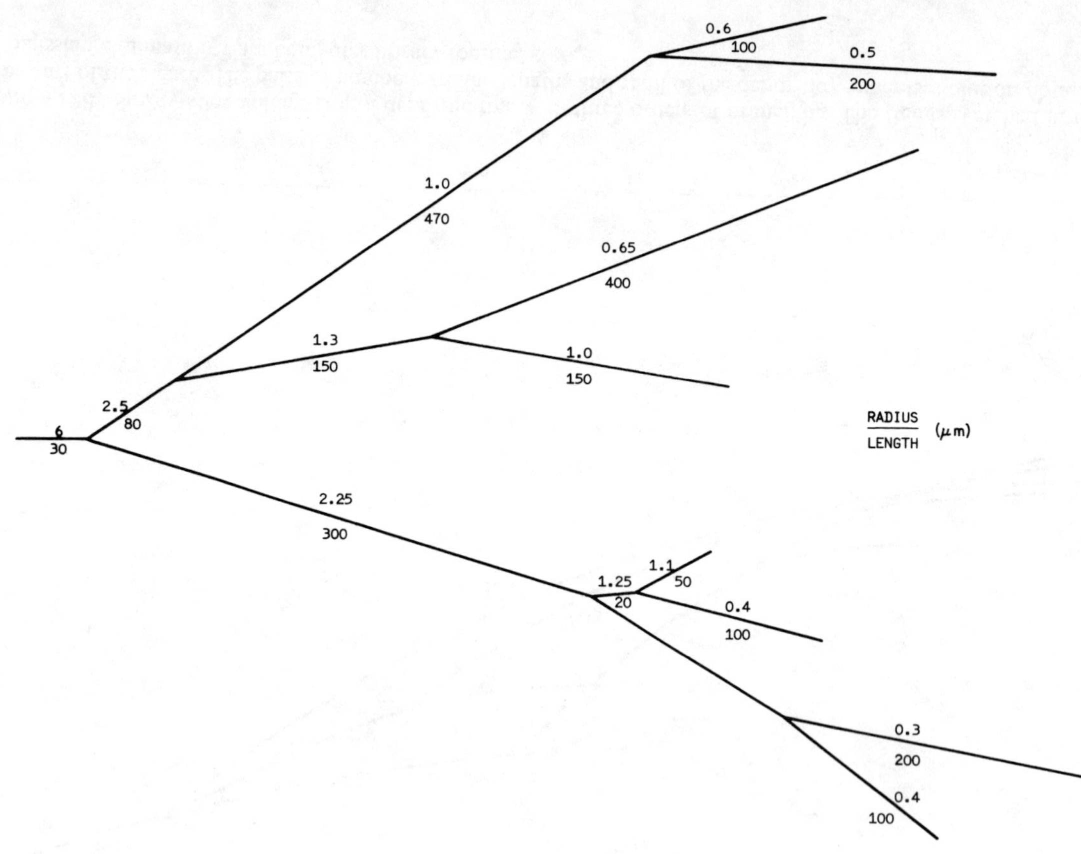

Figure 4.12B. Continued

We now make an initial guess at each x_i, $i=1,\ldots,14$, which we denote by $x_i^{(0)}$. The first iteration calculates $x_i^{(1)}$ by substituting the values of $x_i^{(0)}$ in the right-hand sides of Equations (4.165). Thus the first equation yields

$$x_1^{(1)}=\left(\gamma_{11}-\beta_{11}x_8^{(0)}\right)/\alpha_{11}. \tag{4.166}$$

The procedure is repeated (for each equation) and on the kth iteration we have, for example,

$$x_1^{(k)}=\left(\gamma_{11}-\beta_{11}x_8^{(k-1)}\right)/\alpha_{11}. \tag{4.167}$$

The iterative procedure is continued until the $x_i^{(k)}$'s differ by a prescribed amount from the $x_i^{(k-1)}$'s. A sufficient, but not necessary, condition for the iterature procedure to converge to the solution is that the coefficient matrix be *diagonally dominant*. That is, the absolute value of each principal diagonal element is greater than the sum of the absolute values of the remaining elements in the corresponding row. This condition may sometimes be fulfilled by suitable rearrangement of the system of equations. However, the system (4.164) is not diagonally dominant even when rearranged to make all principal diagonal elements nonzero.

(iii) *Symmetric trees.* When all the daughter cylinders at a given order of branching are identical and boundary conditions at all the terminals are the same, a great reduction in the number of unknown constants and equations occurs. In general, for a tree with n orders of branching, the number of constants to be determined is only $2[n+1]$. For example, when $n=2$, the only constants that need be determined are c_{11}, d_{11}, c_{21}, d_{21}, c_{31} and d_{31}. The system to be solved has only the six equations

$$\begin{bmatrix} \alpha_{11} & 0 & 0 & \beta_{11} & 0 & 0 \\ C_{11} & -1 & 0 & 0 & S_{11} & 0 \\ \kappa_{11} & 0 & 0 & \kappa_{11}^{*} & -2\sigma_{21} & 0 \\ 0 & C_{21} & -1 & 0 & S_{21} & 0 \\ 0 & \kappa_{21} & 0 & 0 & \kappa_{21}^{*} & -2\sigma_{31} \\ 0 & 0 & \alpha_{31}^{*} & 0 & 0 & \beta_{31}^{*} \end{bmatrix} \begin{bmatrix} c_{11} \\ c_{21} \\ c_{31} \\ d_{11} \\ d_{21} \\ d_{31} \end{bmatrix} = \begin{bmatrix} \gamma_{11} \\ 0 \\ 0 \\ 0 \\ 0 \\ \gamma_{31} \end{bmatrix}. \tag{4.168}$$

For the case of current injection of magnitude I_0 at the origin and killed-end conditions at the terminals this becomes

$$\begin{bmatrix} 0 & 0 & 0 & 1 & 0 & 0 \\ C_{11} & -1 & 0 & S_{11} & 0 & 0 \\ \kappa_{11} & 0 & 0 & \kappa_{11}^{*} & -2\sigma_{21} & 0 \\ 0 & C_{21} & -1 & 0 & S_{21} & 0 \\ 0 & \kappa_{21} & 0 & 0 & \kappa_{21}^{*} & -2\sigma_{31} \\ 0 & 0 & C_{31} & 0 & 0 & S_{31} \end{bmatrix} \begin{bmatrix} c_{11} \\ c_{21} \\ c_{31} \\ d_{11} \\ d_{21} \\ d_{31} \end{bmatrix} = \begin{bmatrix} \gamma_{11} \\ 0 \\ 0 \\ 0 \\ 0 \\ 0 \end{bmatrix}, \tag{4.169}$$

where $\gamma_{11} = -\bar{r}_{11} I_0$. This system is solved by Gaussian elimination to yield the solution

$$d_{31} = \frac{-\bar{r}_{31} I_0 \cosh(L_{11})\cosh(L_{21})\left[1 - \tanh^2(L_{11})\right]\left[1 - \tanh^2(L_{21})\right]}{\left[2 + \dfrac{\bar{r}_{21}}{\bar{r}_{11}}\tanh(L_{11})\tanh(L_{21})\right]\left[2 + \dfrac{\bar{r}_{31}}{\bar{r}_{21}}\tanh(L_{21})\tanh(L_{31})\right]} \tag{4.170}$$

$$d_{21} = \frac{-\bar{r}_{21} I_0 \cosh(L_{11})(1 - \tanh^2(L_{11}))}{\left[2 + \dfrac{\bar{r}_{21}}{\bar{r}_{11}}\tanh(L_{11})\tanh(L_{21})\right]}, \tag{4.171}$$

$$d_{11} = -\bar{r}_{11} I_0, \tag{4.172}$$

$$c_{31} = -\tanh L_{31} d_{31}, \tag{4.173}$$

$$c_{21} = \operatorname{sech} L_{21} c_{31} - \tanh L_{21} d_{21}, \tag{4.174}$$

$$c_{11} = \operatorname{sech} L_{11} c_{21} - \tanh {}_{11} d_{11}. \tag{4.175}$$

The input resistance is then c_{11}/I_0, which can be found from the above formulas.

4.11 Complete trees with inputs along the cylinders and/or at the terminals and origin

If inputs occur along the cylinders that comprise the dendritic tree, then Equations (4.135) on cylinder (j, k) become nonhomogeneous; that is,

$$-\frac{d^2 V_{jk}}{dx_{jk}^2} + V_{jk} = I_{jk}(x_{jk}), \qquad 0 < x_{jk} < L_{jk}, \tag{4.176}$$

where $I_{jk}(x_{jk})$ represents the steady-state current density. We saw in Section 4.6 that the general solution of (4.176) can be written as

$$V_{jk}(x_{jk}) = c_{jk}\cosh x_{jk} + d_{jk}\sinh x_{jk} + P_{jk}(x_{jk}), \tag{4.177}$$

where the particular solution is

$$P_{jk}(x_{jk}) = \tfrac{1}{2}\left[e^{-x_{jk}}\int^{x_{jk}} e^{t} I_{jk}(t)\,dt - e^{x_{jk}}\int^{x_{jk}} e^{-t} I_{jk}(t)\,dt\right]. \tag{4.178}$$

For the complete tree with n orders of branching, $j = 1, 2, \ldots, n$ and $k = 1, 2, \ldots, 2^{j-1}$. We use the notation

$$\begin{aligned} P_{jk}(0) &= p_{jk}, \\ P'_{jk}(0) &= q_{jk}, \\ P_{jk}(L_{jk}) &= p^*_{jk}, \\ P'_{jk}(L_{jk}) &= q^*_{jk}. \end{aligned} \tag{4.179}$$

Then the boundary conditions all have extra terms because now

$$\begin{aligned} V_{jk}(0) &= c_{jk} + p_{jk}, \\ V'_{jk}(0) &= d_{jk} + q_{jk}, \\ V_{jk}(L_{jk}) &= c_{jk}\cosh L_{jk} + d_{jk}\sinh L_{jk} + p^*_{jk}, \\ V'_{jk}(L_{jk}) &= c_{jk}\sinh L_{jk} + d_{jk}\cosh L_{jk} + q^*_{jk}. \end{aligned} \tag{4.180}$$

We therefore have slightly modified equations resulting from the boundary conditions.

Boundary condition at the origin

We use the same form for the boundary condition, namely (4.138),

$$\alpha_{11} V_{11}(0) + \beta_{11} V'_{11}(0) = \gamma_{11}, \tag{4.181}$$

to get

$$\alpha_{11} c_{11} + \beta_{11} d_{11} = \gamma_{11} - \alpha_{11} p_{11} - \beta_{11} q_{11}. \tag{4.182}$$

Boundary conditions at the branch points

Continuity of the depolarization at the node N_{jk} gives, as before,

$$\begin{aligned} V_{jk}(L_{jk}) &= V_{j+1,2k-1}(0), \\ V_{jk}(L_{jk}) &= V_{j+1,2k}(0), \end{aligned} \tag{4.183}$$

or now

$$\begin{aligned} c_{jk}\cosh L_{jk} + d_{jk}\sinh L_{jk} + p^*_{jk} &= c_{j+1,2k-1} + p_{j+1,2k-1}, \\ c_{jk}\cosh L_{jk} + d_{jk}\sinh L_{jk} + p^*_{jk} &= c_{j+1,2k} + p_{j+1,2k}. \end{aligned} \tag{4.184}$$

These may be written as

$$\begin{aligned} c_{jk}\cosh L_{jk} + d_{jk}\sinh L_{jk} - c_{j+1,2k-1} &= p_{j+1,2k-1} - p^*_{jk}, \\ c_{jk}\cosh L_{jk} + d_{jk}\sinh L_{jk} - c_{j+1,2k} &= p_{j+1,2k} - p^*_{jk}, \end{aligned} \tag{4.185}$$

where $j = 1, 2, \ldots, n$ and $k = 1, 2, \ldots, 2^{j-1}$.

The conservation of axial current at branch points yields

$$\begin{aligned} \left[c_{jk}\sinh L_{jk} + d_{jk}\cosh L_{jk} + q^*_{jk}\right]/\bar{r}_{jk} &= \left[d_{j+1,2k-1} + q_{j+1,2k-1}\right]/\bar{r}_{j+1,2k-1} \\ &\quad + \left[d_{j+1,2k} + q_{j+1,2k}\right]/\bar{r}_{j+1,2k}. \end{aligned} \tag{4.186}$$

These equations may be written, using the same definitions of the κ's and σ's [(4.146)–(4.149)],

$$\begin{aligned} \kappa_{jk}c_{jk} + \kappa^*_{jk}d_{jk} - \sigma_{j+1,2k-1}d_{j+1,2k-1} - \sigma_{j+1,2k}d_{j+1,2k} \\ = \bar{q}_{j+1,2k-1} + \bar{q}_{j+1,2k} - \bar{q}^*_{jk}, \end{aligned} \tag{4.187}$$

where again $j = 1, 2, \ldots, n$ and $k = 1, 2, \ldots, 2^{j-1}$, and where

$$\bar{q}_{j+1,2k-1} = q_{j+1,2k-1}/\bar{r}_{j+1,2k-1}, \tag{4.188}$$

$$\bar{q}_{j+1,2k} = q_{j+1,2k}/\bar{r}_{j+1,2k}, \tag{4.189}$$

$$\bar{q}^*_{jk} = q^*_{jk}/\bar{r}_{jk}. \tag{4.189A}$$

Boundary conditions at the terminals

If we employ the previous boundary conditions

$$\alpha_{n+1}V_{n+1,k}(L_{n+1,k}) + \beta_{n+1,k}V'_{n+1,k}(L_{n+1,k}) = \gamma_{n+1,k}, \tag{4.190}$$

we will now have

$$\alpha^*_{n+1,k}c_{n+1,k} + \beta^*_{n+1,k}d_{n+1,k} = \gamma^*_{n+1,k}, \tag{4.191}$$

where

$$\begin{aligned} \alpha^*_{n+1,k} &= \alpha_{n+1,k}\cosh L_{n+1,k} + \beta_{n+1,k}\sinh L_{n+1,k}, \\ \beta^*_{n+1,k} &= \alpha_{n+1,k}\sinh L_{n+1,k} + \beta_{n+1,k}\cosh L_{n+1,k}, \\ \gamma^*_{n+1,k} &= \gamma_{n+1,k} - \alpha_{n+1,k}p^*_{n+1,k} - \beta_{n+1,k}q^*_{n+1,k}, \end{aligned} \tag{4.192}$$

and where $k = 1, 2, \ldots, 2^n$.

The linear system for the coefficients c_{jk}, d_{jk}, for the tree with inputs only at the terminals or origin, is thus modified slightly when inputs occur on the cylinders. The coefficient matrix **A** in (4.155) is unchanged, but the vector **b** is modified by the possible additional inputs.

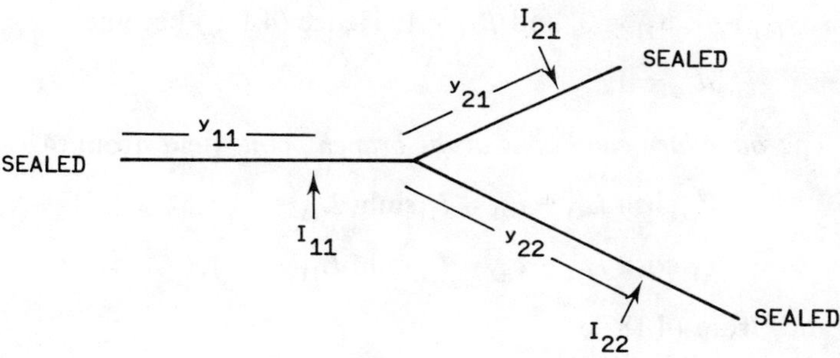

Figure 4.13.

Examples

(i) *One order of branching, current injection on each cylinder: explicit solution with sealed ends.* The arrangement to be considered is sketched in Figure 4.13. Here a current of strength I_{jk} is injected at a point y_{jk} on cylinder (j, k). The equations to be solved are

$$-\frac{d^2V_{jk}}{dx_{jk}^2} + V_{jk} = I_{jk}\delta(x_{jk} - y_{jk}), \qquad 0 < x_{jk} < L_{jk}, \tag{4.193}$$

for (j, k) values (1,1), (2,1), and (2,2). From (4.178) particular solutions are

$$P_{jk}(x_{jk}) = \frac{I_{jk}}{2}\left[e^{-x_{jk}}\int^{x_{jk}} e^{t}\,\delta(t - y_{jk})\,dt \right.$$
$$\left. - e^{x_{jk}}\int^{x_{jk}} e^{-t}\,\delta(t - y_{jk})\,dt\right]. \tag{4.194}$$

Performing the integrations gives

$$P_{jk}(x_{jk}) = \begin{cases} 0, & 0 < x_{jk} < y_{jk}, \\ -I_{jk}\sinh(x_{jk} - y_{jk}), & y_{jk} < x_{jk} < L_{jk}. \end{cases} \tag{4.195}$$

Hence, from (4.179),

$$\begin{aligned} p_{jk} &= 0, \\ q_{jk} &= 0, \\ p_{jk}^* &= -I_{jk}\sinh(L_{jk} - y_{jk}), \\ q_{jk}^* &= -I_{jk}\cosh(L_{jk} - y_{jk}). \end{aligned} \tag{4.196}$$

The *boundary condition at the origin* is

$$V_{11}'(0) = 0, \tag{4.197}$$

so $\gamma_{11}=0$, $\alpha_{11}=0$, and $\beta_{11}=1$. Hence (4.182) becomes

$$d_{11}=0. \tag{4.198}$$

The *boundary conditions at the branch point* yield, from (4.185),

$$\begin{aligned} c_{11}\cosh L_{11}-c_{21}&=I_{11}\sinh(L_{11}-y_{11}),\\ c_{11}\cosh L_{11}-c_{22}&=I_{11}\sinh(L_{11}-y_{11}), \end{aligned} \tag{4.199}$$

and, from (4.187),

$$\kappa_{11}c_{11}-\sigma_{21}d_{21}-\sigma_{22}d_{22}=0. \tag{4.200}$$

Finally, the conditions at the two terminals are

$$\begin{aligned} V_{21}'(L_{21})&=0,\\ V_{22}'(L_{22})&=0, \end{aligned} \tag{4.201}$$

so that in (4.190), $\alpha_{21}=\gamma_{21}=\alpha_{22}=\gamma_{22}=0$ and $\beta_{21}=\beta_{22}=1$. Hence

$$\begin{aligned} \alpha_{21}^*&=\sinh L_{21},\\ \alpha_{22}^*&=\sinh L_{22},\\ \beta_{21}^*&=\cosh L_{21},\\ \beta_{22}^*&=\cosh L_{22},\\ \gamma_{21}^*&=I_{21}\cosh(L_{21}-y_{21}),\\ \gamma_{22}^*&=I_{22}\cosh(L_{22}-y_{22}). \end{aligned} \tag{4.202}$$

The form of the linear system to be solved is

$$\begin{bmatrix} 0 & 0 & 0 & 1 & 0 & 0\\ C_{11} & -1 & 0 & 0 & 0 & 0\\ C_{11} & 0 & -1 & 0 & 0 & 0\\ \kappa_{11} & 0 & 0 & 0 & -\sigma_{21} & -\sigma_{22}\\ 0 & S_{21} & 0 & 0 & C_{21} & 0\\ 0 & 0 & S_{22} & 0 & 0 & C_{22} \end{bmatrix}\begin{bmatrix} c_{11}\\ c_{21}\\ c_{22}\\ d_{11}\\ d_{21}\\ d_{22}\end{bmatrix}$$

$$=\begin{bmatrix} 0\\ I_{11}\sinh(L_{11}-y_{11})\\ I_{11}\sinh(L_{11}-y_{11})\\ I_{11}\cosh(L_{11}-y_{11})/\bar{r}_{11}\\ I_{21}\cosh(L_{21}-y_{21})\\ I_{22}\cosh(L_{22}-y_{22}) \end{bmatrix}. \tag{4.203}$$

This system can be solved by elimination methods to yield

$$c_{11} = \left(I_{11}\sinh(L_{11} - y_{11})\left[\frac{\tanh L_{21}}{\bar{r}_{21}} + \frac{\tanh L_{22}}{\bar{r}_{22}}\right] + \frac{I_{21}\cosh(L_{21} - y_{21})}{\bar{r}_{21}\cosh L_{21}} + \frac{I_{22}\cosh(L_{22} - y_{22})}{\bar{r}_{22}\cosh L_{22}} + \frac{I_{11}\cosh(L_{11} - y_{11})}{\bar{r}_{11}} \right) \Bigg/ \left(\frac{\sinh L_{11}}{\bar{r}_{11}} + \cosh L_{11}\left[\frac{\tanh L_{21}}{\bar{r}_{21}} + \frac{\tanh L_{22}}{\bar{r}_{22}}\right]\right), \tag{4.204}$$

$$d_{11} = 0, \tag{4.205}$$

$$c_{21} = \cosh L_{11}c_{11} - I_{11}\sinh(L_{11} - y_{11}), \tag{4.206}$$

$$d_{21} = \frac{I_{21}\cosh(L_{21} - y_{21})}{\cosh L_{21}} - \tanh L_{21}c_{21}, \tag{4.207}$$

$$c_{22} = c_{21}, \tag{4.208}$$

$$d_{22} = \frac{I_{22}\cosh(L_{22} - y_{22})}{\cosh L_{22}} - \tanh L_{22}c_{21}. \tag{4.209}$$

The value of the depolarization at the origin is

$$V_{11}(0) = c_{11}, \tag{4.210}$$

which is given by (4.204). Hence the effects of various stimulus positions and various strengths can be found by inserting appropriate parameter values.

An example is shown in Figure 4.14. The steady-state voltage over the tree is shown when there is an input on each segment of the tree of Figure 4.11A. The input currents are $I_{11} = I_{21} = I_{22} = 10^{-3}$ and sealed ends are assumed.

(ii) *Extension to multiple inputs on each cylinder.* We consider the last example modified such that there are many points of application of current of various strengths on each cylinder. On segment (j, k) the depolarization satisfies

$$-\frac{d^2V_{jk}}{dx_{jk}^2} + V_{jk} = \sum_{m=1}^{m_{jk}} I_{jk}^m \delta(x_{jk} - y_{jk}^m), \qquad 0 < x_{jk} < L_{jk}, \tag{4.211}$$

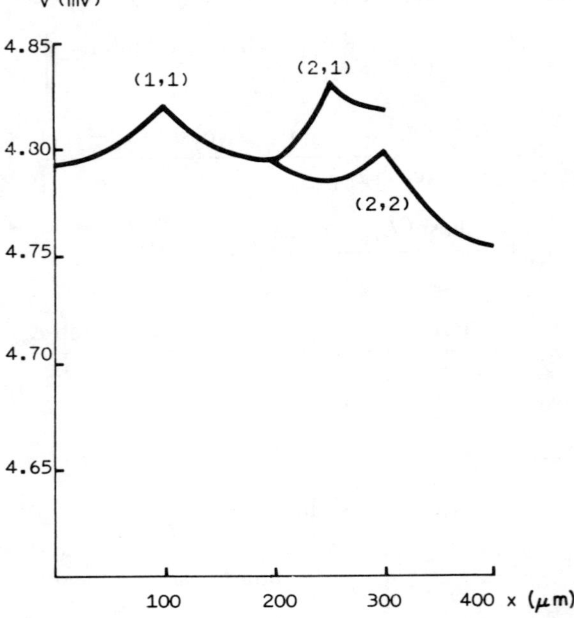

Figure 4.14. Steady-state distribution of depolarization over a dendritic tree with one order of branching when there is current injected on each segment.

where $m = 1, 2, \ldots, m_{jk}$. The I_{jk}^m's are the current strengths and the y_{jk}^m's are their positions. Thus we are considering a situation such as that shown in Figure 4.15, where each arrow represents an input.

The particular solution P_{jk} on cylinder (j, k) is the sum of the particular solutions for each current source considered separately,

$$P_{jk}(x_{jk}) = \sum_{m=1}^{m_{jk}} P_{jk}^m(x_{jk}). \tag{4.212}$$

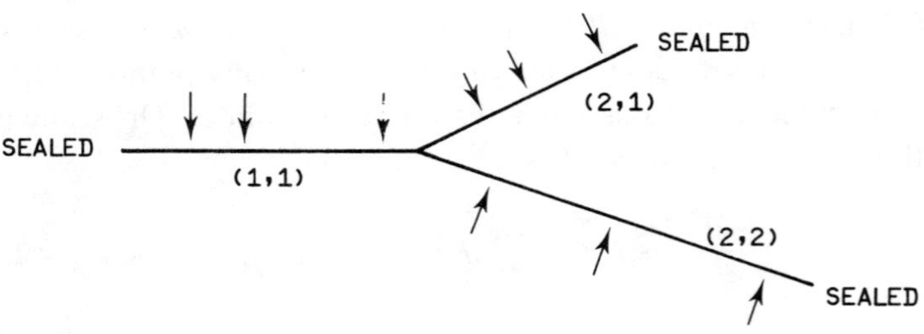

Figure 4.15.

Since we have seen already that

$$P_{jk}^m(x_{jk}) = \begin{cases} 0, & 0 < x_{jk} < y_{jk}^m, \\ -I_{jk}^m \sinh(x_{jk} - y_{jk}^m), & y_{jk}^m < x_{jk} < L_{jk}, \end{cases} \tag{4.213}$$

then $P_{jk}(x_{jk})$ is known. Furthermore,

$$\begin{aligned} p_{jk} &= \sum_{m=1}^{m_{jk}} p_{jk}^m = 0, \\ q_{jk} &= 0, \\ p_{jk}^* &= -\sum_{m=1}^{m_{jk}} I_{jk}^m \sinh(L_{jk} - y_{jk}^m), \\ q_{jk}^* &= -\sum_{m=1}^{m_{jk}} I_{jk}^m \cosh(L_{jk} - y_{jk}^m). \end{aligned} \tag{4.214}$$

If we again employ sealed-end boundary conditions, the linear system to be solved is the same as (4.203) except that each term in the vector **b** that involves a current is replaced by the corresponding sum over inputs. Thus we can immediately obtain the solution. We have therefore, for the depolarization at the origin, which is c_{11},

$$\begin{aligned} V_{11}(0) = a_1 \sum_{m=1}^{m_{11}} I_{11}^m \sinh(L_{11} - y_{11}^m) + a_2 \sum_{m=1}^{m_{21}} I_{21}^m \cosh(L_{21} - y_{21}^m) \\ + a_3 \sum_{m=1}^{m_{22}} I_{22}^m \cosh(L_{22} - y_{22}^m) + a_4 \sum_{m=1}^{m_{11}} I_{11}^m \cosh(L_{11} - y_{11}^m), \end{aligned} \tag{4.215}$$

where

$$a_1 = \frac{1}{\alpha}\left[\frac{\tanh L_{21}}{\bar{r}_{21}} + \frac{\tanh L_{22}}{\bar{r}_{22}}\right], \tag{4.216}$$

$$a_2 = \frac{1}{\alpha \bar{r}_{21} \cosh L_{21}}, \tag{4.217}$$

$$a_3 = \frac{1}{\alpha \bar{r}_{22} \cosh L_{22}}, \tag{4.218}$$

$$a_4 = \frac{1}{\alpha \bar{r}_{11}}, \tag{4.218A}$$

$$\alpha = \frac{\sinh L_{11}}{\bar{r}_{11}} + \cosh L_{11}\left[\frac{\tanh L_{21}}{\bar{r}_{21}} + \frac{\tanh L_{22}}{\bar{r}_{22}}\right]. \tag{4.219}$$

(iii) *Effect of branching on the effect of a stimulus and the effects of damage.* An important question concerns how branching alters the effect that a stimulus of a given strength and given *electrotonic distance* (i.e., distance measured in characteristic lengths) from the origin (presumed location of soma or possibly trigger zone) has on the response. With the above formulas we can answer this question in some simple cases – a more elaborate treatment is given in the next chapter.

We wish to compare the values of the steady-state depolarization at the origin in the cases illustrated in Figure 4.16 (labeled "no branching" and "branching"). It can be seen that we can get the response at the origin in both cases from formula (4.204), where in the case of "no branching" we set $L_{22}=0$. For a current of strength I_{21} at distance y_{21} from the end of cylinder (1,1), we have

$$V^{\text{branching}}(0)=\frac{\dfrac{I_{21}\cosh(L_{21}-y_{21})}{\bar{r}_{21}\cosh L_{21}}}{\dfrac{\sinh L_{11}}{\bar{r}_{11}}+\cosh L_{11}\left[\dfrac{\tanh L_{21}}{\bar{r}_{21}}+\dfrac{\tanh L_{22}}{\bar{r}_{22}}\right]}, \tag{4.220}$$

whereas

$$V^{\text{no branching}}(0)=\frac{\dfrac{I_{21}\cosh(L_{21}-y_{21})}{\bar{r}_{21}\cosh L_{21}}}{\dfrac{\sinh L_{11}}{\bar{r}_{11}}+\dfrac{\cosh L_{11}\tanh L_{21}}{\bar{r}_{21}}}. \tag{4.221}$$

The ratio of the responses is

$$\frac{V^{\text{no branching}}(0)}{V^{\text{branching}}(0)}=1+\frac{\dfrac{\cosh L_{11}\tanh L_{22}}{\bar{r}_{22}}}{\dfrac{\sinh L_{11}}{\bar{r}_{11}}+\dfrac{\cosh L_{11}\tanh L_{21}}{\bar{r}_{21}}}$$

$$=1+\frac{\dfrac{\tanh L_{22}}{\bar{r}_{22}}}{\dfrac{\tanh L_{11}}{\bar{r}_{11}}+\dfrac{\tanh L_{21}}{\bar{r}_{21}}}. \tag{4.222}$$

Recall now that the input resistance of a finite cylinder of electrotonic length L and with sealed end is

$$R_{\text{sealed}}=\bar{r}\coth L, \tag{4.223}$$

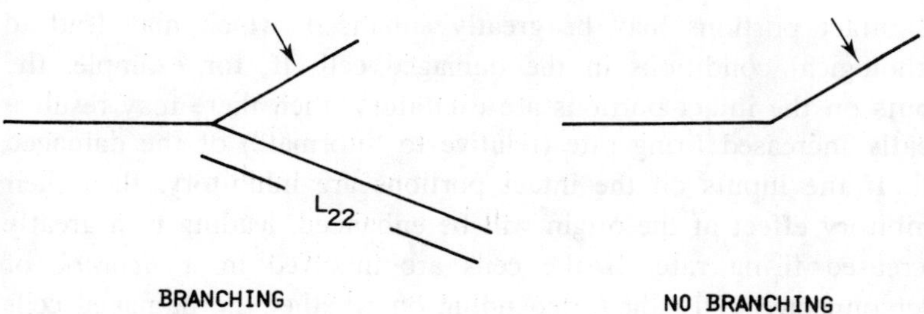

Figure 4.16.

where $\bar{r}$ is the internal resistance for a characteristic length. Also, the *input conductance*, which is the reciprocal of the input resistance, is

$$G_{\text{sealed}} = \tanh L/\bar{r}. \tag{4.224}$$

Thus, if we let g_{jk} be the input conductance of cylinder (j, k), then (4.222) becomes

$$\frac{V^{\text{no branching}}(0)}{V^{\text{branching}}(0)} = 1 + \frac{g_{22}}{g_{11} + g_{21}}. \tag{4.225}$$

Let us define the quantities ρ_{jk} as the relative decrease in the voltage at the origin due to the introduction of cylinder (j, k), where $(j, k) = (2,1)$ or $(2,2)$ and the input occurs on the cylinder of the pair which is already present. Thus

$$\rho_{jk} = \frac{V^{\text{no branching}}(0) - V^{\text{branching}}(0)}{V^{\text{no branching}}(0)}, \tag{4.226}$$

where "branching" means including cylinder (j, k). Then

$$\rho_{21} = \frac{g_{21}}{g_{11} + g_{21} + g_{22}}, \tag{4.227}$$

$$\rho_{22} = \frac{g_{22}}{g_{11} + g_{21} + g_{22}}. \tag{4.228}$$

To obtain a rough guide, *if all the conductances* g_{11}, g_{21}, *and* g_{22} *are equal, then the response at the origin on introducing the cylinders* (2,1) *or* (2,2) (*with an input on the other cylinder of the pair*) *is two-thirds of that without the added cylinder; or, the effect of removing cylinder* (2,1) *or* (2,2) *is to multiply the response by* 1.5.

It may be concluded that if portions of a dendritic tree are *damaged* or lost in some way (e.g., due to aging), then the effects of inputs on the intact portions may be greatly enhanced, which may lead to pathological conditions in the damaged cell. If, for example, the inputs on the intact portions are excitatory, then there may result a greatly increased firing rate (relative to "normal") of the damaged cell. If the inputs on the intact portions are inhibitory, then their inhibitory effect at the origin will be enhanced, leading to a greatly decreased firing rate. If the cells are involved in a *network* of interconnected cells, then, depending on whether the damaged cells are excitatory or inhibitory, these excitatory effects will be enhanced if the damaged portions contain excitatory synapses but decreased if the damaged portions contain inhibitory synapses, whereas the inhibitory effects will be enhanced if the damaged portions contain excitatory synapses but decreased if the damaged portions contain inhibitory synapses. These conclusions are based on an extremely simplified picture but the same quantitative methods can be employed to study more complicated situations.

4.12 Incomplete trees

The equations developed in the preceding sections were for complete trees as defined at the beginning of this chapter. When there are missing cylinders the tree is called *incomplete*. Then one considers the smaller system of equations that results from applying the appropriate boundary conditions to the cylinders of the incomplete tree per se. Since a general formalism will be extremely unwieldy if we attempt to describe all manners of incomplete trees, we will illustrate with an example.

Let us consider the incomplete tree sketched in Figure 4.17. Here there are no inputs on any of the cylinders. There are two orders of branching and the notation for the various component cylinders is marked on the sketch. The previous system of 14 equations (4.164) for

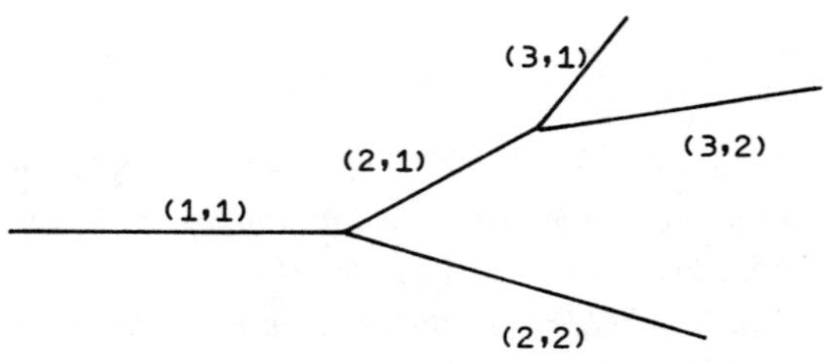

Figure 4.17.

the complete tree becomes one of 10 equations as follows.

$$\begin{bmatrix} \alpha_{11} & 0 & 0 & 0 & 0 & \beta_{11} & 0 & 0 & 0 & 0 \\ C_{11} & -1 & 0 & 0 & 0 & S_{11} & 0 & 0 & 0 & 0 \\ C_{11} & 0 & -1 & 0 & 0 & S_{11} & 0 & 0 & 0 & 0 \\ \kappa_{11} & 0 & 0 & 0 & 0 & \kappa_{11}^{*} & -\sigma_{21} & -\sigma_{22} & 0 & 0 \\ 0 & C_{21} & 0 & -1 & 0 & 0 & S_{21} & 0 & 0 & 0 \\ 0 & C_{21} & 0 & 0 & -1 & 0 & S_{21} & 0 & 0 & 0 \\ 0 & \kappa_{21} & 0 & 0 & 0 & 0 & \kappa_{21}^{*} & 0 & -\sigma_{31} & -\sigma_{32} \\ 0 & 0 & \alpha_{22}^{*} & 0 & 0 & 0 & 0 & \beta_{22}^{*} & 0 & 0 \\ 0 & 0 & 0 & \alpha_{31}^{*} & 0 & 0 & 0 & 0 & \beta_{31}^{*} & 0 \\ 0 & 0 & 0 & 0 & \alpha_{32}^{*} & 0 & 0 & 0 & 0 & \beta_{32}^{*} \end{bmatrix}$$

$$\times \begin{bmatrix} c_{11} \\ c_{21} \\ c_{22} \\ c_{31} \\ c_{32} \\ d_{11} \\ d_{21} \\ d_{22} \\ d_{31} \\ d_{32} \end{bmatrix} = \begin{bmatrix} \gamma_{11} \\ 0 \\ 0 \\ 0 \\ 0 \\ 0 \\ 0 \\ \gamma_{22} \\ \gamma_{31} \\ \gamma_{32} \end{bmatrix}, \tag{4.229}$$

where again $C_{ij} = \cosh L_{ij}$ and $S_{ij} = \sinh L_{ij}$. The equations, when there are inputs along the cylinders, can be obtained by adding the appropriate source terms to the vector on the right-hand side of (4.229).

The problem of determining the steady-state depolarization over a dendritic tree, using the steady-state cable equation on each segment and appropriate boundary conditions at the branch points and terminals, can lead to large linear systems of equations for the unknown coefficients of $\cosh x_{jk}$ and $\sinh x_{jk}$ on cylinder (j, k). In certain cases the necessity for solving such large linear systems can be avoided by a method introduced by Rall (1962). In this method, points on a dendritic tree are mapped onto a single nerve cylinder so that the problem is reduced to finding just two unknown constants. The resulting cylinder is called an *equivalent cylinder*. The theory of the equivalent cylinder is developed in the next chapter.

5

Time-dependent cable theory for nerve cylinders and dendritic trees

5.1 Introduction

In the preceding chapter we showed how, in the framework of cable theory, the steady-state depolarization could be found either in a nerve cylinder or a whole neuron composed of soma, dendritic tree, and axon. To achieve that, we had to solve ordinary differential equations with given boundary conditions at terminals and branch points.

The steady-state solutions are of interest when they can be related to experimental results and can provide some relatively quick insights into the effects of various input patterns and various neuronal geometries. In the natural activity of a nerve cell, however, steady-state conditions will never prevail. In order to understand the dynamic behavior of nerve cells, we must therefore examine the time-dependent solutions.

To obtain the time-dependent solutions, we must solve the partial differential equation (4.25), or the dimensionless version (4.59), for the depolarization $V(x, t)$ at position x at time t. The inclusion of the time variable makes obtaining solutions more difficult, though it may be said that if we are using the linear cable model, there is no problem that we cannot, in principle, solve.

We will begin by showing the usefulness of the Green's function method of solution for nerve cylinders. This extends the approach we used in the previous chapter, and the technique can be used for any spatio–temporal input pattern. We will then look at the case of a nerve cylinder with time-dependent current injection at a terminal. The boundary-value problems that result from considering the time-dependent solutions for dendritic trees are very complicated and we will find that the mapping procedures introduced in Sections 5.9 and 5.10 are often extremely useful.

5.2 The Green's function method of solution for nerve cylinders

The Green's function method of solution we used to solve the steady-state cable equation for nerve cylinders in the previous chapter can be extended to the time-dependent case. We are now interested in the partial differential equation

$$V_t = V_{xx} - V + I, \qquad 0 < x < L,\ t > 0, \tag{5.1}$$

where $I(x, t)$ is the input-current density and the variables are dimensionless. To solve (5.1) we will need *boundary conditions* at the endpoints. To cover the cases of sealed or killed ends we write these as

$$\begin{aligned} \alpha_1 V(0, t) + \beta_1 V_x(0, t) &= 0, \\ \alpha_2 V(L, t) + \beta_2 V_x(L, t) &= 0. \end{aligned} \tag{5.2}$$

We also need to specify the initial value of the depolarization

$$V(x, 0) = v(x), \qquad 0 \le x \le L. \tag{5.3}$$

Definition of the Green's function

The Green's function $G(x, y; t)$ for Equation (5.1) with boundary conditions (5.2) is the solution of

$$G_t = G_{xx} - G + \delta(x - y)\,\delta(t), \qquad 0 < x < L,\ 0 < y < L, \tag{5.4}$$

which satisfies the same boundary conditions and has the value $G = 0$ for $t < 0$. Thus G is the depolarization that results when a unit charge is delivered instantaneously at $t = 0$ at the point y. An alternative and equivalent definition is that the Green's function satisfies the homogeneous equation for $t > 0$,

$$G_t = G_{xx} - G, \tag{5.5}$$

with the initial condition

$$G(x, y; 0) = \delta(x - y). \tag{5.6}$$

Solution of (5.1)

The main result is that once $G(x, y; t)$ is known, the depolarization $V(x, t)$ may be found from the following formula for any suitable current density $I(x, t)$,

$$V(x, t) = \int_0^L G(x, y; t)\, v(y)\, dy + \int_0^L \int_0^t G(x, y; t - s)\, I(y, s)\, ds\, dy. \tag{5.7}$$

Thus one may find $V(x, t)$ for various initial depolarizations $v(x)$ and various current densities by performing the integrations indicated in (5.7).

Proof. There are three things to check. We have to show that the solution given by (5.7) satisfies (a) the initial condition, (b) the boundary conditions, and (c) the partial differential equation.

To show $V(x, t)$ satisfies the *initial condition* we note that contributions from the double integral in (5.7) will be zero for $t = 0$. This leaves the single integral over y and on account of (5.6) we have

$$\begin{aligned} V(x,0) &= \int_0^L G(x, y; 0) v(y)\, dy \\ &= \int_0^L \delta(x-y) v(y)\, dy \\ &= v(x), \end{aligned} \tag{5.8}$$

where we have used the substitution property of the delta function (see Chapter 2, Section 2.16). Thus $V(x, t)$ has the correct initial value.

The boundary conditions are checked as follows, for example, at $x = 0$. We have, on carrying out the partial differentiation with respect to x,

$$\begin{aligned} &\alpha_1 V(0, t) + \beta_1 V_x(0, t) \\ &\quad = \alpha_1\left[\int_0^L G(0, y; t) v(y)\, dy \right. \\ &\qquad\quad \left. + \int_0^L \int_0^t G(0, y; t-s) I(y, s)\, ds\, dy\right] \\ &\qquad + \beta_1\left[\int_0^L G_x(0, y; t) v(y)\, dy \right. \\ &\qquad\quad \left. + \int_0^L \int_0^t G_x(0, y; t-s) I(y, s)\, ds\, dy\right] \\ &\quad = \int_0^L [\alpha_1 G(0, y; t) + \beta_1 G_x(0, y; t)] v(y)\, dy \\ &\qquad + \int_0^L \int_0^t [\alpha_1 G(0, y; t-s) + \beta_1 G_x(0, y; t-s)] \\ &\qquad \times I(y, s)\, dy\, ds. \end{aligned} \tag{5.9}$$

Since G, by definition, satisfies the same boundary conditions as V, it follows that $\alpha_1 G + \beta_1 G_x$ vanishes at $x = 0$ so the integrands in the last two integrals are identically zero. Hence the boundary condition at $x = 0$ is satisfied and a similar procedure shows that the boundary condition at $x = L$ is also satisfied.

There remains to check that $V(x,t)$, as given by (5.7), satisfies Equation (5.1) for $t > 0$. We need the following result. Let $f(s,t)$ be a suitable function of s and t. Then

$$\frac{\partial}{\partial t}\int_0^t f(s,t)\,ds = f(t,t) + \int_0^t \frac{\partial f(s,t)}{\partial t}\,ds, \tag{5.10}$$

a proof of which can be found in many calculus texts [see, for example, Greenspan and Benney (1973), Chapter 6]. Applying this formula, we find

$$\begin{aligned}
\frac{\partial V}{\partial t} &= \int_0^L \frac{\partial}{\partial t}\int_0^t G(x,y;t-s)I(y,s)\,ds\,dy + \int_0^L \frac{\partial G}{\partial t}(x,y;t)v(y)\,dy \\
&= \int_0^L G(x,y;0)I(y,t)\,dy + \int_0^L\int_0^t \frac{\partial G}{\partial t}(x,y;t-s)I(y,s)\,ds\,dy \\
&\quad + \int_0^L \frac{\partial G}{\partial t}(x,y;t)v(y)\,dy \\
&= I(x,t) + \int_0^L\int_0^t \frac{\partial G}{\partial t}(x,y;t-s)I(y,s)\,ds\,dy \\
&\quad + \int_0^L \frac{\partial G}{\partial t}(x,y;t)v(y)\,dy,
\end{aligned} \tag{5.11}$$

where we have again used (5.6) and the substitution property of the delta function. The right-hand side of (5.1) becomes, from (5.7),

$$\begin{aligned}
&I(x,t) + \int_0^L \left[\frac{\partial^2 G}{\partial x^2}(x,y;t) - G(x,y;t)\right]v(y)\,dy \\
&\quad + \int_0^L\int_0^t \left[\frac{\partial^2 G}{\partial x^2}(x,y;t-s) - G(x,y;t-s)\right]I(y,s)\,ds\,dy.
\end{aligned} \tag{5.12}$$

The $I(x,t)$'s on each side cancel and when we collect terms with the same integrations we get an identity because

$$\begin{aligned}
&\int_0^L \left[\frac{\partial G}{\partial t}(x,y;t) - \frac{\partial^2 G}{\partial x^2}(x,y;t) + G(x,y;t)\right]v(y)\,dy \\
&\quad + \int_0^L\int_0^t \left[\frac{\partial G}{\partial t}(x,y;t-s) - \frac{\partial^2 G}{\partial x^2}(x,y;t-s)\right. \\
&\qquad\qquad \left. + G(x,y;t-s)\right]I(y,s)\,ds\,dy \\
&\quad = 0,
\end{aligned} \tag{5.13}$$

by virtue of (5.5). This completes the proof.

5.3 Methods of determining the Green's function

There are three different methods of finding the Green's function and each method has its own merits. Namely,

(a) the separation of variables;
(b) the method of images; and
(c) the Laplace transform.

In this section we will illustrate methods (a) and (b). The Laplace transform method will be used later in particular problems. In all cases we are finding solutions $G(x, y; t)$ of

$$G_t = -G + G_{xx}, \qquad t > 0, 0 < x < L, \tag{5.14}$$

with specified boundary conditions of the form of (5.2). By definition $G = 0$ for $t < 0$ and the initial condition for (5.14) is

$$G(x, y; 0) = \delta(x - y). \tag{5.15}$$

5.3.1 Separation of variables

The solutions are written as the product of a function of x and a function of t and then the variables are *separated*. Thus we look for particular solutions of the form

$$G_n(x, y; t) = \phi_n(x) T_n(t), \tag{5.16}$$

where n is an index over the various particular solutions and y is suppressed because it is regarded as a constant, not a variable. If we substitute (5.16) in the differential equation (5.14), we get

$$\phi_n \frac{dT_n}{dt} = \frac{d^2\phi}{dx^2} T_n - \phi_n T_n. \tag{5.17}$$

We divide by $\phi_n T_n$,

$$\frac{1}{T_n(t)} \frac{dT_n(t)}{dt} = \frac{1}{\phi_n(x)} \frac{d^2\phi_n(x)}{dx^2} - 1. \tag{5.18}$$

The variables are now separated since the left-hand side is a function of t and the right-hand side is a function of x. It is now claimed that the only way a function of t could be equal to a function of x for all x and all t is if both functions are constant. Let us call the corresponding constants for (5.18) k_n. This leads to the *ordinary differential equations*

$$\frac{1}{T_n} \frac{dT_n}{dt} = k_n, \tag{5.19A}$$

and

$$\frac{1}{\phi_n}\frac{d^2\phi_n}{dx^2} - 1 = k_n. \tag{5.19B}$$

The solution of (5.19A) is

$$T_n(t) = c_n e^{k_n t},$$

where c_n is a constant to be found. The solution of (5.19B) must satisfy the boundary conditions, from (5.2), at $x = 0$ and $x = L$,

$$\alpha_1\phi_n(0) + \beta_1\phi_n'(0) = 0,$$
$$\alpha_2\phi_n(L) + \beta_2\phi_n'(L) = 0,$$

where the prime denotes derivative. The imposition of these constraints on the ϕ_n's leads to the requirement that the k_n's take on particular values – the so-called *eigenvalues*. The ϕ_n for each value of k_n is called a *spatial eigenfunction*. It can be shown (as will be verifiable later when examples are considered) that the spatial eigenfunctions with different eigenvalues are *orthogonal*; that is,

$$\int_0^L \phi_n(x)\phi_m(x)\,dx = 0, \qquad n \neq m. \tag{5.20}$$

It is convenient to incorporate in the ϕ_n's multiplicative constants to ensure that each ϕ_n is *normalized* in the sense that

$$\int_0^L \phi_n^2(x)\,dx = 1. \tag{5.21}$$

The solution we seek is obtained as a sum over the particular solutions

$$G(x, y; t) = \sum_n c_n\phi_n(x)e^{k_n t}, \qquad t \geq 0. \tag{5.22}$$

This sum satisfies the same equation as its individual terms. Finally, the constants c_n are found as follows. From (5.15) we have

$$G(x, y; 0) = \sum_n c_n\phi_n(x) = \delta(x - y).$$

Multiplying by $\phi_m(x)$ and integrating with respect to x,

$$\sum_n \int_0^L c_n\phi_n(x)\phi_m(x)\,dx = \int_0^L \phi_m(x)\,\delta(x - y)\,dx. \tag{5.23}$$

From (5.20), the orthogonality relation, the only contribution to the sum on the left-hand side of (5.23), is from the term with $n = m$. Because of the normalization condition (5.21) this term is c_m. By the substitution property of the delta function, the right-hand side of

(5.23) is $\phi_m(y)$. Hence we find

$$c_m = \phi_m(y).$$

The requirement that $G = 0$ for $t < 0$ can be incorporated by multiplying by the step function

$$H(t) = \begin{cases} 1, & t \geq 0, \\ 0, & t < 0. \end{cases}$$

We thus finally have

$$G(x, y; t) = H(t) \sum_n \phi_n(y) \phi_n(x) e^{k_n t}. \tag{5.24}$$

The solution of (5.1) may now be written, using (5.7),

$$V(x, t) = \sum_n \phi_n(x) e^{k_n t} \int_0^L \phi_n(y) v(y)\, dy + \sum_n \phi_n(x) e^{k_n t} \int_0^L \int_0^t \phi_n(y) e^{-k_n s} I(y, s)\, ds\, dy, \qquad t \geq 0. \tag{5.25}$$

An example: a finite nerve cylinder with sealed ends

As an example of determining the Green's function in a particular case we will find $G(x, y; t)$ for the case of a nerve cylinder with sealed ends at $x = 0$ and $x = L$. Thus the boundary conditions satisfied by the Green's function are

$$G_x(0, y; t) = G_x(L, y; t) = 0.$$

From (5.19B) we must solve

$$\phi_n'' + \mu_n \phi_n = 0, \tag{5.26}$$

with $\phi_n'(0) = \phi_n'(L) = 0$, where we have set

$$\mu_n = -(1 + k_n). \tag{5.27}$$

The general solution of (5.26) is

$$\phi_n(x) = c_n \cos\sqrt{\mu_n}\, x + d_n \sin\sqrt{\mu_n}\, x, \tag{5.28}$$

where c_n and d_n are constants. Since

$$\phi_n'(x) = -c_n\sqrt{\mu_n} \sin\sqrt{\mu_n}\, x + d_n\sqrt{\mu_n} \cos\sqrt{\mu_n}\, x, \tag{5.29}$$

we have from the condition at $x = 0$ that $d_n = 0$. Furthermore, the condition at $x = L$ will be met if $L\sqrt{\mu_n}$ is $0, \pm\pi, \pm 2\pi \ldots$, or, in general,

$$L\sqrt{\mu_n} = n\pi, \qquad n = 0, 1, 2, \ldots. \tag{5.30}$$

[We do not need to consider the negative multiples of π because

$\cos x = \cos(-x)$.] From (5.30) we have

$$\mu_n = n^2\pi^2/L^2, \qquad n = 0, 1, 2, \ldots. \tag{5.31}$$

Thus the unnormalized spatial eigenfunctions are

$$\phi_n(x) = c_n \cos(n\pi x/L). \tag{5.32}$$

To normalize the ϕ_n's according to (5.21), we note that

$$\int_0^L \cos^2(n\pi x/L)\,dx = L/2, \qquad n = 1, 2, \ldots. \tag{5.33}$$

Thus, since the normalization condition requires

$$c_n^2 \int_0^L \cos^2(n\pi x/L)\,dx = 1, \tag{5.34}$$

we have $c_n^2 = 2/L$, or

$$c_n = \sqrt{2/L}, \qquad n = 1, 2, \ldots. \tag{5.35}$$

For $n = 0$ we have

$$c_0^2 \int_0^L 1\,dx = 1, \tag{5.36}$$

which leads to

$$c_0 = \sqrt{1/L}. \tag{5.37}$$

In summary the normalized spatial eigenfunctions are

$$\phi_n(x) = \begin{cases} \sqrt{1/L}, & n = 0, \\ \sqrt{2/L}\cos(n\pi x/L), & n = 1, 2, \ldots. \end{cases} \tag{5.38}$$

The Green's function may now be written down from (5.24) if we note that $k_n = -1 - \mu_n$,

$$G(x, y; t) = \frac{\exp(-t)}{L} + \frac{2}{L}\sum_{n=1}^{\infty} \cos\left(\frac{n\pi x}{L}\right)\cos\left(\frac{n\pi y}{L}\right) \times \exp\left[-\left(1 + \frac{n^2\pi^2}{L^2}\right)t\right], \qquad t \geq 0. \tag{5.39}$$

5.3.2 Method of images

It can be seen in Equation (5.39) that for large t, the factors $\exp[-(1 + n^2\pi^2 t/L^2)]$ will be very small. Hence for large t the series will converge very rapidly. However, when t is small these same exponential factors will be close to unity and the convergence will be very slow. If one seeks to find the solution accurately, it is better to avoid the calculation of a large number of terms because of the possibility of a large accumulated error. Fortunately, whereas the eigenfunction expansion obtained by separation of variables converges rapidly for large t, the series representations of the Green's function we will obtain by the method of images converge rapidly for small t. This means that one may compute the Green's function accurately and efficiently for all t.

To employ the *method of images* we need the Green's function $G^*(x, y; t)$ for the cable equation on the infinite interval $(-\infty, \infty)$. This is the solution of (5.5) with initial condition (5.6) and boundary conditions $\lim_{|x| \to \infty} G(x, y; t) = 0$. It can be shown, by Laplace transforms, for example, that this Green's function is

$$G^*(x, y; t) = \frac{\exp[-t]\exp\left[-(x-y)^2/4t\right]}{\sqrt{4\pi t}}, \qquad t > 0. \tag{5.40}$$

The method of images consists of finding suitable linear combinations of G^*'s with different source locations (values of y) that satisfy the required boundary conditions.

An example: a finite nerve cylinder with sealed ends

We wish to find a linear combination of functions G^* that will have zero space derivatives at $x = 0$ and $x = L$. It is important to first realize that a function of x, which has even symmetry about a point x_1, will have a zero space derivative there. For example, the function $f(x) = \cos x$ has even symmetry about $x = 0$ [since $f(x) = f(-x)$] and has zero derivative at $x = 0$.

To ensure that the initial condition is satisfied, namely a delta function concentrated at $x = y$, we must have a term $G^*(x, y; t)$ in the linear combination. The placement of this source at $x = y$ must be compensated to maintain symmetry about $x = 0$ and $x = L$. It will be seen that the placement of sources of unit strength at the points $2nL - y$ and $2nL + y$, $n = 0, \pm 1, \pm 2, \ldots$, will lead to symmetry about $x = 0$ and $x = L$. Thus the Green's function obtained by the method of images is

$$G(x, y; t) = \sum_{n=-\infty}^{\infty} \left[G^*(x, 2nL - y; t) + G^*(x, 2nL + y; t)\right], \tag{5.41}$$

or

$$G(x, y; t) = \frac{\exp[-t]}{\sqrt{4\pi t}} \sum_{n=-\infty}^{\infty} \left[\exp\left\{-\frac{(x-2nL-y)^2}{4t}\right\} + \exp\left\{-\frac{(x-2nL+y)^2}{4t}\right\}\right], \qquad t > 0. \tag{5.42}$$

The function G given by (5.42) satisfies the cable equation because it is a linear combination of terms, each of which is itself a solution. Furthermore at $t=0$, on the interval $[0, L]$, G is equal to $\delta(x-y)$ and satisfies the required boundary conditions. Hence G as given by (5.42) is the Green's function for the problem.

Effect of delivering an impulse at various locations

Although the Green's function is primarily useful for determining the responses to many different kinds of inputs, we may use it directly to find the response to an instantaneous current pulse. In particular, we can see how the response to such a stimulus changes as the distance from the location of the input changes.

In the case of a nerve cylinder with sealed ends at $x=0$ and $x=L$, the response at $x=0$ to an impulse being delivered at $t=0$ at the point y is, from (5.39),

$$G(0, y; t) = \frac{\exp(-t)}{L}\left[1 + 2\sum_{n=1}^{\infty} \cos\left(\frac{n\pi y}{L}\right)\exp\left\{-\frac{n^2\pi^2 t}{L^2}\right\}\right], \quad t>0, \quad (5.43)$$

or, from (5.42),

$$G(0, y; t) = \frac{\exp(-t)}{\sqrt{4\pi t}}\sum_{n=-\infty}^{\infty}\left[\exp\left\{-\frac{(2nL+y)^2}{4t}\right\} + \exp\left\{-\frac{(2nL-y)^2}{4t}\right\}\right], \quad t>0. \quad (5.44)$$

We find $G(0, y; t)$ by summing sufficient terms to meet some criterion of accuracy such as $|(S_{n+1}-S_n)/S_n| < \varepsilon$, where S_n is the sum of n terms and ε is a small positive number. This was done for $L=1.5$, a value appropriate for cylinders representing dendritic trees of cat spinal motoneurons, and with $y=0.1$, $y=0.75$, and $y=1.4$; that is, for an input close to, at an intermediate distance, and remote from the recording point $x=0$. The results are shown in Figure 5.1

Two observations are immediate. First, the amplitude (i.e., maximum value) of the response depends strongly on the distance between input location and recording point. The maximum of the depolariza-

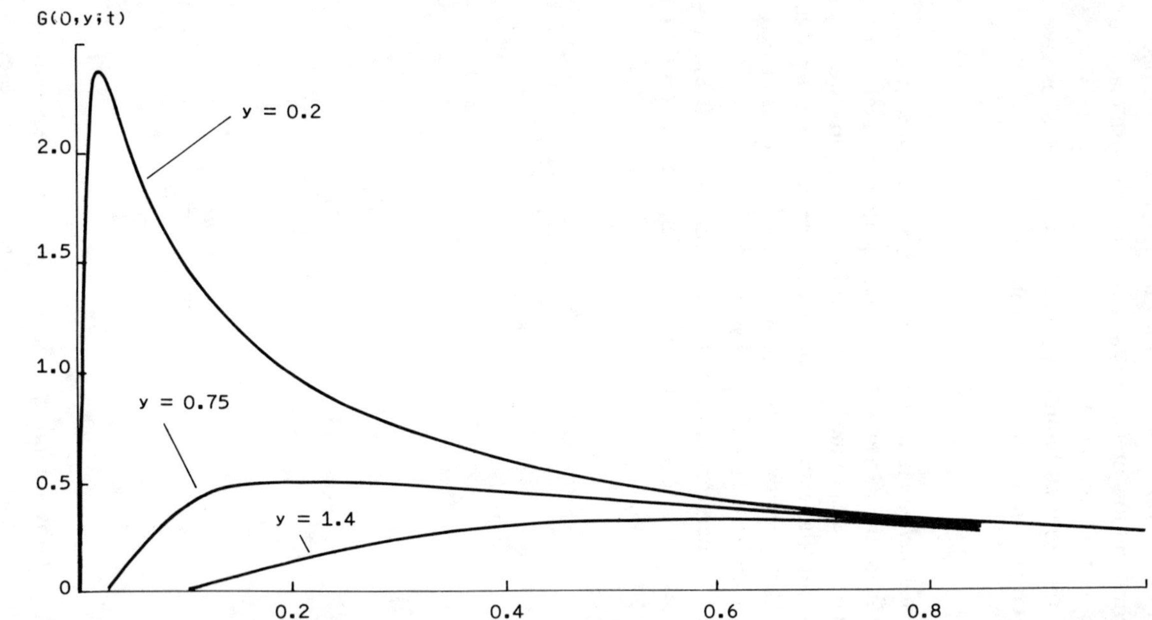

Figure 5.1. Response (depolarization) at $x = 0$ to impulse currents of the same strength delivered at $t = 0$ at various locations y on a nerve cylinder whose total electrotonic length is $L = 1.5$ in the case of sealed ends at $x = 0$ and $x = L$.

tion at $x=0$ with the stimulus close by ($y=0.1$) is about 20 times that when the stimulus is located at $y=1.4$. This indicates that a synaptic input located near the soma (the possible location of the trigger zone) is much more influential on the level of excitation there than one on distal parts of the dendritic tree. This statement, however, may only be partly correct because, as can be seen from Figure 5.1, when the stimulus is more remote the response takes longer to attain its maximum and longer to decay. A reasonable quantitative index on the effect of a stimulus might be the integral from $t=0$ to $t=\infty$ of $V(x,t)$ at fixed x. If this is used as a measure instead of the maximum value of the response, then the difference in the effectivenesses of distal and proximal stimuli is not as pronounced. A calculation shows that a stimulus close to the origin is then only about twice as effective as one near the remote end of the cable structure.

The half-width is defined as the time interval between the two time points on either side of the maximum at which the depolarization is one-half of its maximum (for an IPSP, minimum). Plots of "half-width" versus "time-to-peak" are made on experimental EPSPs to obtain clues as to the possible location of synaptic inputs (see Chapter 6).

The advantages of linearity

Though the cable equation does not provide an accurate quantitative picture of membrane-potential dynamics during action-potential generation, what it lacks in reality is made up for by its tremendous mathematical advantage of *linearity*. This linearity may be fully exploited (it is implicit in the use of Green's functions) by virtue of the following. If V_1 is the solution of the equation

$$V_t = -V + V_{xx} + I_1,$$

with initial value $v_1(x)$, and V_2 is the solution of

$$V_t = -V + V_{xx} + I_2,$$

with initial value $v_2(x)$ and the same boundary conditions, then the solution of

$$V_t = -V + V_{xx} + I_1 + I_2,$$

with the same boundary conditions and initial value $v_1 + v_2$, is

$$V(x,t) = V_1(x,t) + V_2(x,t).$$

This can be verified by substituting in the cable equation. For example, suppose an impulse of magnitude a_1 is delivered at $t=t_1$ at position y_1 and an impulse of magnitude a_2 is delivered at t_2 at position y_2, then the total response is

$$V(x,t) = a_1 G(x, y_1; t - t_1) + a_2 G(x, y_2; t - t_2). \qquad (5.45)$$

5.4 Response to a rectangular current wave

To illustrate the usefulness of the Green's function method and result (5.7), we will find the depolarization $V(x,t)$ that results when a current is injected at the point $x = x_0$ between the times t_1 and t_2 where $0 \le t_1 < t_2 \le \infty$. Without loss of generality we can take $t_1 = 0$. Thus the depolarization satisfies

$$V_t = -V + V_{xx} + \alpha\,\delta(x - x_0)[H(t) - H(t - t_2)], \qquad t > 0,$$

where $H(t)$ is the unit step function [Equation (2.145)], $0 < x_0 < L$, and α is a constant. *We assume the initial depolarization is zero and the ends at $x = 0$ and $x = L$ are sealed.*

We have already found the Green's function for this problem – it is given by (5.39) or (5.42). The applied current density is

$$I(x,t) = \alpha\,\delta(x - x_0)[H(t) - H(t - t_2)].$$

Substituting this in (5.7), we have

$$V(x,t) = \int_0^L \int_0^t G(x, y; t - s) I(y, s)\,ds\,dy$$

$$= \alpha \int_0^L \int_0^t G(x, y; t - s)\,\delta(y - x_0)[H(s) - H(s - t_2)]\,ds\,dy.$$

Performing the y-integration gives, on using the substitution property of the delta function,

$$V(x,t) = \alpha \int_0^t G(x, x_0; t - s)[H(s) - H(s - t_2)]\,ds. \qquad (5.46)$$

The resulting form of V depends on t. First, we have

$$V(x,t) = 0, \qquad t < 0,$$

since the applied current is switched on at $t = 0$. Second, for those times for which the stimulus is on, $0 \le t < t_2$, we have

$$V(x,t) = \alpha \int_0^t G(x, x_0; t - s)[H(s) - H(s - t_2)]\,ds$$

$$= \alpha \int_0^t G(x, x_0; t - s)\,ds,$$

since the integrand in (5.46) is zero for $s < 0$. The third possibility is

$t \geq t_2$, in which case,

$$V(x,t) = \alpha \int_0^{t_2} G(x, x_0; t-s)\, ds, \qquad t \geq t_2.$$

Solution suitable for calculations at small times

The Green's function series representation, which converges rapidly for small times, is that obtained by the method of images in formula (5.42). To evaluate the integrals that arise with this representation, we introduce the *complementary error function* erfc(x), defined by

$$\operatorname{erfc}(x) = \frac{2}{\sqrt{\pi}} \int_x^{\infty} e^{-z^2}\, dz. \tag{5.47}$$

Particular values of this function are $\operatorname{erfc}(-\infty) = 2$, $\operatorname{erfc}(0) = 1$, and $\operatorname{erfc}(\infty) = 0$. This is a well-tabulated function and is available as a library routine on most large computers. The determination of the response to a rectangular current wave is facilitated by use of the following identity, proof of which is left as an exercise:

$$\int_0^t \frac{\exp[(-x^2/4T) - T]}{\sqrt{T}}\, dT = \frac{\sqrt{\pi}}{2} \{\exp(-|x|)\operatorname{erfc}[(|x| - 2t)/2\sqrt{t}] \\ - \exp|x| \operatorname{erfc}[(|x| + 2t)/2\sqrt{t}]\}. \tag{5.48}$$

It is convenient to introduce

$$\alpha_n = x - 2nL - x_0, \qquad n = 0, \pm 1, \pm 2,, \ldots, \tag{5.49A}$$

$$\beta_n = x - 2nL + x_0, \qquad n = 0, \pm 1, \pm 2, \ldots, \tag{5.49B}$$

whereupon

$$G(x, x_0; t) = \frac{\exp(-t)}{\sqrt{4\pi t}} \sum_{n=-\infty}^{\infty} \{\exp[-\alpha_n^2/4t] + \exp[-\beta_n^2/4t]\}.$$

(i) $0 \leq t < t_2$. While the current is switched on, we find that the depolarization at x at time t for a stimulus location x_0 is

$$V(x,t) \\ = \frac{\alpha}{4} \sum_{n=-\infty}^{\infty} \left[e^{-|\alpha_n|} \operatorname{erfc}\left(\frac{|\alpha_n| - 2t}{2\sqrt{t}} \right) - e^{|\alpha_n|} \operatorname{erfc}\left(\frac{|\alpha_n| + 2t}{2\sqrt{t}} \right) \right. \\ \left. + e^{-|\beta_n|} \operatorname{erfc}\left(\frac{|\beta_n| - 2t}{2\sqrt{t}} \right) - e^{|\beta_n|} \operatorname{erfc}\left(\frac{|\beta_n| + 2t}{2\sqrt{t}} \right) \right]. \tag{5.50A}$$

(ii) $t > t_2$. After the current is switched off we obtain

$$
\begin{aligned}
V(x,t) \\
= \frac{\alpha}{4} \sum_{n=-\infty}^{\infty} \Bigg[& e^{-|\alpha_n|}\left\{\operatorname{erfc}\left(\frac{|\alpha_n| - 2t}{2\sqrt{t}}\right) - \operatorname{erfc}\left(\frac{|\alpha_n| - 2(t-t_2)}{2\sqrt{t-t_2}}\right)\right\} \\
& - e^{|\alpha_n|}\left\{\operatorname{erfc}\left(\frac{|\alpha_n| + 2t}{2\sqrt{t}}\right) - \operatorname{erfc}\left(\frac{|\alpha_n| + 2(t-t_2)}{2\sqrt{t-t_2}}\right)\right\} \\
& + e^{-|\beta_n|}\left\{\operatorname{erfc}\left(\frac{|\beta_n| - 2t}{2\sqrt{t}}\right) - \operatorname{erfc}\left(\frac{|\beta_n| - 2(t-t_2)}{2\sqrt{t-t_2}}\right)\right\} \\
& - e^{|\beta_n|}\left\{\operatorname{erfc}\left(\frac{|\beta_n| + 2t}{2\sqrt{t}}\right) - \operatorname{erfc}\left(\frac{|\beta_n| + 2(t-t_2)}{2\sqrt{t-t_2}}\right)\right\}\Bigg]. \quad (5.50B)
\end{aligned}
$$

Solution suitable for calculation at large times

For large times we have seen that we should employ the Green's function in the form (5.39) obtained by separation of variables.

(i) $0 \le t < t_2$. Evaluating the integrals gives, while the current is on,

$$V(x,t) = \frac{\alpha}{L}\left[1 - e^{-t} + 2\sum_{n=1}^{\infty} \frac{\phi_n(x)\phi_n(x_0)}{\mu_n}\{1 - e^{-\mu_n t}\}\right], \tag{5.50C}$$

where $\mu_n = 1 + n^2\pi^2/L^2$ and $\phi_n(x) = \cos(n\pi x/L)$, $n = 1, 2, \dots$.

(ii) $t \ge t_2$. After the current is turned off we obtain

$$V(x,t) = \frac{\alpha}{L}\left[e^{-t}\{e^{t_2} - 1\} + 2\sum_{n=1}^{\infty} \frac{\phi_n(x)\phi_n(x_0)}{\mu_n}\{e^{-\mu_n(t-t_2)} - e^{-\mu_n t}\}\right]. \tag{5.50D}$$

We now consider the following two cases.

(A) *The current is applied indefinitely.* Here $t_2 = \infty$. It can be seen from expression (5.50C) that as $t \to \infty$, the exponential terms go to zero and leave the constant terms corresponding to the *steady state*

$$\tilde{V}(x) = \frac{\alpha}{L}\left[1 + 2\sum_{n=1}^{\infty} \frac{\phi_n(x)\phi_n(x_0)}{\mu_n}\right], \tag{5.51}$$

which must correspond to the steady-state solution given in Table 4.3, case 4, so

$$\tilde{V}(x)=\begin{cases}\dfrac{\alpha\cosh x\cosh(L-x_0)}{\sinh L}, & x\le x_0,\\[2ex] \dfrac{\alpha\cosh x_0\cosh(L-x)}{\sinh L}, & x\ge x_0.\end{cases} \tag{5.52}$$

[That (5.52) is equivalent to (5.51) means that (5.51) is the Fourier expansion of (5.52) as can be verified by writing

$$\tilde{V}(x)=\sum\left[a_n\cos\left(\frac{n\pi x}{L}\right)+b_n\sin\left(\frac{n\pi x}{L}\right)\right]$$

and determining the coefficients a_n and b_n.]

This observation enables rapid convergence to be obtained in (5.50C) by separating the steady-state solution $\tilde{V}(x)$ and the transient solution $V_T(x,t)$; that is,

$$V(x,t)=\tilde{V}(x)-V_T(x,t),$$

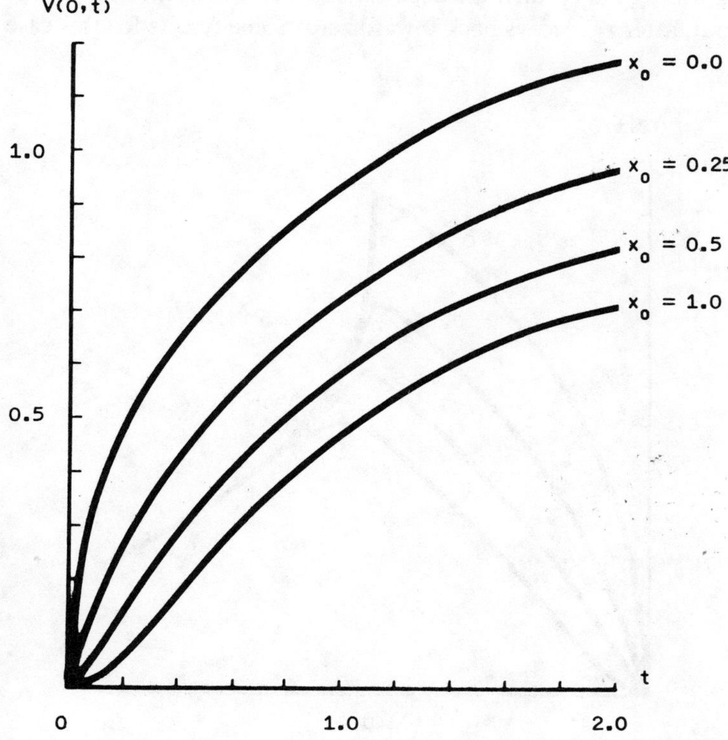

Figure 5.2. Depolarization at the origin as a function of time when a current is applied at the point x_0 for $t\ge 0$. These solutions are computed from (5.50) with $\alpha=1$ and $L=1$.

where $\tilde{V}(x)$ is given by (5.52), and

$$V_T(x,t) = \frac{\alpha}{L}\left[e^{-t} + 2\sum_{n=1}^{\infty} \frac{\phi_n(x)\phi_n(x_0)}{\mu_n} e^{-\mu_n t}\right].$$

It can be seen, since $\mu_n = 1 + n^2\pi^2/L^2$, that the series for V_T will converge rapidly for even quite small values of t.

Of interest is the approach to the steady state. Unfortunately, closed-form expressions for $V(x,t)$ given by (5.50) are not known, so the solution must be computed from the series directly. This has been done for $x = 0$ and various values of t to produce the voltage-versus-time curves at the origin given in Figure 5.2. The constant α has been set at unity and the total electrotonic length of the cylinder set at $L = 1$. The results are given for various x_0, and again indicate how a stimulus close to the recording point is more rapidly effective than one remote from it.

(B) *The current is switched off at* t_2. If the applied current is switched off at a finite time t_2, then the depolarization will not reach the steady state, but, after t_2, decays back toward zero. Some results for this case

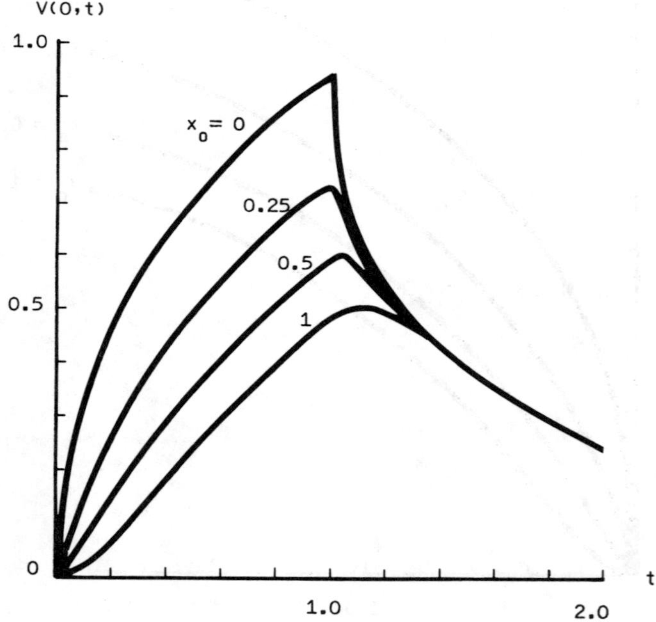

Figure 5.3. Depolarization at the origin when a constant current is applied at the point x_0 for $0 \le t \le 1$ and switched off at $t = 1$. These solutions are computed from (5.50) with $\alpha = L = 1$.

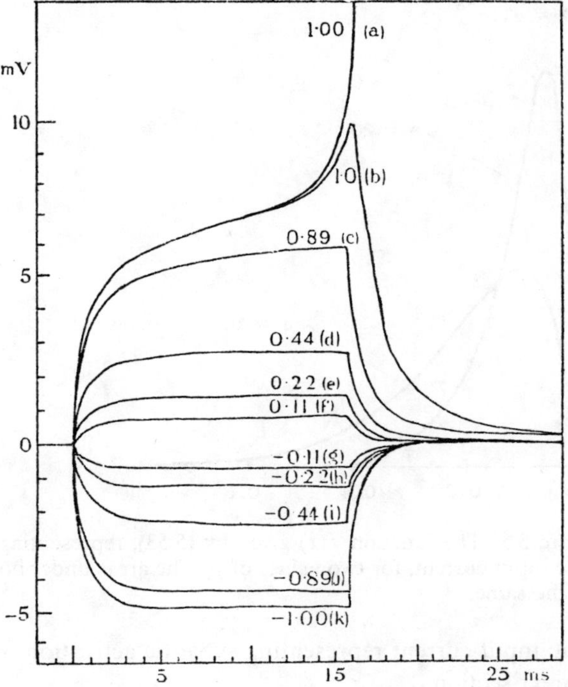

Figure 5.4. Response (membrane depolarization positive) of a *Carcinus* axon to rectangular current wave of duration 15 ms for various current strengths relative to threshold for action-potential generation. Hyperpolarizing as well as depolarizing currents were applied. [From Hodgkin and Rushton (1946). Reproduced with the permission of The Royal Society and the authors.]

are shown in Figure 5.3 with $t_2 = 1$, again with $\alpha = L = 1$ and for various values of x_0.

Figure 5.4 shows the depolarization that resulted in a *Carcinus* (crab) axon when rectangular waves of current were applied through a narrow (200-μm) electrode. As indicated, the current strengths were of various values relative to the threshold value, taken as unity, required to elicit an action potential. A threshold current did not always elicit an action potential as indicated by the two results marked 1b and 1a. The results indicate the usefulness of cable theory for subthreshold responses.

The application of a rectangular current wave of short duration may mimic the input current derived from synaptic activation so these results can be used as an approximation for postsynaptic potentials seen at the soma when a synapse at $x = x_0$ is activated. A somewhat

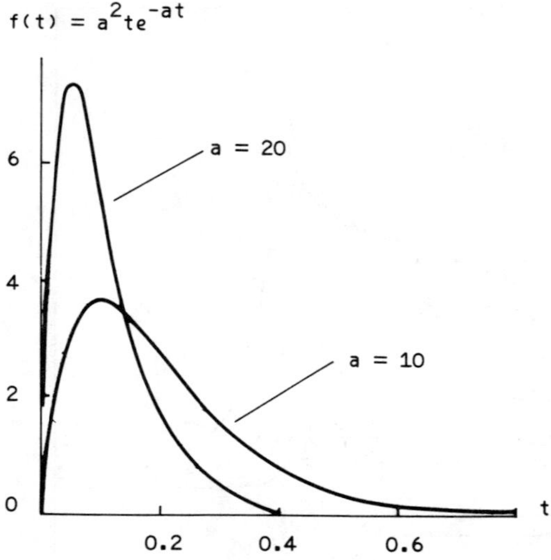

Figure 5.5. The function $f(t)$ given by (5.53), representing a synaptic input current, for two values of a. The areas under both curve are the same.

more realistic input current representing synaptic activation is considered in the next section.

5.5 Response to an approximation to synaptic input current (alpha function)

There is evidence [cited by Jack and Redman (1971a)] that input currents, which arise when certain synapses are active, may be well approximated by the function, often called an *alpha-function*,

$$f(t) = bte^{-at}, \qquad t > 0, \tag{5.53}$$

where a is positive and b may be positive or negative corresponding to the cases of excitation and inhibition, respectively. The function $f(t)$ rises smoothly as t increases to a maximum at $t = 1/a$ and then decays smoothly to zero. The decay eventually becomes close to exponential according to be^{-at}. The larger the value of a, the more rapid is the rise and fall of the current. In fact, in the limit $a \to \infty$, the function $f(t)$ approaches a multiple of $\delta(t)$. The function is illustrated for two values of a in Figure 5.5.

If the current, whose form is (5.53), arises at the point x_0, then the input-current density function is

$$I(x, t) = \delta(x - x_0)bte^{-at}, \qquad t > 0, 0 < x < L. \tag{5.54}$$

The usefulness of the Green's function method is now apparent as we insert (5.54) in formula (5.7) with the appropriate G for the specified boundary conditions. If we assume $V(x, t)$ is initially zero, then we get

$$\begin{aligned} V(x,t) &= \int_0^L \int_0^t G(x, y; t-s) I(y, s)\, dy\, ds \\ &= \int_0^L \int_0^t G(x, y; t-s)\, \delta(y - x_0) bse^{-as}\, ds\, dy \\ &= b\int_0^t G(x, x_0; t-s) se^{-as}\, ds, \qquad t > 0. \end{aligned} \tag{5.55}$$

If we employ the sealed-end boundary conditions at $x = 0$ and $x = L$, then the appropriate Green's function is again (5.39) or (5.42).

Solution suitable for calculation at small times

Proceeding as in the previous section, we have, employing (5.42) for the Green's function (i.e., assuming sealed-end conditions at $x = 0$ and $x = L$),

$$\begin{aligned} V(x,t) &= \frac{b}{2\sqrt{\pi}} \int_0^t \frac{s\exp(-as)\exp[-(t-s)]}{\sqrt{t-s}} \\ &\quad \times \sum_{n=-\infty}^{\infty} \left[\exp\left\{ -\frac{\alpha_n^2}{4(t-s)} \right\} + \exp\left\{ -\frac{\beta_n^2}{4(t-s)} \right\} \right] ds. \end{aligned} \tag{5.56}$$

Putting $T = t - s$ at fixed t, we get

$$\begin{aligned} V(x,t) = \frac{b\exp(-at)}{2\sqrt{\pi}} & \sum_{n=-\infty}^{\infty} \int_0^t \\ & \times \left[\frac{t\exp[-T(1-a)]}{\sqrt{T}} - \exp[-T(1-a)]\sqrt{T} \right] \\ & \times \left[\exp\left\{ \frac{-\alpha_n^2}{4T} \right\} + \exp\left\{ \frac{-\beta_n^2}{4T} \right\} \right] dT. \end{aligned}$$

To evaluate this, we need integrals of the form

$$I_1(x) = \int_0^t \frac{e^{-T(1-a)} e^{-x^2/4T}}{\sqrt{T}}\, dT,$$

and

$$I_2(x) = \int_0^t e^{-T(1-a)} e^{-x^2/4T} \sqrt{T}\, dT.$$

The observation that, when I_1 is regarded as a function of a, differentiation gives

$$I_2 = dI_1/da,$$

means that I_2 can be found from I_1. Assuming $a < 1$, a change of variable to $T' = T(1-a)$ yields, with $t' = t(1-a)$, $x' = x\sqrt{(1-a)}$, the integral we evaluated in the previous section,

$$\begin{aligned} I_1(x) &= \frac{1}{\sqrt{1-a}} \int_0^{t'} \frac{e^{-T'} e^{-x'^2/4T'}}{\sqrt{T'}} dT' \\ &= \frac{1}{2}\left(\frac{\pi}{1-a}\right)^{1/2} \left[e^{-|x|\sqrt{1-a}} \operatorname{erfc}\left(\frac{|x|}{2\sqrt{t}} - \sqrt{t(1-a)}\right) \right. \\ &\qquad \left. - e^{|x|\sqrt{1-a}} \operatorname{erfc}\left(\frac{|x|}{2\sqrt{t}} + \sqrt{t(1-a)}\right) \right], \end{aligned} \tag{5.57}$$

from (5.48).

Differentiation gives

$$\begin{aligned} I_2(x) = \frac{\sqrt{\pi}}{2(1-a)} \Bigg[& e^{-|x|\sqrt{1-a}} \frac{\operatorname{erfc}(R)}{2} \left\{ |x| + \frac{1}{\sqrt{1-a}} \right\} \\ & + e^{|x|\sqrt{1-a}} \operatorname{erfc}(S) \left\{ |x| - \frac{1}{\sqrt{1-a}} \right\} \\ & - \sqrt{\frac{t}{\pi}} \left\{ e^{-|x|\sqrt{1-a}} P + e^{|x|\sqrt{1-a}} Q \right\} \Bigg], \end{aligned} \tag{5.58}$$

where

$$R = \frac{|x|}{2\sqrt{t}} - \sqrt{t(1-a)},$$

$$S = \frac{|x|}{2\sqrt{t}} + \sqrt{t(1-a)},$$

$$P = e^{-R^2},$$

$$Q = e^{-S^2}.$$

In terms of I_1 and I_2 we obtain

$$V(x,t) = \frac{be^{-at}}{2\sqrt{\pi}} \sum_{n=-\infty}^{\infty} \left[t\{I_1(\alpha_n) + I_1(\beta_n)\} - \{I_2(\alpha_n) + I_2(\beta_n)\} \right]. \tag{5.59}$$

It must be remembered that these formulas are derived on the assumption $a < 1$. When $a > 1$ the various arguments become complex and a numerical integration of the integrals in (5.56) seems desirable.

Solution suitable for calculation at large times

Since other than sealed-end conditions may be of interest, we will first find an expression for $V(x, t)$ using the general formula

$$G(x, y; t) = \sum_n \phi_n(x)\phi_n(y)e^{k_n t}, \qquad t > 0. \tag{5.60}$$

Substituting (5.60) in (5.55) gives

$$V(x, t) = b\sum_n \phi_n(x_0)\phi_n(x)e^{k_n t}\int_0^t se^{-(k_n+a)s}\,ds. \tag{5.61}$$

The integration over s is done by parts

$$\int_0^t se^{-(k_n+a)s}\,ds = \left[\frac{se^{-(k_n+a)s}}{-(k_n+a)}\right]_0^t + \frac{1}{(k_n+a)}\int_0^t e^{-(k_n+a)s}\,ds.$$

Evaluating this expression gives

$$\int_0^t se^{-(k_n+a)s}\,ds = \frac{1}{(k_n+a)}\left[\frac{1}{(k_n+a)} - e^{-(k_n+a)t}\left\{t + \frac{1}{(k_n+a)}\right\}\right],$$

$$a \neq -k_n.$$

Evaluating this and substituting in (5.61), we obtain

$$V(x, t) = b\sum_n \frac{\phi_n(x_0)\phi_n(x)}{(k_n+a)}\left[\frac{e^{k_n t}}{(k_n+a)} - e^{at}\left\{t + \frac{1}{(k_n+a)}\right\}\right],$$

$$a \neq -k_n. \tag{5.62}$$

When the ends are sealed, for example, we may use this formula with

$$k_n = -\left(1 + \frac{n^2\pi^2}{L^2}\right), \tag{5.63}$$

and ϕ_n given by (5.38). Note that care is needed less the value of a be equal to $-k_n$ for some n. For example, if $a = -1$, the $n = 0$ term in (5.63) is $(b/2L)t^2e^{-t}$.

The responses at the origin when the input current of form (5.53) is employed for two values of a are shown in Figure 5.6. Again we set $L = 1.5$. Furthermore, the constant b in bte^{-at} is chosen such that

$$\int_0^\infty f(t)\,dt = 1, \tag{5.64}$$

which requires

$$b = a^2, \tag{5.65}$$

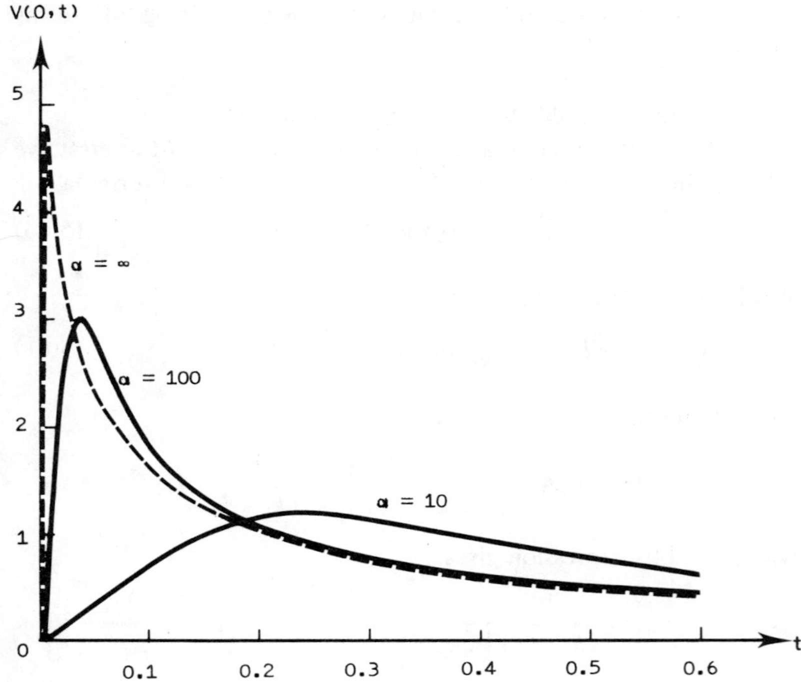

Figure 5.6. Response at the origin to the (synaptic) input current a^2te^{-at} for various indicated values of a. The dashed curve corresponds to a delta-function input at the same location $x_0 = 0.1$. The total electrotonic length of the nerve cylinder is $L = 1.5$.

as seen in Figure 5.5. Thus the total charge delivered is the same as that with a delta function.

The effects of applying a current of form (5.53), where the charge is delivered over an extended period rather than instantaneously, can be seen by examining Figure 5.6. The solid line denotes the results for (5.53), whereas the dashed line corresponds to the results for the delta function. The more abruptly the charge is delivered, the faster the voltage change and the larger the maximum response attained. We also mention that Ashida (1985) has performed calculations of the response of a cable-model neuron with both ends sealed.

5.6 The Green's function for various cable lengths and boundary conditions

In the previous three sections, a nerve cylinder of length L, with sealed ends at $x = 0$ and $x = L$, was considered for various forms of input current. To find the voltage in each case, the Green's function given by (5.39) was employed. We will now summarize the Green's functions corresponding to the cases given in Table 4.3. Some have been given before in this context [e.g., Rall (1969a) and Jack and Redman (1971a, b)]. For more on the properties of Green's functions, see Duff and Naylor (1966) and Roach (1970). In the following formulas, the unit step function $H(t)$ is omitted. When available, both eigenfunction expansion and method-of-images representation are given.

(i) *Infinite cable*:

$$G(x, y; t) = \frac{\exp(-t)}{\sqrt{4\pi t}} \exp\left(-\frac{(x-y)^2}{4t}\right), \qquad -\infty < x < \infty. \tag{5.66}$$

(ii) *Semiinfinite cable, sealed at* $x = 0$:

$$G(x, y; t) = \frac{\exp(-t)}{\sqrt{4\pi t}} \left[\exp\left(-\frac{(x-y)^2}{4t}\right) + \exp\left(-\frac{(x+y)^2}{4t}\right)\right], \quad x \geq 0. \tag{5.67}$$

(iii) *Semiinfinite cable, killed at* $x = 0$:

$$G(x, y; t) = \frac{\exp(-t)}{\sqrt{4\pi t}} \left[\exp\left(-\frac{(x-y)^2}{4t}\right) - \exp\left(-\frac{(x+y)^2}{4t}\right)\right], \quad x \geq 0. \tag{5.68}$$

(iv) *Finite cable, sealed at* $x = 0$ *and* $x = L$:

$$G(x, y; t) = \frac{\exp(-t)}{L} \left[1 + 2 \sum_{n=1}^{\infty} \cos\left(\frac{n\pi x}{L}\right) \cos\left(\frac{n\pi y}{L}\right) \times \exp\left(-\frac{n^2\pi^2}{L^2} t\right)\right], \tag{5.69A}$$

or

$$G(x, y; t) = \frac{\exp(-t)}{\sqrt{4\pi t}} \sum_{n=-\infty}^{\infty} \left[\exp\left(-\frac{(x-2nL-y)^2}{4t}\right) + \exp\left(-\frac{(x-2nL+y)^2}{4t}\right)\right], \quad 0 \leq x \leq L. \tag{5.69B}$$

(v) *Finite cable, killed at $x=0$, sealed at $x=L$:*

$$G(x, y; t) = \frac{2\exp(-t)}{L} \sum_{n=0}^{\infty} \sin\left(\frac{(2n+1)\pi x}{2L}\right) \times \sin\left(\frac{(2n+1)\pi y}{2L}\right) \exp\left(-\frac{(2n+1)^2\pi^2}{4L^2} t\right), \tag{5.70A}$$

or

$$G(x, y; t) = \frac{\exp(-t)}{\sqrt{4\pi t}} \sum_{n=-\infty}^{\infty} (-)^n \left[\exp\left(-\frac{(x-2nL-y)^2}{4t}\right) - \exp\left(-\frac{(x-2nL+y)^2}{4t}\right)\right], \quad 0 \le x \le L. \tag{5.70B}$$

(vi) *Finite cable, sealed at $x=0$, killed at $x=L$:*

$$G(x, y; t) = \frac{2\exp(-t)}{L} \sum_{n=0}^{\infty} \cos\left(\frac{(2n+1)\pi x}{L}\right) \times \cos\left(\frac{(2n+1)\pi y}{L}\right) \exp\left(-\frac{(2n+1)^2\pi^2}{4L^2} t\right), \tag{5.71A}$$

or

$$G(x, y; t) = \frac{\exp(-t)}{\sqrt{4\pi t}} \sum_{n=-\infty}^{\infty} (-)^n \left[\exp\left(-\frac{(x-2nL-y)^2}{4t}\right) + \exp\left(-\frac{(x-2nL+y)^2}{4t}\right)\right], \quad 0 \le x \le L. \tag{5.71B}$$

(vii) *Finite cable, killed at $x=0$ and $x=L$:*

$$G(x, y; t) = \frac{2\exp(-t)}{L} \sum_{n=1}^{\infty} \sin\left(\frac{n\pi x}{L}\right) \sin\left(\frac{n\pi y}{L}\right) \exp\left(-\frac{n^2\pi^2}{L^2} t\right), \tag{5.72A}$$

or

$$G(x, y; t) = \frac{\exp(-t)}{\sqrt{4\pi t}} \sum_{n=-\infty}^{\infty} \left[\exp\left(-\frac{(x-2nL-y)^2}{4t}\right) - \exp\left(-\frac{(x-2nL+y)^2}{4t}\right)\right], \quad 0 \le x \le L. \tag{5.72B}$$

5.7 Current injection at a terminal

It is desirable to be able to obtain expressions for the depolarization over a cylinder as a function of space and time when there is current injected at a terminal (e.g., the origin or soma in the equivalent cylinder picture) as opposed to the case when the current is injected along the cylinder (Rall and Rinzel 1973; Rinzel and Rall 1974; Rinzel 1975). The injected current then becomes part of the boundary conditions rather than the nonhomogeneous term in the partial differential equation for $V(x, t)$.

We here consider a finite cylinder of length L on which the depolarization satisfies

$$V_t = -V + V_{xx}, \qquad 0 < x < L,\ t > 0, \tag{5.73}$$

with initially zero depolarization everywhere,

$$V(x,0) = 0, \qquad 0 \le x \le L. \tag{5.74}$$

We assume a sealed end at $x = L$,

$$V_x(L, t) = 0, \tag{5.75}$$

and a time-dependent longitudinal current $I(t)$ injected at $x = 0$ so that

$$V_x(0, t) = -\bar{r}I(t), \tag{5.76}$$

where $\bar{r}$ is the internal resistance of a characteristic length.

Solution by the Green's function method

We can define a Green's function for the above problem as follows. The Green's function is the response when $I(t)$ in (5.76) is a delta function $\delta(t)$, so that a charge is instantaneously delivered into the terminal at $x = 0$. Thus $G(x, t)$, the Green's function, satisfies

$$G_t = -G + G_{xx}, \qquad 0 < x < L,\ t > 0, \tag{5.77}$$

with zero initial data on the cylinder

$$G(x,0) = 0, \qquad 0 \le x \le L, \tag{5.78}$$

a sealed-end condition at $x = L$,

$$G_x(L, t) = 0, \tag{5.79}$$

and the following boundary condition at $x = 0$,

$$G_x(0, t) = -\bar{r}\delta(t). \tag{5.80}$$

Note that the last condition implies that $G_x(0, t) = 0$ for $t > 0$, so that after the impulse is delivered, the origin has a sealed-end condition.

Solution of (5.73)–(5.76) *via* $G(x, t)$

The solution of the general problem described by the differential equation (5.73) and the boundary conditions (5.74)–(5.76) is

$$V(x, t) = \int_0^t G(x, t-s) I(s)\, ds. \tag{5.81}$$

Proof. We have, from (5.10),

$$\begin{aligned} V_t &= G(x,0) I(t) + \int_0^t G_t(x, t-s) I(s)\, ds \\ &= \int_0^t G_t(x, t-s) I(s)\, ds, \end{aligned} \tag{5.82}$$

using (5.78). Differentiating (5.81) with respect to x twice gives

$$V_{xx} = \int_0^t G_{xx}(x, t-s) I(s)\, ds. \tag{5.83}$$

Hence

$$V_t - V_{xx} + V = \int_0^t (G_t - G_{xx} + G) I(s)\, ds = 0, \tag{5.84}$$

by (5.77). Since G satisfies the condition $G_x(L, t) = 0$, V does also from (5.81). Furthermore,

$$\begin{aligned} V_x(0, t) &= \int_0^t G_x(0, t-s) I(s)\, ds \\ &= -\bar{r} \int_0^t \delta(t-s) I(s)\, ds \\ &= -\bar{r} I(t). \end{aligned} \tag{5.85}$$

From (5.81) $V(x, t)$ also satisfies the correct initial condition since at $t = 0$ the integral is zero. Hence V, given by (5.81), is the required solution.

Determining the Green's function–Laplace transform method

We will show how the Green's function may be found by first finding its Laplace transform. The Laplace transform of $G(x, t)$ is defined as

$$G_L(x; s) = \int_0^\infty G(x, t) e^{-st}\, dt. \tag{5.86}$$

The idea is to Laplace transform the differential equation (5.77) and the boundary conditions to give, for fixed s, an ordinary differential equation for G_L. The Laplace transform of G_t is

$$\begin{aligned} \int_0^\infty G_t(x, t) e^{-st}\, dt &= G(x, t) e^{-st} \Big|_0^\infty + s \int_0^\infty G(x, t) e^{-st}\, dt \\ &= s G_L(x; s), \end{aligned} \tag{5.87}$$

where we have integrated by parts and used the fact that $G(x, 0) = 0$.

Also, the transform of G_{xx} is

$$\int_0^\infty G_{xx}(x, t)e^{-st}\,dt = \frac{\partial^2}{\partial x^2}\int_0^\infty G(x, t)e^{-st}\,dt = G_L''. \tag{5.88}$$

The boundary condition (5.79) becomes

$$G_L'(L) = 0. \tag{5.89}$$

(Primes denote differentiation with respect to x with s held fixed.) We also Laplace transform the boundary condition (5.80) at $x = 0$ to give

$$\begin{aligned} G_L'(0) &= -\bar{r}\int_0^\infty e^{-st}\,\delta(t)\,dt \\ &= -\bar{r}. \end{aligned} \tag{5.90}$$

To find G_L we now need to solve

$$-G_L'' + (1+s)G_L = 0, \tag{5.91}$$

which comes from transforming (5.77), subject to the conditions (5.89) and (5.90). As we saw in the last chapter, the general solution of (5.91) is

$$G_L(x; s) = c(s)\cosh\sqrt{(1+s)}\,x + d(s)\sinh\sqrt{(1+s)}\,x, \tag{5.92}$$

where c and d are constants which depend on s. Since

$$G_L' = \sqrt{(1+s)}\left[c(s)\sinh\sqrt{(1+s)}\,x + d(s)\cosh\sqrt{(1+s)}\,x\right], \tag{5.93}$$

the boundary condition at $x = 0$ gives

$$-\bar{r} = d(s)\sqrt{(1+s)}\,, \tag{5.94}$$

or

$$d(s) = -\bar{r}/\sqrt{(1+s)}\,. \tag{5.95}$$

At $x = L$ we now have

$$\left[c(s)\sinh\sqrt{(1+s)}\,L - \frac{\bar{r}\cosh\sqrt{(1+s)}\,L}{\sqrt{(1+s)}}\right] = 0, \tag{5.96}$$

or

$$c(s) = \frac{\bar{r}\coth\sqrt{(1+s)}\,L}{\sqrt{(1+s)}}. \tag{5.97}$$

Thus the Laplace transform of the Green's function is

$$\begin{aligned} G_L(x; s) = \bar{r}\big[&\coth\sqrt{(1+s)}\,L\cosh\sqrt{(1+s)}\,x \\ &\times - \sinh\sqrt{(1+s)}\,x\big]/\sqrt{(1+s)}\,. \end{aligned} \tag{5.98}$$

To find $G(x, t)$ we need to invert this transform, which is not given in the table of transforms (Table 3.2). By making use of formula (5.52), it can be shown that

$$\mathscr{L}\left[\exp(-t)\left\{1+2\sum_{n=1}^{\infty}\cos\left(\frac{n\pi x}{L}\right)\exp\left(-\frac{n^2\pi^2}{L^2}t\right)\right\}\right]$$
$$=\frac{L}{\sqrt{1+s}}\{\coth(\sqrt{1+s}\,L)\cosh(\sqrt{1+s}\,x)-\sinh(\sqrt{1+s}\,x)\},$$
$$(5.99)$$

where $\mathscr{L}$ denotes the Laplace transform operation. Thus the inverse transform of (5.98) is

$$G(x,t)=\frac{\bar{r}\exp(-t)}{L}\left[1+2\sum_{n=1}^{\infty}\cos\left(\frac{n\pi x}{L}\right)\exp\left(-\frac{n^2\pi^2}{L^2}t\right)\right]. \qquad (5.100)$$

As can be seen by comparing (5.100) with (5.39) evaluated with $y=0$, the Green's function with an impulse applied at a terminal (i.e., as a longitudinal current at $x=0$) is $\bar{r}$ times the Green's function for the previous problem (cf. the note on input conductances in Chapter 4). Thus an expansion that converges rapidly for small t is given by $\bar{r}$ times expression (5.42) for $G(x, y; t)$ with $y=0$.

With the Green's function available, it is a matter of substituting in (5.81) for the injected current of interest and performing the time integration to obtain the response.

Application of a constant current

Suppose a constant current of magnitude I_0 is applied at $x=0$ so that

$$I(t)=I_0H(t), \qquad (5.101)$$

where $H(t)$ is the unit step function. Then, from (5.81), the response is

$$V(x,t)=I_0\int_0^t G(x,t-s)H(s)\,ds$$
$$=I_0\int_0^t G(x,t-s)\,ds$$
$$=\frac{I_0\bar{r}}{L}\int_0^t\left[\exp[-(t-s)]+2\sum_{n=1}^{\infty}\cos\left(\frac{n\pi x}{L}\right)\right.$$
$$\left.\times\exp\left(-\left(1+\frac{n^2\pi^2}{L^2}\right)(t-s)\right)\right]ds. \qquad (5.102)$$

The integrals are the same as in Section 5.4. We obtain

$$V(x,t)=\frac{I_0\bar{r}}{L}\left[1-\exp(-t)+2\sum_{n=1}^{\infty}\frac{\cos\left(\frac{n\pi x}{L}\right)}{1+\frac{n^2\pi^2}{L^2}}\times\left\{1-\exp\left(-\left(1+\frac{n^2\pi^2}{L^2}\right)t\right)\right\}\right]. \quad (5.103)$$

Accelerated convergence is obtained by separating the steady-state solution $\tilde{V}(x)$ and the transient solution $V_T(x,t)$,

$$V(x,t)=\tilde{V}(x)-V_T(x,t), \quad (5.104)$$

where, from (5.52),

$$\tilde{V}(x)=\frac{I_0\bar{r}\cosh(L-x)}{\sinh L}, \quad (5.104A)$$

and

$$V_T(x,t)=\frac{I_0\bar{r}}{L}\left[\exp(-t)+2\sum_{n=1}^{\infty}\frac{\cos\left(\frac{n\pi x}{L}\right)}{1+\frac{n^2\pi^2}{L^2}}\times\exp\left[-\left(1+\frac{n^2\pi^2}{L^2}\right)t\right]\right]. \quad (5.104B)$$

An alternative representation of the solution is, from that used in Equation (5.50A),

$$V(x,t)=\frac{I_0\bar{r}}{2}\sum_{n=-\infty}^{\infty}\left[e^{-|x-2nL|}\operatorname{erfc}\left(\frac{|x-2nL|-2t}{2\sqrt{t}}\right)-e^{|x-2nL|}\operatorname{erfc}\left(\frac{|x-2nL|+2t}{2\sqrt{t}}\right)\right].$$

5.8 Time-dependent potential over dendritic trees

The steady-state potential over complete or incomplete dendritic trees could be found by the methods outlined in Chapter 4. When the inputs are time dependent and the time-dependent potential distribution is sought, the resulting boundary-value problems become extremely difficult.

We use the same notation for labeling the cylinders of the dendritic tree as in Chapter 4. Consider a complete tree with n orders of branching. On cylinder (j, k) we have, with dimensionless space and time variables,

$$\frac{\partial V_{jk}}{\partial t} = -V_{jk} + \frac{\partial^2 V_{jk}}{\partial x_{jk}^2} + I_{jk}, \qquad 0 < x_{jk} < L_{jk},\ t > 0, \tag{5.105}$$

where L_{jk} is the electrotonic length (physical length l_{jk} divided by space constant λ_{jk}), and we have the initial conditions

$$V_{jk}(x_{jk}, 0) = v_{jk}(x_{jk}), \qquad 0 \le x_{jk} \le L_{jk}. \tag{5.106}$$

Boundary condition at the origin

The various possible boundary conditions at the origin, including sealed or killed end, voltage clamp (fixed), can be incorporated in the equation

$$\alpha_{11} V_{11}(0, t) + \beta_{11} V_{11,x}(0, t) = \gamma_{11}(t), \tag{5.107}$$

where $V_{11,x}$ denotes the partial derivative of V_{11} with respect to x_{11}. [Equation (5.107) could be further generalized by making either, or both, α_{11} and β_{11} functions of time and including a term in $V_{11,t}$, but many cases of interest are covered by (5.107) as it is.]

Boundary conditions at the branch points

We again have the physical constraints of continuity of potential and conservation of axial current at branch points. At the branch point N_{jk} we have

$$\begin{aligned} V_{jk}(L_{jk}, t) &= V_{j+1,2k-1}(0, t), \\ V_{jk}(L_{jk}, t) &= V_{j+1,2k}(0, t), \end{aligned} \tag{5.108}$$

and

$$\frac{1}{\bar{r}_{jk}} V_{jk,x}(L_{jk}, t) = \frac{1}{\bar{r}_{j+1,2k-1}} V_{j+1,2k-1;x}(0, t) + \frac{1}{\bar{r}_{j+1,2k}} V_{j+1,2k;x}(0, t). \tag{5.109}$$

Boundary conditions at terminals

To allow for flexibility, the boundary conditions at the 2^n terminals are written as

$$\alpha_{n+1,k}V_{n+1,k}(L_{n+1,k}, t) + \beta_{n+1,k}V_{n+1,k;x}(L_{n+1,k}, t) = \gamma_{n+1,k}(t). \tag{5.110}$$

Laplace transforms

Equations (5.105)–(5.110) present a difficult problem for which an analytic solution is probably only obtainable under simplifying assumptions. One case has been solved recently (Tuckwell 1984), where an exact solution was obtained at the origin of a tree with an arbitrary number of cylinders emanating therefrom provided the cylinders had equal electrotonic lengths. It is desirable, in general, to turn to Laplace transforms to convert the coupled system of partial differential equations to a coupled system of ordinary differential equations. We define the following Laplace transforms:

$$\bar{V}_{jk}(x_{jk}; s) = \int_0^\infty e^{-st}V_{jk}(x_{jk}, t)\, dt, \tag{5.111}$$

$$\bar{I}_{jk}(x_{jk}; s) = \int_0^\infty e^{-st}I_{jk}(x_{jk}, t)\, dt, \tag{5.112}$$

$$\bar{\gamma}_{jk}(s) = \int_0^\infty e^{-st}\gamma_{jk}(t)\, dt. \tag{5.113}$$

We make one simplifying assumption – that the initial depolarizations $v_{jk}(x_{jk})$ are all zero. Then using primes to denote differentiation with respect to space variables, we have, from the same calculation as in (5.87) applied to (5.105), that the Laplace transforms of the depolarizations on the various cylinders of the tree satisfy the ordinary differential equations,

$$-\bar{V}_{jk}'' + (1+s)\bar{V}_{jk} = \bar{I}_{jk}, \qquad 0 < x_{jk} < L_{jk}, \tag{5.114}$$

whereas the boundary conditions transform to

$$\alpha_{11}\bar{V}_{11}(0; s) + \beta_{11}\bar{V}_{11}'(0; s) = \bar{\gamma}_{11}(s), \tag{5.115}$$

$$\bar{V}_{jk}(L_{jk}; s) = \bar{V}_{j+1,2k-1}(0; s), \tag{5.116}$$

$$\bar{V}_{jk}(L_{jk}; s) = \bar{V}_{j+1,2k}(0; s), \tag{5.117}$$

$$\frac{1}{\bar{r}_{jk}}\bar{V}_{jk}'(L_{jk}; s) = \frac{1}{\bar{r}_{j+1,2k-1}}\bar{V}_{j+1,2k-1}'(0; s) + \frac{1}{\bar{r}_{j+1,2k}}\bar{V}_{j+1,2k}'(0; s), \tag{5.118}$$

$$\alpha_{n+1,k}\bar{V}_{n+1,k}(L_{n+1,k}; s) + \beta_{n+1,k}\bar{V}_{n+1,k}'(L_{n+1,k}; s) = \bar{\gamma}_{n+1,k}(s). \tag{5.119}$$

The similarity of these equations to those of the steady-state problem suggests changing variables and defining new constants as follows:

$$y_{jk} = \sqrt{(1+s)}\, x_{jk}, \tag{5.120}$$

$$\overline{W}_{jk}(y_{jk}; s) = \overline{V}(x_{jk}; s), \tag{5.121}$$

$$\bar{J}_{jk}(y_{jk}; s) = \frac{\bar{I}_{jk}(x_{jk}; s)}{(1+s)}, \tag{5.122}$$

$$M_{jk} = \sqrt{(1+s)}\, L_{jk}, \tag{5.123}$$

$$\bar{\alpha}_{jk} = \alpha_{jk}, \tag{5.124}$$

$$\bar{\beta}_{jk} = \sqrt{(1+s)}\, \beta_{jk}. \tag{5.125}$$

Then the differential equations (5.114) become

$$-\overline{W}_{jk}'' + \overline{W}_{jk} = \bar{J}_{jk}, \qquad 0 < y_{jk} < M_{jk}, \tag{5.126}$$

whereas the boundary conditions are now

$$\bar{\alpha}_{11}\overline{W}_{11}(0; s) + \bar{\beta}_{11}\overline{W}_{11}'(0; s) = \bar{\gamma}_{11}(s), \tag{5.127}$$

$$\overline{W}_{jk}(M_{jk}; s) = \overline{W}_{j+1,2k-1}(0; s), \tag{5.128}$$

$$\overline{W}_{jk}(M_{jk}; s) = \overline{W}_{j+1,2k}(0; s), \tag{5.129}$$

$$\frac{1}{\bar{r}_{jk}}\overline{W}_{jk}'(M_{jk}; s) = \frac{1}{\bar{r}_{j+1,2k-1}}\overline{W}_{j+1,2k-1}'(0; s) + \frac{1}{\bar{r}_{j+1,2k}}\overline{W}_{j+1,2k}'(0; s), \tag{5.130}$$

$$\bar{\alpha}_{n+1,k}\overline{W}_{n+1,k}(M_{n+1,k}; s) + \bar{\beta}_{n+1,k}\overline{W}_{n+1,k}'(M_{n+1,k}; s) = \bar{\gamma}_{n+1,k}(s). \tag{5.131}$$

Equations (5.126)–(5.131) are of the same form as the steady-state equations. The solutions of (5.126) can be written as

$$\overline{W}_{jk}(y_{jk}; s) = \bar{c}_{jk}\cosh y_{jk} + \bar{d}_{jk}\sinh y_{jk} + \overline{P}_{jk}(y_{jk}; s), \tag{5.132}$$

where the $\overline{P}_{jk}$'s are particular solutions. Equation (5.132) is the same in form as (4.177) and the boundary conditions also have the same form. Hence solutions for the steady-state potential over dendritic trees can be used to find the Laplace transform of the time-dependent potential over the same tree.

No inputs along the cylinders

When the only possible sources of current are at the origin or terminals, we have the following equations for the constants $\bar{c}_{jk}$ and

$\bar{d}_{jk}$:

$$\bar{\alpha}_{11}\bar{c}_{11} + \bar{\beta}_{11}\bar{d}_{11} = \bar{\gamma}_{11}, \tag{5.133}$$

$$\bar{C}_{jk}\bar{c}_{jk} + \bar{S}_{jk}\bar{d}_{jk} - \bar{c}_{j+1,2k-1} = 0, \tag{5.134}$$

$$\bar{C}_{jk}\bar{c}_{jk} + \bar{S}_{jk}\bar{d}_{jk} - \bar{c}_{j+1,2k} = 0, \tag{5.135}$$

$$\bar{\kappa}_{jk}\bar{c}_{jk} + \bar{\kappa}^*_{jk}\bar{d}_{jk} - \bar{\sigma}_{j+1,2k-1}\bar{d}_{j+1,2k-1} - \bar{\sigma}_{j+1,2k}\bar{d}_{j+1,2k} = 0, \tag{5.136}$$

$$\bar{\alpha}^*_{n+1,k}\bar{c}_{n+1,k} + \bar{\beta}^*_{n+1,k}\bar{d}_{n+1,k} = \bar{\gamma}_{n+1,k}, \tag{5.137}$$

where in equations (5.134)–(5.136), $j = 1, 2, \ldots, n$ and $k = 1, 2, \ldots, 2^{j-1}$, and in (5.137) $k = 1, 2, \ldots, 2^n$. The constants are defined thusly:

$$\begin{aligned}
\bar{C}_{jk} &= \cosh(M_{jk}), \\
\bar{S}_{jk} &= \sinh(M_{jk}), \\
\bar{\kappa}_{jk} &= \bar{S}_{jk}/\bar{r}_{jk}, \\
\bar{\kappa}^*_{jk} &= \bar{C}_{jk}/\bar{r}_{jk}, \\
\bar{\sigma}_{jk} &= 1/\bar{r}_{jk}, \\
\bar{\alpha}^*_{n+1,k} &= \bar{\alpha}_{n+1,k}\bar{C}_{n+1,k} + \bar{\beta}_{n+1,k}\bar{S}_{n+1,k}, \\
\bar{\beta}^*_{n+1,k} &= \bar{\alpha}_{n+1,k}\bar{S}_{n+1,k} + \bar{\beta}_{n+1,k}\bar{C}_{n+1,k}.
\end{aligned} \tag{5.138}$$

An example

We consider a complete tree with one order of branching, with current $I_0(t)$ injected at the origin, and killed ends at the dendritic terminals. The linear system to be solved is

$$\begin{bmatrix} 0 & 0 & 0 & 1 & 0 & 0 \\ \bar{C}_{11} & -1 & 0 & \bar{S}_{11} & 0 & 0 \\ \bar{C}_{11} & 0 & -1 & \bar{S}_{11} & 0 & 0 \\ \bar{\kappa}_{11} & 0 & 0 & \kappa^*_{11} & -\bar{\sigma}_{21} & -\bar{\sigma}_{22} \\ 0 & \bar{C}_{21} & 0 & 0 & \bar{S}_{21} & 0 \\ 0 & 0 & \bar{C}_{22} & 0 & 0 & \bar{S}_{22} \end{bmatrix} \begin{bmatrix} \bar{c}_{11} \\ \bar{c}_{21} \\ \bar{c}_{22} \\ \bar{d}_{11} \\ \bar{d}_{21} \\ \bar{d}_{22} \end{bmatrix} = \begin{bmatrix} \bar{\gamma}^*_{11} \\ 0 \\ 0 \\ 0 \\ 0 \\ 0 \end{bmatrix}, \tag{5.139}$$

where

$$\bar{\gamma}^*_{11} = \frac{-\bar{r}_{11}\bar{I}_0(s)}{\sqrt{(1+s)}}, \tag{5.140}$$

and $\bar{I}_0(s)$ is the Laplace transform of $I_0(t)$. From (4.160) we see that

$$\bar{c}_{11} = \frac{\dfrac{\bar{r}_{11}\bar{I}_0(s)}{\sqrt{1+s}}\left[\dfrac{1}{\bar{r}_{11}} + \tanh(M_{11})\left\{\dfrac{\coth(M_{21})}{\bar{r}_{21}} + \dfrac{\coth(M_{22})}{\bar{r}_{22}}\right\}\right]}{\dfrac{\tanh(M_{11})}{\bar{r}_{11}} + \dfrac{\coth(M_{21})}{\bar{r}_{21}} + \dfrac{\coth(M_{22})}{\bar{r}_{22}}}. \tag{5.141}$$

Thus the Laplace transform of the depolarization at the origin is

$$\begin{aligned}\bar{V}_{11}(0;s) = {} & \left(\frac{\bar{r}_{11}\bar{I}_0(s)}{\sqrt{1+s}}\left[\frac{1}{\bar{r}_{11}} + \tanh\left(L_{11}\sqrt{1+s}\right)\right.\right.\\ & \left.\left.\times\left\{\frac{\coth\left(L_{21}\sqrt{1+s}\right)}{\bar{r}_{21}} + \frac{\coth\left(L_{22}\sqrt{1+s}\right)}{\bar{r}_{22}}\right\}\right]\right)\Bigg/\\ & \left(\frac{\tanh\left(L_{11}\sqrt{1+s}\right)}{\bar{r}_{11}} + \frac{\coth\left(L_{21}\sqrt{1+s}\right)}{\bar{r}_{21}}\right.\\ & \left. + \frac{\coth\left(L_{22}\sqrt{1+s}\right)}{\bar{r}_{22}}\right).\end{aligned} \tag{5.142}$$

As a check on this result, when the equivalent cylinder constraints (see Section 5.9),

$$\frac{1}{\bar{r}_{11}} = \frac{1}{\bar{r}_{21}} + \frac{1}{\bar{r}_{22}}, \tag{5.143}$$

and

$$L_{21} = L_{22}, \tag{5.143A}$$

are met, (5.142) reduces to

$$\bar{V}_{11}(0;s) = \frac{\bar{r}_{11}\bar{I}_0(s)\tanh\left(L\sqrt{1+s}\right)}{\sqrt{1+s}}, \tag{5.144}$$

where $L = L_{11} + L_{21}$. Formula (5.144) is the Laplace transform of the depolarization on a single cylinder of length L.

The graphical calculus of Butz and Cowan

Butz and Cowan (1974) have given a useful method for quickly obtaining the Laplace transform of the potential over dendritic trees with arbitrary geometries. This enables one to solve complicated linear systems of equations for the coefficients in the solution on each dendritic segment by inspection. A brief summary of the underlying observations and rules for obtaining the Laplace transform will be given.

Consider first a single cable of length L with current injection $I(t)$ at y, with say killed ends at $x = 0$ and $x = L$. Thus V satisfies

$$V_t = -V + V_{xx} + I(t)\,\delta(x-y),$$
$$V(0,t) = V(L,t) = 0,$$
$$V(x,0) = 0.$$

The Laplace transform $\bar{V}(x;s)$ of V satisfies the ordinary differential equation

$$-\bar{V}'' + (1+s)\bar{V} = \bar{I}(s)\,\delta(x-y),$$

where $\bar{I}$ is the Laplace transform of $I(t)$. Solving for $\bar{V}$ gives

$$\bar{V}(x;s) = \frac{Z\sinh(\gamma(L-y))\sinh(\gamma x)}{\sinh(\gamma L)}\,\frac{\bar{I}(s)}{\bar{r}}, \tag{5.144A}$$

where

$$\gamma = \sqrt{1+s},$$

$\bar{r}$ is the axial resistance per characteristic length and

$$Z = \bar{r}/\gamma,$$

is the *characteristic impedance* of the cable.

Now consider a tree with one order of branching and suppose, in the notation employed previously, there is an input current $I(t)$ on the (2,1) cylinder. Assuming killed-end conditions, the coupled boundary-value problems are

$$V_{11,t} = -V_{11} + V_{11,xx},$$
$$V_{21,t} = -V_{21} + V_{21,x} + \delta(x_{21} - y_{21})I(t),$$
$$V_{22,t} = -V_{22} + V_{22,x},$$
$$V_{11}(0,t) = V_{21}(L_{21},t) = V_{22}(L_{22},t) = 0,$$
$$V_{11}(L_{11},t) = V_{21}(0,t) = V_{22}(0,t),$$
$$V_{11,x}(L_{11},t)/\bar{r}_{11} = V_{21,x}(0,t)/\bar{r}_{21} + V_{22,x}(0,t)/\bar{r}_{22}.$$

Upon Laplace transforming this problem, the following expression is obtained for the Laplace transform of the voltage on the primary branch:

$$\bar{V}(x_{11};s) = \frac{Z_{11}Z_{21}Z_{22}\sinh(\gamma L_{22})\sinh(\gamma(L_{21}-y_{21}))\sinh(\gamma x_{11})(\bar{I}(s)/\bar{r}_{21})}{\begin{bmatrix} Z_{11}Z_{21}\sinh(\gamma L_{11})\sinh(\gamma L_{21})\cosh(\gamma L_{22}) \\ +Z_{11}Z_{22}\sinh(\gamma L_{11})\cosh(\gamma L_{21})\sinh(\gamma L_{22}) \\ +Z_{21}Z_{22}\cosh(\gamma L_{11})\sinh(\gamma L_{21})\sinh(\gamma L_{22}) \end{bmatrix}}. \tag{5.144B}$$

Butz and Cowan represent expressions such as (5.144A) and (5.144B) as the quotient of two factors. The numerator contains the geometrical relation between the input location and the recording electrode [x_{11} in (5.144B)], whereas the denominator depends topologically on the entire tree. Then (5.144A) and (5.144B) are represented by

$$\bar{V}(x;s) = \frac{[\underline{x}][\underline{L-y}]}{[\underline{L}]}(\bar{I}(s)/\bar{r}), \tag{5.144A$'$}$$

and

$$\bar{V}_{11}(x_{11};s) = \frac{Z_{11}Z_{21}Z_{22}[\underline{x_{11}}][\underline{L_{21}-y_{21}}][\underline{L_{22}}](\bar{I}(s)/\bar{r}_{21})}{\left[\ \underline{L_{11}}\ \begin{smallmatrix} \diagup\, L_{21} \\ \diagdown\, L_{22} \end{smallmatrix}\ \right]}. \tag{5.144B$'$}$$

The *numerator* is the product of factors obtained as follows. *Delete* the direct path from the recording electrode to the input location. This breaks the tree. The numerator contains products of hyperbolic sines (for killed ends) of γ times the lengths of the remaining segments as well as the factor $\sinh(\gamma x)$ and the input current term $\bar{I}(s)/\bar{r}$. In addition, for each branch point encountered along the deleted path, there appears a product of the three characteristic impedances of the cylinders emanating from the branch point.

The *denominator* in (5.144B) has cyclic symmetry. Each term therein is of the form $Z_A Z_B \sinh(\gamma A)\sinh(\gamma B)\cosh(\gamma C)$, where A, B, and C are $(1,1)$, $(2,1)$, and $(2,2)$, respectively, without repetition.

For the tree with two orders of branching and input on cylinder $(3,1)$,

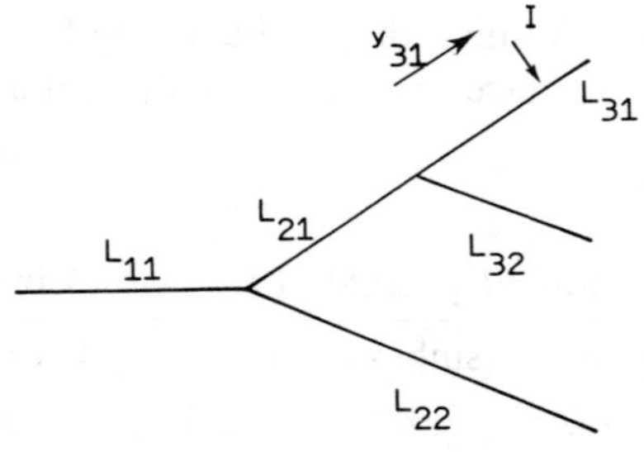

we obtain by using the *deletion rule* (which leaves segments $L_{31}-y_{31}$,

L_{32}, and L_{22}),

$$\overline{V}_{11}(x_{11};s)=\frac{Z_{11}Z_{21}^{2}Z_{22}Z_{31}Z_{32}\left[x_{11}\right]\left[L_{31}-y_{31}\right]\left[L_{32}\right]\left[L_{22}\right]\left(\bar{I}(s)/\bar{r}_{31}\right)}{\left[\text{tree: } L_{11},\ L_{21},\ L_{31},\ L_{32},\ L_{22}\right]},\qquad(5.144\text{C})$$

where

L_{11} L_{21} L_{31} L_{32} L_{22} = L_{11} L_{21} L_{22} ⊗ L_{21} L_{31} L_{32}

⊗ means multiplication in the ordinary sense, except that products of terms involving the same segment obey the following contraction rules:

$$\sinh(\gamma L)\sinh(\gamma L)\rightarrow\sinh(\gamma L),$$

$$\cosh(\gamma L)\sinh(\gamma L)\rightarrow\cosh(\gamma L),$$

$$\cosh(\gamma L)\cosh(\gamma L)\rightarrow\sinh(\gamma L).$$

It is apparent that the graphical calculus provides an expeditious method of solution of some of the complicated linear systems encountered. Butz and Cowan have proved their rules are valid in general and have covered the cases of multiple branching and general linear boundary conditions including sealed ends and lumped-soma terminations. Further details and an interesting extension to include spine-bearing dendrites are given by Kawato and Tsukahara (1983).

There remains the problem of inverting the Laplace transforms. Norman (1972) gives a computer program for numerical inversion of the Laplace transforms of solutions of the cable equation with various boundary conditions using the fast Fourier transform. Horwitz (1981, 1983) has pursued the inversion of the transforms for dendritic trees. It was noted that the transforms obtained by the *Butz–Cowan al-*

gorithm were factorizable as follows:

$$\bar{V}(x;s) = \bar{R}(x,s)\bar{A}(D,s)\bar{F}(G,s)\bar{I}(s),$$

where $\bar{R}$ is the factor describing location of the recording electrode; $\bar{A}$ is the factor describing input location; $\bar{F}$ is the factor containing the geometry of the dendritic tree; and $\bar{I}$ is the input-current transform.

Using the convolution theorem for Laplace transforms the inversion may be written as:

$$V(x,t) = \int_0^t dt_2 \int_0^{t_2} dt_3 \int_0^{t_3} dt_4\, R(x,t_4) F(G, t_3 - t_4) \times A(D, t_2 - t_3) I(t - t_2). \quad (5.145)$$

Horwitz has utilized the results for certain special symmetric dendritic trees to obtain results in general of the form of what he calls primitive integrals plus correction terms. A few cases have been worked out thoroughly – for example, the effects of departures from Rall's three-halves power law on the response to delta-function inputs in trees with one order of branching.

Compartmental analysis

Thus far we have let the potential over dendritic segments be governed by cable equations. An alternative is to compartmentalize the neuron into a set of lumped circuits of the Lapicque type, the assumption being that the potential is approximately constant over such local regions. With this approach, the voltage in each small segment satisfies an ordinary differential equation and thus one has to deal with a set of coupled such equations. Given that ordinary differential equations are so much easier to handle than partial differential equations, this approach is not without usefulness (Rall 1964, 1967; Perkel and Mulloney 1978; Perkel, Mulloney, and Budelli 1981; Edwards and Mulloney 1984).

5.9 The equivalent cylinder

In the previous section we saw that, in general, the determination of the Laplace transform of the depolarization over a dendritic tree is straightforward but the inversion of the transform is usually difficult. It is a great advantage to be able to reduce the many cable equations for the various segments of a dendritic tree to a single cable equation that can, in many cases of interest, be solved explicitly (cf. Sections 5.3–5.5). Under some circumstances this is possible using the equivalent cylinder concept of Rall. We will deduce the conditions under which a dendritic tree with one order of branching can be

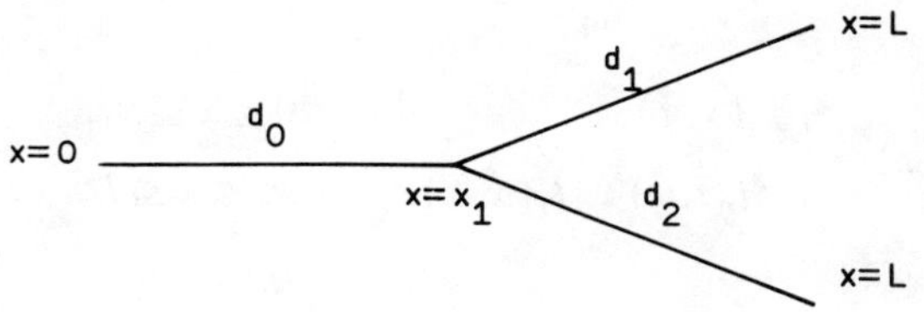

Figure 5.7. Tree with one order of branching (Theorem 5.1).

reduced to an equivalent cylinder. The general case can be treated by successive reductions.

Recall that when distance is in space constants λ and time is in time constants τ, the cable equation on a cylinder of diameter d has the form, from Equation (4.57),

$$V_t = -V + V_{xx} + k/d^{3/2} I, \tag{5.146}$$

where k is a constant and $I(x,t)$ is the current density. In the following we will always work with dimensionless space and time variables (except that diameters are in centimeters). We now establish the following.

Theorem 5.1 (Reduction of a tree with one order of branching to an equivalent cylinder)

Consider a dendritic tree of the kind sketched in Figure 5.7. Let the diameter of the parent cylinder be d_0 and let those of the daughter cylinders be d_1 and d_2. Let the corresponding current densities be I_0, I_1, and I_2, and let the initial depolarizations be $v_0(x)$, $v_1(x)$, and $v_2(x)$. If (i) $d_0^{3/2} = d_1^{3/2} + d_2^{3/2}$, (ii) *the terminals of the daughter cylinders are the same distance from* $x = 0$, (iii) $v_1(x) = v_2(x)$, (iv) *the boundary conditions at the terminals of the daughter cylinders are the same,* (v) $I_1(x,t)/d_1^{3/2} = I_2(x,t)/d_2^{3/2}$, *then* (a) *the depolarization is the same at points on the daughter cylinders, which are the same distance from* $x = 0$, (b) *the depolarization at any point on the tree, which is a distance x from $x = 0$, can be obtained from the solution of the cable equation*

$$V_t = -V + V_{xx} + \frac{k}{d_0^{3/2}} I, \qquad 0 < x < L, \tag{5.147}$$

with the given boundary conditions at $x = 0$ and $x = L$ and with initial value

$$v(x) = \begin{cases} v_0(x), & 0 \le x \le x_1, \\ v_1(x), & x_1 \le x \le L, \end{cases} \tag{5.148}$$

where

$$I(x,t)=\begin{cases} I_0(x,t), & 0\le x\le x_1,\\ (d_0/d_1)^{3/2}I_1(x,t), & x_1\le x\le L. \end{cases} \tag{5.149}$$

Note that we are only considering the possibility of a common bifurcation at $x = x_1$. The case of three or more daughter cylinders can be handled in the same way. The three-halves power law for diameters is approximately satisfied in some cases (Rall 1977) but in others it is clearly violated (Hillman 1979).

Proof. (a) Because of the assumed continuity of the potential at the branch point, the potential at the left-hand endpoints of the daughter cylinders is the same. Conditions (ii), (iii), (iv), and (v) imply that the potentials $V_1(x,t)$ and $V_2(x,t)$ on the daughter cylinders satisfy the same cable equation on the same space interval, with the same boundary conditions at the ends of the interval and the same initial condition. Hence, by existence and uniqueness theorems for linear partial differential equations (John 1982), $V_1(x,t) = V_2(x,t)$ for all $t \ge 0$ and $x_1 \le x \le L$. This establishes (a).

(b) Since $V_1 = V_2$ we need only consider the equations

$$V_{0,t} = -V_0 + V_{0,xx} + \frac{k}{d_0^{3/2}}I_0, \qquad 0<x<x_1, \tag{5.150}$$

$$V_{1,t} = -V_1 + V_{1,xx} + \frac{k}{d_1^{3/2}}I_1, \qquad x_1<x<L. \tag{5.151}$$

Denote the Laplace transform of a quantity by a bar [e.g., the Laplace transform of $V_0(x,t)$ is $\overline{V}_0(x;s)$, where s is the transform variable]. When we take the Laplace transforms of (5.150) and (5.151) we get, with primes denoting differentiation with respect to x,

$$-\overline{V}_0'' + (1+s)\overline{V}_0 = v_0 + \frac{k}{d_0^{3/2}}\bar{I}_0, \qquad 0<x<x_1, \tag{5.152}$$

$$-\overline{V}_1'' + (1+s)\overline{V}_1 = v_1 + \frac{k}{d_1^{3/2}}\bar{I}_1, \qquad x_1<x<L, \tag{5.153}$$

with a given (transformed) boundary condition for $\overline{V}_0$ at $x=0$ and a given (transformed) boundary condition for $\overline{V}_1$ at $x = L$. It is sufficient to show the equivalence of (5.152) and (5.153) to an equation for a single cylinder and then by inversion of the Laplace transforms to obtain the result for the time-dependent potentials. We define

$$\bar{V}(x; s) = \begin{cases} \bar{V}_0(x; s), & 0 \le x \le x_1, \\ \bar{V}_1(x; s), & x_1 \le x \le L, \end{cases} \tag{5.154}$$

$$\bar{I}(x; s) = \begin{cases} \bar{I}_0(x; s), & 0 \le x \le x_1, \\ (d_0/d_1)^{3/2}\bar{I}_1(x; s), & x_1 \le x \le L, \end{cases} \tag{5.155}$$

$$v(x) = \begin{cases} v_0(x), & 0 \le x \le x_1, \\ v_1(x), & x_1 \le x \le L. \end{cases} \tag{5.156}$$

Then

$$-\bar{V}'' + (1+s)\bar{V} = v + \frac{k}{d_0^{3/2}}\bar{I}, \qquad 0 < x < x_1, \tag{5.157}$$

$$-\bar{V}'' + (1+s)\bar{V} = v + \frac{k}{d_0^{3/2}}\bar{I}, \qquad x_1 < x < L, \tag{5.158}$$

and $\bar{V}$ satisfies the given boundary conditions at $x = 0$ and $x = L$.

We will show that if

$$d_0^{3/2} = d_1^{3/2} + d_2^{3/2},$$

then $\bar{V}(x; s) = \bar{U}(x; s)$, where $\bar{U}$ is the solution of the single equation

$$-\bar{U}'' + (1+s)\bar{U} = v + \frac{k}{d_0^{3/2}}\bar{I}, \qquad 0 < x < L, \tag{5.159}$$

with the given boundary conditions at $x = 0$ and $x = L$.

We observe that continuity of potential at x_1 gives

$$\bar{V}_0(x_1; s) = \bar{V}_1(x_1; s) \doteq a(s), \tag{5.160}$$

and conservation of axial current gives

$$\frac{\bar{V}_0'(x_1; s)}{\bar{r}_0} = \frac{\bar{V}_1'(x_1; s)}{\bar{r}_1} + \frac{\bar{V}_2'(x_1; s)}{\bar{r}_2}$$

$$= \bar{V}_1'(x_1; s)\left[\frac{1}{\bar{r}_1} + \frac{1}{\bar{r}_2}\right], \tag{5.161}$$

from part (a). We have seen in Section 4.13 that $1/\bar{r}_0 = 1/\bar{r}_1 + 1/\bar{r}_2$ implies $d_0^{3/2} = d_1^{3/2} + d_2^{3/2}$. Hence the latter condition implies

$$\bar{V}_0'(x_1; s) = \bar{V}_1'(x_1; s) \doteq b(s). \tag{5.162}$$

Now, Equation (5.157) with the "initial" values, $\bar{V}(x_1; s) = a(s)$ and $\bar{V}'(x_1; s) = b(s)$, completely and uniquely determines $\bar{V}(x; s)$ for $0 \le x \le x_1$. Similarly, Equation (5.158) with the "initial" values,

$\overline{V}(x_1; s) = a(s)$ and $V'(x_1; s) = b(s)$, completely and uniquely determines $\overline{V}(x; s)$ for $x_1 \le x \le L$. Furthermore, Equation (5.159) with "initial" values, $\overline{U}(x_1; s) = a(s)$ and $\overline{U}'(x_1; s) = b(s)$, completely and uniquely determines $\overline{U}(x; s)$ for $0 \le x \le L$. Thus

$$\overline{U}(x; s) = \overline{V}(x; s), \quad \text{for } 0 \le x \le x_1, \tag{5.163}$$

and

$$\overline{U}(x; s) = \overline{V}(x; s), \quad \text{for } x_1 \le x \le L. \tag{5.164}$$

But $\overline{V}$, by definition, satisfies the given boundary conditions at $x = 0$ and $x = L$. Thus, since $\overline{U} = \overline{V}$, it follows that $\overline{U}$ also satisfies these boundary conditions at $x = 0$ and $x = L$. But Equation (5.159) with these boundary conditions completely and uniquely determines $\overline{U}$ on $0 \le x \le L$. It follows that $\overline{V}$ is the solution of the same equation with the same boundary conditions. Inverting the Laplace transforms, we see that Equation (5.147) is established, which completes the proof.

The above theorem generalizes as follows.

Theorem 5.2 (Equivalent cylinder)

Consider a general dendritic tree. If (i) *the three-halves power of the parent cylinder at each branch point is the sum of the three-halves powers of the diameters of the daughter cylinders*, (ii) *the dendritic terminals are all the same distance from the origin*, (iii) *the boundary conditions at all dendritic terminals are the same*, (iv) *the initial depolarization is the same at all points equidistant from the origin*, (v) *on each segment alive at a given distance from the origin, the current density divided by the three-halves power of the diameter of the segment is the same, then* (a) *the depolarization is the same at all points which are the same distance from the origin*, (b) *the depolarization at any point on the tree, which is a distance x from the origin, can be obtained from the solution of the cable equation* (5.147), *with the given boundary conitions at $x = 0$ and $x = L$, with initial depolarization at each point the same as on any segment at that distance from the origin, and with $I(x, t) = (d_0/d_i)^{3/2} I_i(x, t)$, where d_i is the diameter, $I_i(x, t)$ is the current density on any segment (the ith) alive at x and d_0 is the trunk diameter.*

Proof. The proof follows by a reduction argument whereby subtrees as in Theorem 5.1 are reduced to equivalent cylinders. Theorem 5.2 is a special case of the following result.

5.10 Determining the potential over the soma–dendritic surface for arbitrary input current: the role of dendrites

In the previous section we saw that under certain symmetry conditions the dendritic tree could be transformed to an equivalent cylinder. Solving the cable equation for the equivalent cylinder immediately gives the potential at all points on the dendritic tree. The symmetry requirements on the input-current density and boundary conditions are highly restrictive, however. In real physiological situations the symmetry requirement on the input current will hardly ever be met, except in experiments where current is injected into the soma.

In this section we give a method of determining the potential over the soma–dendritic surface when the input-current density is *arbitrary*. First, for the case of a neuron with a single dendritic tree, the method relies on a mapping theorem (Theorem 5.3). A transformation is made from tree to (nonequivalent) cylinder and the solution of the cable equation for the cylinder immediately gives the potential at the soma and on the primary trunk. The potential over the rest of the nerve cell can be found by solving a sequence of cable equations with known boundary conditions. Second, a theorem, which applies to neurons with several dendritic trees emanating from a common soma, enables the potential at the soma to be found from the solution of a single cable equation. Again the potential over the rest of the neuronal surface can be found by solving a sequence of cable equations with known boundary conditions. More details than appear here are given in Walsh and Tuckwell (1985).

5.10.1 A neuron with a single dendritic tree

We consider a dendritic tree such as that sketched in Figure 5.8. The origin of the tree is marked 0, representing the soma, and the right-hand extremities are labeled $T_1, \ldots, T_n$, representing the dendritic terminals. Branching occurs at electrotonic distances $x_1, x_2, \ldots, x_m$, from the origin, and with $x_0 = 0$, we have $x_0 < x_1 < \cdots < x_m$ There may be two or more daughter cylinders at each branch point but the following condition must be met. If d_0 is the diameter of the parent cylinder and $d_1, d_2, \ldots$ are the diameters of the daughter cylinders, then

$$d_0^{3/2} = d_1^{3/2} + d_2^{3/2} + \cdots . \tag{5.165}$$

Each segment is assumed to have the same diameter throughout its length.

On each segment of the dendritic tree it is assumed that the depolarization satisfies a cable equation. With each point on the

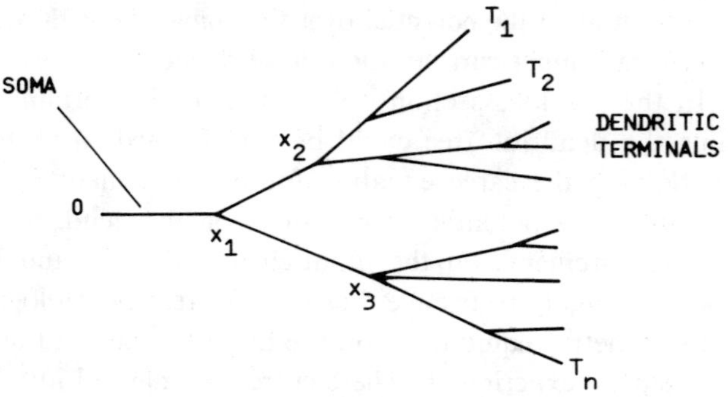

Figure 5.8. Schematic representation of a dendritic tree of the kind considered in the theorem of Section 5.10.1. The soma is labelled 0, the dendritic terminals $T_1, \ldots, T_n$ and successive branch points occur at distances $x_1, x_2, \ldots, x_m$ from the origin.

dendritic tree we can associate an electrotonic distance x from the origin. We suppose there are n_i cylinders alive between x_i and x_{i+1} and let their corresponding depolarizations, current densities, and diameters be V_{ij}, I_{ij}, and d_{ij}, respectively, where $j = 1, 2, \ldots, n_i$ and $i = 0, 1, \ldots, m$. Thus, from (4.57),

$$V_{ij,t} = -V_{ij} + V_{ij,xx} + \frac{k}{d_{ij}^{3/2}} I_{ij}, \qquad x_i < x < x_{i+1}. \qquad (5.166)$$

The distances of the dendritic terminals from the origin must all have the same value l, in any practical application. We let the boundary conditions at the terminals be

$$\alpha V_{mj}(l,t) + \beta V_{mj,x}(l,t) = \gamma_j(t), \qquad j = 1, 2, \ldots, n, \qquad (5.167)$$

where α and β are the same for each terminal. Also, suppose there are given the initial values for the depolarization

$$V_{ij}(x,0) = v_{ij}(x), \qquad x_i < x < x_{i+1}, \qquad (5.168)$$

and a given boundary condition at the soma. Then we have the following result.

Theorem 5.3

The weighted sum of the depolarizations

$$V(x,t) = \sum_{j=1}^{n_i} \left(\frac{d_{ij}}{d_{01}} \right)^{3/2} V_{ij}(x,t), \qquad x_i < x < x_{i+1},\ i = 0, 1, \ldots, m, \qquad (5.169)$$

satisfies the cable equation

$$V_t = -V + V_{xx} + \frac{k}{d_{01}^{3/2}} I(x,t), \qquad 0 < x < l,\ t > 0, \qquad (5.170)$$

where

$$I(x,t) = \sum_{j=1}^{n_i} I_{ij}(x,t), \qquad x_i < x < x_{i+1},\ i = 0,1,\ldots,m, \qquad (5.171)$$

with initial value

$$V(x,0) = \sum_{j=1}^{n_i} \left(\frac{d_{ij}}{d_{01}}\right)^{3/2} v_{ij}(x), \qquad x_i < x < x_{i+1}, \qquad (5.172)$$

boundary condition

$$\alpha V(l,t) + \beta V_x(l,t) = \sum_{j=1}^{n} \left(\frac{d_{mj}}{d_{01}}\right)^{3/2} \gamma_j(t), \qquad (5.173)$$

and the given boundary condition at the origin.

The proof of this theorem is given in Walsh and Tuckwell (1985). Theorem 5.3 enables the depolarization at the soma and on the trunk of the dendritic tree to be found by solving a single cable equation. The reason for this is that the mapping (5.169) is an identity on the trunk

$$V(x,t) = V_{01}(x,t), \qquad 0 \le x \le x_1. \qquad (5.174)$$

Thus by solving (5.170) with boundary conditions at $x = 0$ and $x = l$, one immediately obtains the depolarization on the trunk.

This result is very useful because usually, either in the analysis of experimental results or in theoretical studies of neuronal integration, one is more interested in the response at and near the soma (the presumed site of implantation of a recording electrode or possibly the trigger zone) than at points elsewhere on the dendritic tree. The depolarization may be found at points not on the trunk, but more cable equations, again with two-point boundary conditions need to be solved.

Effect of location of an input

Suppose an input current with density $i(t)\,\delta(x-y)$ arises anywhere on a single segment of the dendritic tree at an electrotonic distance y from the soma. By Theorem 5.3 the depolarization on the trunk is the solution of

$$V_t = -V + V_{xx} + \frac{k}{d_{01}^{3/2}} \delta(x-y) i(t), \qquad 0 < x < l,\ t > 0, \tag{5.175}$$

restricted to $0 \leq x \leq x_1$. Note that Equation (5.175) is precisely the equation obtained if the trunk is extended to l and the same input current is applied on the extended cylinder at the same distance y from the origin. We therefore have the following principle.

Principle of independence of response on geometry

For a single dendritic tree, satisfying the requirements of Theorem 5.3, *the depolarization on the trunk and at the soma in response to a given input current occurring at any point on the dendritic tree depends only on the electrotonic distance of the input from the soma and not on the details of the geometry of the dendritic tree.*

Notice that this principle applies not only to the steady-state response but also the time-dependent response.

Response to an arbitrary input current. Given arbitrary (but reasonably well mathematically behaved) input-current densities I_{ij}, $i = 0, 1, \ldots, m$ and $j = 1, 2, \ldots, n_i$, on the various segments of the dendritic tree, the cable equation (5.170) satisfied by the weighted sum of potentials (5.169) can be written down using formula (5.171) for $I(x, t)$. Suppose, with suitable given boundary conditions at $x = 0$ and $x = l$, the Green's function for (5.170) is $G(x, y; t)$. Then, assuming an initially zero depolarization, the solution of the cable equation is

$$V(x, t) = \int_0^l \int_0^t G(x, y; t-s) I(y, s)\, ds\, dy, \tag{5.176}$$

so that the depolarization at the soma and trunk can be found immediately for an arbitrary input configuration.

Role of dendrites

It can be seen from the above principle of independence of response on geometry and the result for an arbitrary input current, that the neuron, with a dendritic tree of the class described, loses nothing by attenuation due to branching as far as the effect of stimuli at the soma and trunk is concerned. No input is discriminated by virtue of its position on the dendritic tree except with regard to electrotonic distance from the origin. Branching does not serve to attenuate, only to create, the possibility of a greater number of inputs (synapses) at a given distance from the soma. This enables the neuron to achieve the integration and discrimination (by means of a sequence of output spikes) of more complex input patterns. In fact, by virtue of branching, more inputs are possible at a given distance from the soma. If these inputs are from a common source, the branching serves to increase the response at the soma and trunk when these inputs are active.

Finding the depolarization over the entire dendritic tree

Theorem 5.3 enables the depolarization at the soma and trunk to be found immediately by solving a cable equation with boundary conditions at $x = 0$ and $x = l$. The depolarization over the remainder of the dendritic tree can be found by solving a sequence of cable equations with two-point boundary conditions. The procedure, in outline, is as follows.

Solving Equation (5.170) gives the depolarization for $0 \le x \le x_1$, and, in particular, it gives $V(x_1, t)$. Suppose at x_1, the first branch point, there are just two daughter cylinders. Now apply Theorem 5.3 to the "upper" dendritic tree extending from $x = x_1$ to $x = l$. This enables the potential to be found on the trunk of this subtree from x_1 to the next branch point, since the value of the potential at its origin $(x = x_1)$ is known. The lower subtree can be treated in the same way, or, if both upper and lower subtrees have their first branch points at the same distance from the origin, the potential on one of them can be found by subtraction if it is known on the other. Proceeding in this fashion, one may obtain the potential over the whole dendritic tree.

An example

We will apply the technique just described to the situation depicted in Figure 5.9A. An impulse is delivered at $t = 0$ to a point on the lowermost dendrite at a distance y from the soma. There are

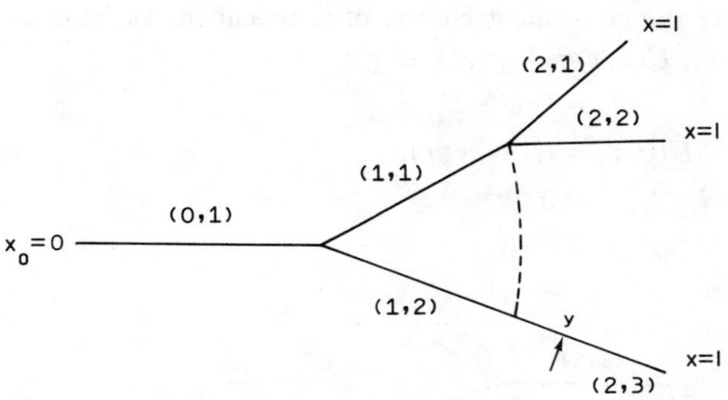

Figure 5.9A. Dendritic tree with input shown for text example.

sealed-end conditions at the soma and all dendritic terminals. For convenience, assume $y > x_2$ and note that in accordance with the notation of Theorem 5.3 the segments (1, 2) and (2, 3) both occur on the same dendrite.

The cable equations for each segment are

$$V_{01,t} = -V_{01} + V_{01,xx}, \tag{5.177A}$$

$$V_{11,t} = -V_{11} + V_{11,xx}, \tag{5.177B}$$

$$V_{12,t} = -V_{12} + V_{12,xx}, \tag{5.177C}$$

$$V_{21,t} = -V_{21} + V_{21,xx}, \tag{5.177D}$$

$$V_{22,t} = -V_{22} + V_{22,xx}, \tag{5.177E}$$

$$V_{23,t} = -V_{23} + V_{23,xx} + \frac{k}{d_{23}^{3/2}}\delta(x-y)\,\delta(t). \tag{5.177F}$$

Then $V(x,t)$ defined by (5.169) satisfies

$$V_t = -V + V_{xx} + \frac{k}{d_{01}^{3/2}}\delta(x-y)\,\delta(t), \qquad 0 < x < l, \tag{5.178}$$

and the solution of this is $G(x, y; t)$ given by either (5.39) or (5.42). Thus the potential on the trunk is $V_{01}(x,t) = G(x, y; t)$, $0 \le x < x_1$, and the value of the depolarization at x_1 is $G(x_1, y; t) = V_{01}(x_1, t)$.

Consider now the upper subtree shown in Figure 5.9B. Applying Theorem 5.3 to this tree is not necessary since we may use Theorem 5.2. That is, the subtree satisfies conditions for the equivalent cylinder. Letting $V_1(x,t)$ be the depolarization at distance x from x_1, we have

$$V_{1,t} = -V_1 + V_{1,xx}, \qquad x_1 < x < l, \tag{5.179}$$

with boundary conditions

$$V_1(x_1, t) = G(x_1, y; t), \tag{5.179A}$$

$$V_{1,x}(l, t) = 0. \tag{5.179B}$$

We make the convenient change of independent variable $z = x - x_1$, $l_1 = l - x_1$, $U(z,t) = V_1(x,t)$, to get

$$U_t = -U + U_{zz}, \tag{5.180}$$

$$U(0,t) = G(x_1, y; t), \tag{5.180A}$$

$$U_z(l_1, t) = 0. \tag{5.180B}$$

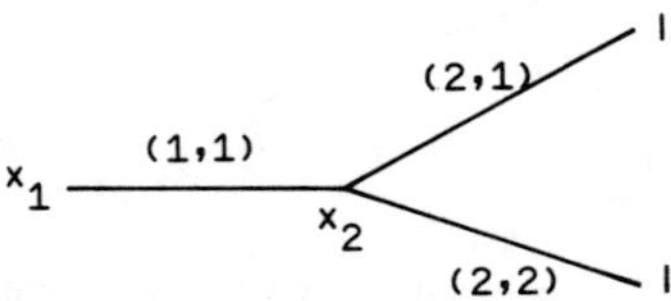

Figure 5.9B. Upper subtree in text example.

Equation (5.180A) has a nonhomogeneous boundary condition and it is desirable to change dependent variables to obtain a homogeneous boundary condition. Thus put

$$W(z,t) = U(z,t) - G(x_1, y; t), \tag{5.181}$$

so that

$$W_t = -W + W_{zz} - [G(x_1, y; t) + G_t(x_1, y; t)], \tag{5.182}$$

with boundary conditions

$$W(0,t) = 0, \tag{5.182A}$$

$$W_z(l_1, t) = 0. \tag{5.182B}$$

Now let $G_1(z, y_1; t)$ be the Green's function for a cable with a sealed end at l_1 and a killed end at 0 (see Section 5.6). Then, from (5.7), we have

$$W(z,t) = -\int_0^{l_1}\int_0^t G_1(z, y_1; t-s)[G(x_1, y; s) + G_s(x_1, y; s)]\, ds\, dy_1. \tag{5.183}$$

If we use the separation of variables form for the Green's functions, we obtain, on substituting in the above formulas,

$$V_1(x,t) = \alpha\exp(-t)\left[1 + 2\sum_{n=1}^{\infty}\phi_n\exp(-\beta_n t)\right] + \frac{8\alpha\exp(-t)}{\pi}$$

$$\times\sum_{n=1}^{\infty}\frac{\psi_n(x)}{2n+1}\sum_{m=0}^{\infty}\beta_m\phi_m\left\{\frac{\exp(-\beta_m t) - \exp(-\mu_n t)}{\mu_n - \beta_m}\right\}, \tag{5.184}$$

where

$$\alpha = \frac{k}{d_{01}^{3/2}l},$$

$$\beta_n = \frac{n^2\pi^2}{l^2},$$

$$\mu_n = \left(\frac{(2n+1)\pi}{2l_1}\right)^2,$$

$$\phi_n = \cos\left(\frac{n\pi x_1}{l}\right)\cos\left(\frac{n\pi y}{l}\right),$$

$$\psi_n(x) = \sin\frac{(2n+1)\pi x}{2l_1}.$$

Note that if $\mu_n = \beta_m$ the factor in curly brackets is replaced by $t\exp(-\mu_n t)$.

In particular, (5.184), for $x_1 \le x \le x_2$, gives the depolarization $V_{11}(x, t)$ on segment (1,1), and it gives $V_{21}(x, t)$ and $V_{22}(x, t)$, these being equal, for $x_2 \le x \le l$. The depolarization on dendrite (1,2) is obtained by subtraction as

$$V_{12}(x, t) = d_{12}^{-3/2}\left[d_{01}^{3/2}V(x, t) - d_{11}^{3/2}V_1(x, t)\right], \qquad x_1 \le x \le x_2, \tag{5.185}$$

where $V(x, t) = G(x, y; t)$ and $V_1(x, t)$ is given by (5.184). Similarly, on the remaining segment (2,3), we find

$$V_{23}(x, t) = d_{23}^{-3/2}\left[d_{01}^{3/2}V(x, t) - d_{11}^{3/2}V_1(x, t)\right], \qquad x_2 \le x \le l. \tag{5.186}$$

Thus the depolarization is found explicitly over the whole dendritic tree. For numerical results see Walsh and Tuckwell (1985).

5.10.2 **A neuron with several dendritic trees**

Many nerve cells, such as spinal motoneurons (see Chapter 1) and pyramidal cells, which are found in the cerebral cortex and hippocampus, have several dendritic trees whose trunks emerge from the cell body. This leads us to consider several cylinders, each of the same electrotonic length l, emanating from a common origin, representing the soma. Each cylinder may represent a reduced dendritic tree in accordance with the procedure given in Theorem 5.3 of the previous subsection. A schematic example is shown in Figure 5.10, which indicates that there are no symmetry requirements.

Suppose there are M cylinders emanating from O ($x = 0$) and that on the ith cylinder the depolarization obeys

$$V_{i,t} = -V_i + V_{i,xx} + kd_i^{-3/2}I_i, \qquad 0 < x < l,\ t > 0, \tag{5.187}$$

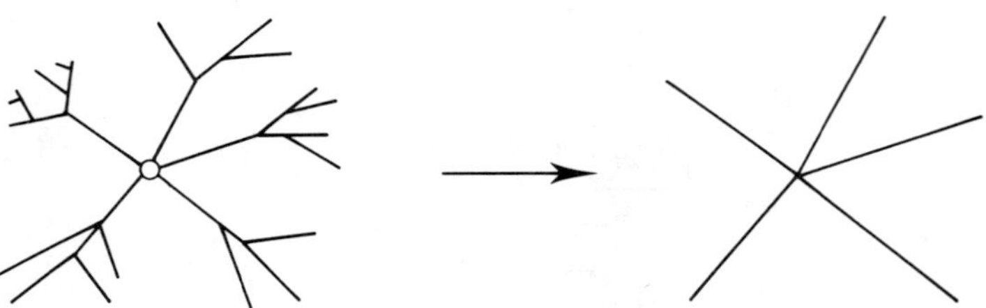

Figure 5.10. A schematic neuron with several dendritic trees emanating from a common soma. One dendritic tree is bifurcated near its terminals to suggest that it may represent an axon with telodendria. In the right part of the figure, the trees are transformed to cylinders with a common origin.

where d_i is the diameter of the ith cylinder. We suppose there are given initial values

$$V_i(x,0) = v_i(x), \qquad 0 \le x \le l,$$

and boundary conditions

$$\alpha V_i(l,t) + \beta V_{i,x}(l,t) = \gamma_i(t).$$

We then have the following result.

Theorem 5.4
The depolarization at the origin (soma) is

$$V(t) = U(0,t) \Big/ \sum_{i=1}^{M} \frac{1}{\bar{r}_i}, \tag{5.188}$$

where $\bar{r}_i$ is the internal resistance per characteristic length of the ith *cylinder and $U(x,t)$ is the solution of the single cable equation*

$$U_t = -U + U_{xx} + k \sum_{i=1}^{M} d_i^{-3/2}(\bar{r}_i)^{-1} I_i, \tag{5.189}$$

with zero space derivative $[U_x(0,t) = 0]$ at the origin, with boundary condition at $x = l$,

$$\alpha U(l,t) + \beta U_x(l,t) = \sum_{i=1}^{M} \frac{\gamma_i(t)}{\bar{r}_i}, \tag{5.190}$$

and with initial value $U(x,0) = \Sigma_{i=1}^{M}[v_i(x)/\bar{r}_i]$.

This theorem is also proved in Walsh and Tuckwell (1985).

Effect of input location

Whereas Theorem 5.3 is useful for a cell with a single dendritic tree, Theorem 5.4 applies, without any constraints at the origin except that there is conservation of current, to cells with several dendritic trees. When Theorem 5.4 is applied, the solution of a single cable equation yields the depolarization at the soma for an arbitrary input.

Suppose a neuron has M dendritic trees, each of which satisfies the conditions for Theorem 5.3 and can therefore be transformed to a cylinder of length l. Suppose an input current $i(t)$ occurs on tree j at a distance y from the origin. Applying Theorem 5.3 the potential on the corresponding transformed cylinder satisfies

$$V_{j,t} = -V_j + V_{j,xx} + \frac{k}{d_j^{3/2}}\,\delta(x-y)i(t),$$

where d_j is the diameter of the trunk of the jth tree. (Note that k is assumed to be a constant of the whole neuron.) Then Equation (5.189) for the quantity U, which has dimensions of a current, becomes

$$U_t = -U + U_{xx} + \frac{k}{\bar{r}_j d_j^{3/2}}\,\delta(x-y)i(t).$$

U satisfies the boundary condition (5.190) at $x = l$ and has zero space derivative at $x = 0$. We note that because $\bar{r}_j = \lambda_j r_{i,j}$, where λ_j is the characteristic length and $r_{i,j}$ is the internal resistance per centimeter of the jth trunk, the quantity $\bar{r}_j d_j^{3/2}$ does not depend on diameter and the equation for U can be written as

$$U_t = -U + U_{xx} + k_1\,\delta(x-y)i(t),$$

where $k_1 = k\pi/2\sqrt{\delta\rho_i\rho_m}$, which is also a constant of the neuron. By Theorem 5.4 the depolarization at the origin of the dendritic trees is

$$V(t) = U(0,t)\Big/\sum_{i=1}^{M}\bar{g}_i,$$

where $\bar{g}_i = \bar{r}_i^{-1}$ is the conductance of a characteristic length of the trunk of the ith tree.

The solution of the equation for $U(x,t)$, for a given input current $i(t)$, depends only on the distance from input location to soma. Hence $U(0,t)$ and also the soma potential $V(t)$ depend only on the distance from input to origin. We may therefore state the following.

Extended principle of independence of response on geometry

The depolarization at the soma in response to a given input current occurring at any point on any dendritic tree, depends, for a given neuron, only on the distance between input and soma. The response is independent of which dendritic tree receives the input and is independent of the geometrical details of the tree that receives the input and the geometrical details of all the other dendritic trees. The magnitude of the response is inversely proportional to the sum of the conductances per characteristic length of the dendritic trunks.

Again we see that the response at the soma can be found for an arbitrary configuration of arbitrary input currents by using the formula

$$U(x,t) = \int_0^l \int_0^t G(x,y;t-s) I(y,s)\, ds\, dy,$$

where G is the Green's function and I is the forcing term in Equation (5.189). The value of $U(0, t)$ gives the potential at the soma using (5.188). The soma potential gives the boundary condition at $x = 0$ for each of the M dendritic trees. The application of Theorem 5.4 to each dendritic tree then gives the depolarization on the trunk of each dendritic tree on solution of the corresponding cable equation. By proceeding as outlined in the previous subsection and in the example considered there, the depolarization may be found over the entire neuron.

6

Rall's model neuron

6.1 Introduction

We have seen that the problem of determining the potential throughout dendritic trees is usually quite difficult. We have also seen that under some conditions the dendritic tree may be reduced to a cylinder. We wish to extend the model to include the soma.

One way to do this is to impose a sealed-end condition at the origin of the cylinder. Then the somatic membrane may be considered as part of the cylinder close to $x = 0$. A more satisfactory approach, introduced by Rall (1960), based on the assumption that the membrane potential is uniform across the soma, is to represent the soma membrane as a single capacitance C_s and resistance R_s in parallel, affixed to the dendritic trunk at $x = 0$. Such a termination is referred to as a *lumped soma* at $x = 0$. The procedure is shown schematically in Figure 6.1.

Consider therefore a cylinder of length l cm with such a lumped soma at $x = 0$. The applied current $I_s(t)$ at $x = 0$ divides into a portion that becomes longitudinal current in the cylinder and another portion that passes into the lumped soma. This leads to the boundary condition at $x = 0$,

$$I_s(t) = \frac{V(0,t)}{R_s} + C_s V_t(0,t) - \frac{1}{r_i} V_x(0,t), \tag{6.1}$$

where x is measured in centimeters, t is in seconds, and I_s is in amperes.

The equation for the potential on the cylinder is (4.56A), where space and time are dimensionless,

$$\bar{V}_T = \bar{V}_{XX} - \bar{V} + \bar{I}/\bar{c}_m. \tag{6.2}$$

Letting $\bar{I}_s(T) = \tau I_s(t)$ be the injected current, measured in coulombs per unit (dimensionless) time, the boundary condition (6.1) can be

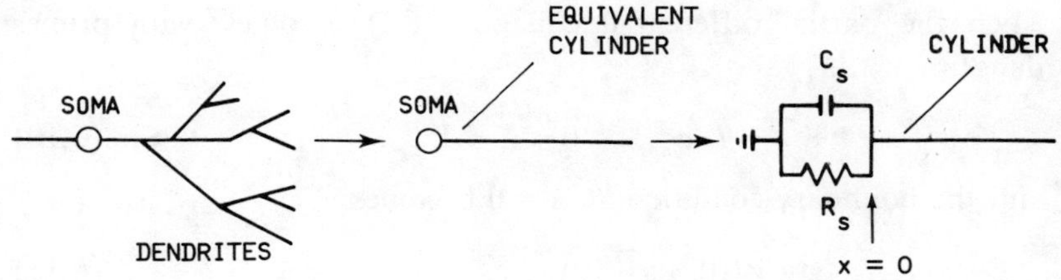

Figure 6.1. The reduction of the soma and dendrites to a cylinder with a lumped-soma attachment.

written as

$$\frac{R_s \bar{I}_s(T)}{\tau} = \bar{V}(0,T) + \frac{\tau_s}{\tau}\bar{V}_T(0,T) - \frac{R_s}{\bar{r}_i}\bar{V}_X(0,T), \qquad (6.3)$$

where the soma time constant is

$$\tau_s = C_s R_s, \qquad (6.4)$$

and $\bar{r}_i$ is the internal resistance of a characteristic length of the cylinder.

We introduce the quantity

$$\gamma = R_s/\bar{r}_i = \bar{g}_i/G_s, \qquad (6.5)$$

which is the ratio of the conductance $\bar{g}_i$ of a characteristic length of cylinder to the conductance of the soma. We also define

$$\sigma = \tau_s/\tau, \qquad (6.6)$$

which is the ratio of the soma time constant to the time constant of the cylinder. Then the boundary condition at the origin is

$$\frac{R_s \bar{I}_s(T)}{\tau} = \bar{V}(0,T) + \sigma\bar{V}_T(0,T) - \gamma\bar{V}_X(0,T). \qquad (6.7)$$

There will be a further boundary condition at $X = L$, such as the sealed-end condition,

$$\bar{V}_X(L,T) = 0. \qquad (6.8)$$

Steady-state equations

We now assume that as $T \to \infty$ a steady state is attained; thus

$$\begin{aligned} \bar{V}(X,T) &\to \tilde{\bar{V}}(X), \\ \bar{V}_T(X,T) &\to 0, \\ \bar{I}(X,T) &\to \tilde{\bar{I}}(X), \\ \bar{I}_s(T) &\to \tilde{\bar{I}}_s. \end{aligned} \qquad (6.9)$$

Then the partial differential equation (6.2) becomes, with primes denoting d/dX,

$$-\bar{\tilde{V}}'' + \bar{\tilde{V}} = \bar{\tilde{I}}/\bar{c}_m, \qquad 0 < X < L, \tag{6.10}$$

and the boundary condition at $X = 0$ becomes

$$R_s \bar{\tilde{I}}_s/\tau = \bar{\tilde{V}}(0) - \gamma \bar{\tilde{V}}'(0), \tag{6.11}$$

whereas the sealed-end condition, for example, becomes

$$\bar{\tilde{V}}'(L) = 0. \tag{6.12}$$

Note that the steady-state problem does not involve the parameter σ.

Steady-state solution with constant current I_0 injected at the soma

We will obtain the steady-state solution when there is a constant current I_0 A injected at the soma and there are no current sources along the cylinder so that $\bar{\tilde{I}} \equiv 0$. We thus solve

$$-\bar{\tilde{V}}'' + \bar{\tilde{V}} = 0, \tag{6.13}$$

subject to the boundary conditions

$$R_s \bar{I}_0/\tau = \bar{\tilde{V}}(0) - \gamma \bar{\tilde{V}}'(0), \tag{6.14}$$

and, assuming a sealed end at L,

$$\bar{\tilde{V}}'(L) = 0. \tag{6.15}$$

The general solution of (6.13) is

$$\bar{\tilde{V}}(X) = A \cosh X + B \sinh X, \tag{6.16}$$

and on imposing the boundary conditions at $X = 0$ and $X = L$, it will be found that

$$\bar{\tilde{V}}(X) = \frac{I_0}{G_s + \dfrac{\tanh L}{\bar{r}_i}} \frac{\cosh(L - X)}{\cosh L}. \tag{6.17}$$

By comparing this expression with (4.100), it is seen that the *form* of the solution is unaltered by including the soma resistance, but the depolarization is reduced uniformly by a multiplicative factor.

Considering the soma and cylinder to be in parallel, with the latter having an input conductance of G_{cyl}, we see that the *whole neuron input conductance* is

$$G_N = G_s + G_{\text{cyl}}. \tag{6.18}$$

Since, from (4.98), a finite cylinder with a sealed end has an input conductance

$$= \tanh L/\bar{r}_i, \tag{6.19}$$

be written as

$$_N \frac{\cosh(L - X)}{\cosh L}, \tag{6.20}$$

euron input resistance is defined through

$$_N = 1/R_s + 1/R_{\text{cyl}}, \tag{6.21}$$

ted since R_s and R_{cyl} are resistances in parallel.

ote that in dimensionless space and time variables, the alternative lumped-soma boundary condition (4.45), becomes [cf. Rall (1960)]

$$\frac{\bar{I}_s(T)}{\tau} = \frac{\bar{V}(0,T)}{R_s} + \frac{\bar{V}_T(0,T)}{R_s} - \frac{\rho}{R_s} V_X(0,T). \tag{6.22}$$

Use of this boundary condition leads to an inconsistent result in the case of a finite cable with a sealed end at $X = L$, but not if the cable is semiinfinite.

6.2 The Green's function for a semiinfinite cylinder with a lumped soma at $x = 0$

The Green's function for a finite cable attached to a lumped soma is quite difficult to obtain and we will first consider the case where the cylinder is semiinfinite, which results in simpler expressions (Rall 1960). Accordingly, we seek $G(x, y; t)$ obeying the equation

$$G_t = -G + G_{xx}, \qquad t > 0, 0 < x < \infty, \tag{6.23}$$

with initial condition

$$G(x, y; 0) = \delta(x - y), \tag{6.24}$$

with boundary condition at infinity

$$|G(\infty, y; t)| < \infty, \tag{6.25}$$

and with boundary condition at the origin

$$G(0, y; t) + G_t(0, y; t) - \gamma G_x(0, y; t) = 0, \tag{6.26}$$

where $\gamma = 1/\bar{r}G_s$. G_s is the soma conductance and $\bar{r}$ is the internal resistance of a characteristic length of the cylinder. It has been assumed that the time constant of the membrane of the soma is the same as that of the dendrites – see Durand (1984) for calculations when this assumption is relaxed.

We proceed by the method of Laplace transforms, defining the transform of the Green's function

$$G_L(x, y; s) = \int_0^\infty e^{-st} G(x, y; t)\, dt. \tag{6.27}$$

Then, denoting the Laplace transform operation by $\mathscr{L}$ and using the initial condition, we find

$$\begin{aligned} \mathscr{L}[G_t] &= sG_L - G(x, y; 0) \\ &= sG_L - \delta(x - y). \end{aligned} \tag{6.28}$$

Thus the differential equation transforms to

$$-G_L'' + (1 + s)G_L = \delta(x - y), \qquad 0 < x < \infty, \tag{6.29}$$

where the primes denote differentiation with respect to x.

Transforming the boundary condition at infinity gives

$$|G_L(\infty, y; s)| < \infty, \tag{6.30}$$

and the boundary condition at $x = 0$ gives

$$G_L(0, y; s)[1 + s] - \gamma G_L'(0, y; s) = 0. \tag{6.31}$$

We utilize the linearly independent solutions of the homogeneous version of (6.29), $\exp(\sqrt{1+s}\,x)$ and $\exp(-\sqrt{1+s}\,x)$, in obtaining a particular solution $G_{L,p}$ for (6.29) using formula (4.73). The Wronskian of these independent solutions is $W = -2\sqrt{1+s}$ and inserting this and the delta function in (4.73) gives

$$G_{L,p} = \begin{cases} 0, & x < y, \\ \dfrac{1}{\sqrt{1+s}} \sinh\sqrt{1+s}\,(y - x), & x \geq y. \end{cases} \tag{6.32}$$

Hence the Laplace transform of the Green's function may be written as

$$\begin{aligned} G_L(x, y; s) = {} & A(s)\exp(\sqrt{1+s}\,x) + B(s)\exp(-\sqrt{1+s}\,x) \\ & + \begin{cases} 0, & x < y, \\ \dfrac{1}{\sqrt{1+s}} \sinh\sqrt{1+s}\,(y - x), & x \geq y, \end{cases} \end{aligned} \tag{6.33}$$

where $A(s)$ and $B(s)$ are constants (at fixed s) to be found from the boundary conditions. Insisting that these are satisfied leads finally to

$$\begin{aligned} G_L(x, y; s) = {} & \frac{1}{2\sqrt{1+s}} e^{-\sqrt{1+s}\,y} \left[e^{\sqrt{1+s}\,x} + \frac{\gamma - \sqrt{1+s}}{\gamma + \sqrt{1+s}} e^{-\sqrt{1+s}\,x} \right] \\ & + \begin{cases} 0, & x < y, \\ \dfrac{1}{\sqrt{1+s}} \sinh\sqrt{1+s}\,(y - x), & x \geq y. \end{cases} \end{aligned} \tag{6.34}$$

To invert the transform and recover $G(x, y; t)$ requires the following three transforms [see, for example, Abramowitz and Stegun (1964), pages 1026 and 1027]:

$$\mathscr{L}\left\{\frac{1}{\sqrt{\pi t}}\exp\left[-\frac{k^2}{4t}\right]\right\} = \frac{1}{\sqrt{s}}\exp(-k\sqrt{s}), \qquad k \geq 0, \tag{6.35}$$

$$\mathscr{L}\left\{\frac{1}{\sqrt{\pi t}}\exp\left[-\frac{k^2}{4t}\right] - a\exp(ak)\exp(a^2 t)\operatorname{erfc}\left[a\sqrt{t} + \frac{k}{2\sqrt{t}}\right]\right\}$$
$$= \frac{\exp(-k\sqrt{s})}{a+\sqrt{s}}, \qquad k \geq 0, \tag{6.36}$$

$$\mathscr{L}\left\{\operatorname{erfc}\left[\frac{k}{2\sqrt{t}}\right] - \exp(ak)\exp(a^2 t)\operatorname{erfc}\left[a\sqrt{t} + \frac{k}{2\sqrt{t}}\right]\right\}$$
$$= \frac{a\exp(-k\sqrt{s})}{s(a+\sqrt{s})}, \qquad k < 0, \tag{6.37}$$

and the general result

$$\mathscr{L}^{-1}\{f_L(s+1)\} = e^{-t}f(t), \tag{6.38}$$

where $f_L(s)$ is the transform of $f(t)$. We then obtain, for $x < y$,

$$G(x, y; t)$$
$$= \frac{\exp(-t)}{2}\left[\frac{1}{\sqrt{\pi t}}\left\{\exp\left(-\frac{(y-x)^2}{4t}\right) - \exp\left(-\frac{(y+x)^2}{4t}\right)\right\}\right.$$
$$\left. + 2\gamma\exp[\gamma(y+x)]\exp(\gamma^2 t)\operatorname{erfc}\left(\gamma\sqrt{t} + \frac{y+x}{2\sqrt{t}}\right)\right]. \tag{6.39}$$

We notice that when $\gamma = 0$, corresponding to an infinite soma conductance, the above expression reduces to

$$G(x, y; t) = \frac{\exp(-t)}{\sqrt{4\pi t}}\left[\exp\left(-\frac{(y-x)^2}{4t}\right) - \exp\left(-\frac{(y+x)^2}{4t}\right)\right], \tag{6.40}$$

which is the result obtained (Section 5.6) for the case of a killed end at $x = 0$. Also, at the soma, $x = 0$, (6.39) reduces to

$$G(0, y; t) = \gamma e^{-t}e^{\gamma y}e^{\gamma^2 t}\operatorname{erfc}\left(\gamma\sqrt{t} + \frac{y}{2\sqrt{t}}\right). \tag{6.41}$$

6.3 The Green's function for a finite cylinder with a lumped soma at $x = 0$

Although the Green's function may be readily obtained for a semiinfinite cylinder with a lumped soma at $x = 0$, in view of the practical applications it is important to have the result for a finite cylinder. It will be seen that a lengthier calculation ensues.

We will confine our attention to a sealed-end condition at $x = L$. Our steps will be as follows. First, we will try the method of separation of variables. We will see that this leads to a partial success because, although the eigenvalues and eigenfunctions can be found, the coefficients of the eigenfunctions in the series solution cannot be found using standard techniques because the eigenfunctions with different eigenvalues are not orthogonal. We will then find the Laplace transform of the Green's function directly from the differential equation. This serves two purposes. By comparing this Laplace transform with the Laplace transform of the series (eigenfunction) expansion, we will be able to find the coefficients that could not be obtained by standard techniques. In addition, by expanding the Laplace transform as a series that converges rapidly for large s (s being the transform variable), we will obtain a representation of the Green's function that converges rapidly for small times, whereas the eigenfunction expansion converges rapidly for large times.

6.3.1 Method of separation of variables

The Green's function of interest is the solution of

$$G_t = -G + G_{xx}, \qquad 0 < x < L,\ t > 0, \tag{6.42}$$

with initial condition

$$G(x, y; 0) = \delta(x - y), \tag{6.43}$$

and with boundary conditions

$$G(0, y; t) + G_t(0, y; t) - \gamma G_x(0, y; t) = 0, \tag{6.44}$$

$$G_x(L, y; t) = 0. \tag{6.45}$$

The boundary condition (6.44) is not of the form covered in Section 5.2, but it will be verified that the solution of

$$V_t = -V + V_{xx} + I, \tag{6.46}$$

$$V(x, 0) = v(x), \tag{6.47}$$

with the lumped-soma condition at $x = 0$ and sealed-end condition at $x = L$, is

$$V(x, t) = \int_0^L G(x, y; t) v(y)\, dy + \int_0^L \int_0^t G(x, y; t - s) I(y, s)\, ds\, dy. \tag{6.48}$$

To find the Green's function by separation of variables, we put, guided by the results of Section 5.3.1,

$$G(x, y; t) = \sum_n A_n \phi_n(x) e^{k_n t}, \qquad t \geq 0, \tag{6.49}$$

where the A_n's are coefficients that depend on y. From (5.20) the ϕ_n's satisfy the equations

$$-\phi_n'' + (1 + k_n)\phi_n = 0. \tag{6.50}$$

One boundary condition is, from (6.45),

$$\phi_n'(L) = 0. \tag{6.51}$$

To obtain the boundary condition for (6.50) at $x = 0$, we substitute each term in (6.49) in Equation (6.44) to get

$$e^{k_n t} A_n [\phi_n(0) + k_n \phi_n(0) - \gamma \phi_n'(0)] = 0,$$

or

$$\phi_n(0)[1 + k_n] - \gamma \phi_n'(0) = 0. \tag{6.52}$$

We see that the boundary-value problem is therefore different from any of those previously encountered in that the eigenvalues k_n appear in the boundary condition. We set

$$\lambda_n^2 = -(1 + k_n), \tag{6.53}$$

and observe that the general solution of (6.50) is

$$\phi_n(x) = c_n \cos(\lambda_n x) + d_n \sin(\lambda_n x),$$

where c_n and d_n are constants to be found. Since

$$\phi_n'(x) = -\lambda_n c_n \sin(\lambda_n x) + \lambda_n d_n \cos(\lambda_n x),$$

application of the boundary conditions leads to the two equations

$$-c_n \sin(\lambda_n L) + d_n \cos(\lambda_n L) = 0, \tag{6.54}$$

$$c_n \lambda_n^2 - \gamma d_n \lambda_n = 0. \tag{6.55}$$

Thus

$$d_n = -c_n \lambda_n / \gamma, \tag{6.56}$$

and, from (6.54), we obtain

$$\gamma \tan(\lambda_n L) + \lambda_n = 0, \tag{6.57}$$

which determines the λ_n's. This equation cannot be solved exactly. We rewrite it as

$$\tan(\lambda_n L) + (\lambda_n L)/\gamma L = 0, \tag{6.58}$$

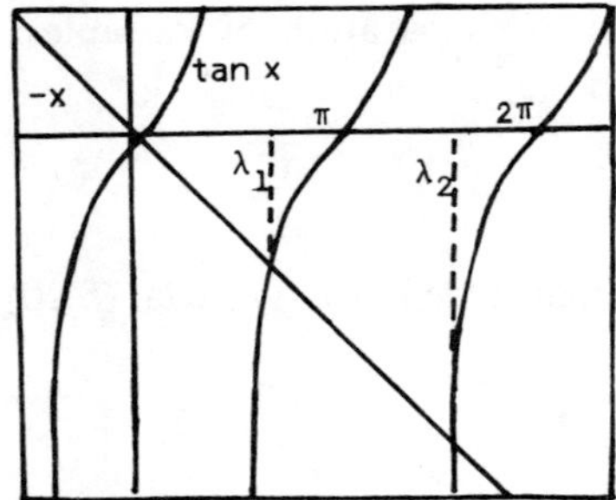

Figure 6.2. Graphical solution of the equation $\tan x = -x$.

so, if the roots of

$$\tan x = -x/\gamma L, \tag{6.59}$$

are denoted $x_0, x_1, \ldots,$ then the eigenvalues are

$$\lambda_n = x_n/L. \tag{6.60}$$

The eigenvalues of interest are all nonnegative.

Figure 6.2 shows the graphical solution of (6.59) for the eigenvalues when $L = \gamma = 1$. The smallest eigenvalue is $\lambda_0 = 0$, and successively larger eigenvalues are designated $\lambda_1, \lambda_2, \ldots$. It can be seen in the case considered that λ_1 is close to 2 and that $\lambda_n \to (2n-1)\pi/2$ as $n \to \infty$. In the general case,

$$x_n \underset{n\to\infty}{\sim} \frac{(2n-1)\pi}{2} \quad \text{so } \lambda_n \underset{n\to\infty}{\sim} \frac{(2n-1)\pi}{2L}.$$

Table 6.1 gives the first 11 values of x_n that satisfy (6.59) for various values of γL. The eigenvalues λ_n must be obtained from (6.60). It can be seen that in all cases considered the error in approximating x_n by $(2n-1)\pi/2$ is very small for $n \geq 10$.

In summary, thus far we have found that the Green's function for a finite cylinder with a lumped soma attached at $x = 0$ and a sealed end at $x = L$ is

$$G(x, y; t) = e^{-t} \sum_{n=0}^{\infty} A_n(y) \left[\cos(\lambda_n x) - \frac{\lambda_n}{\gamma} \sin(\lambda_n x) \right] e^{-\lambda_n^2 t}, \quad t \geq 0. \tag{6.61}$$

The eigenfunctions

$$\phi_n(x) = \cos(\lambda_n x) - \frac{\lambda_n}{\gamma} \sin(\lambda_n x), \qquad n = 0, 1, 2, \ldots, \tag{6.61A}$$

Table 6.1. *Roots of Equation* (6.59) *for various values of* γL

	$\gamma L = 0$	$\gamma L = 0.5$	$\gamma L = 0.75$	$\gamma L = 1$	$\gamma L = 1.5$	$\gamma L = 2$	$\gamma L = 3$	$\gamma L = 14$
x_0	0.00000	0.00000	0.00000	0.00000	0.00000	0.00000	0.00000	0.00000
x_1	1.57080	1.83660	1.93974	2.02876	2.17463	2.28893	2.45564	2.93495
x_2	4.71239	4.81585	4.86533	4.91318	5.00365	5.08698	5.23294	5.88524
x_3	7.85398	7.91706	7.94807	7.97867	8.03846	8.09617	8.20453	8.86052
x_4	10.99557	11.04083	11.06326	11.08554	11.12954	11.17271	11.25604	11.86339
x_5	14.13717	14.17243	14.18998	14.20744	14.24210	14.27635	14.34335	14.89171
x_6	17.27876	17.30764	17.32203	17.33638	17.36493	17.39325	17.44903	17.94138
x_7	20.42035	20.44480	20.45700	20.46917	20.49342	20.51753	20.56521	21.00818
x_8	23.56194	23.58314	23.59372	23.60428	23.62535	23.64632	23.68792	24.08842
x_9	26.70354	26.72224	26.73159	26.74092	26.75953	26.77809	26.81495	27.17919
x_{10}	29.84513	29.86188	29.87024	29.87858	29.89526	29.91189	29.94498	30.27823

are *not* orthogonal with weight factor unity in the sense that

$$\int_0^L \phi_n(x)\phi_m(x)\,dx \neq 0, \qquad n \neq m.$$

Thus the coefficients $A_n(y)$ cannot be found as they were in Section 5.3. [There may exist a weight function $r(x)$ such that $\int \phi_n \phi_m r\,dx = 0$, $n \neq m$, but such a function has not been found.] We therefore turn to the evaluation of the Laplace transform of $G(x, y; t)$ whose knowledge enables us to find the A_n's.

6.3.2 Laplace transform of the Green's function

The differential equation for the Laplace transform of G is the same as in Section 6.2, and the boundary condition at $x = 0$ is unchanged. At $x = L$ we have, with a sealed-end condition,

$$G_L'(L; y; s) = 0. \tag{6.62}$$

It is easier to work with $\overline{G}(x, y; t) = e^{-t}G(x, y; t)$, whose transform is denoted $\overline{G}_L$. From Section 3.9 we see that

$$\overline{G}_L(x, y; s) = G_L(x, y; s+1). \tag{6.63}$$

Furthermore,

$$-\overline{G}_L'' + s\overline{G}_L = \delta(x-y), \tag{6.64}$$

$$s\overline{G}_L(0, y; s) - \gamma \overline{G}_L'(0, y; s) = 0, \tag{6.65}$$

$$\overline{G}_L'(L, y; s) = 0. \tag{6.66}$$

Using the procedure of Section 4.6, we find that the general solution of (6.64) is

$$\overline{G}_L(x, y; s) = A(y, s)e^{x\sqrt{s}} + B(y, s)e^{-x\sqrt{s}} + \begin{cases} 0, & x < y, \\ \dfrac{1}{\sqrt{s}}\sinh(y-x)\sqrt{s}, & x \geq y. \end{cases} \tag{6.67}$$

Applying the two end conditions yields the constants (for fixed y and s),

$$A(y, s) = \frac{\cosh\left[(y-L)\sqrt{s}\right](\gamma + \sqrt{s})}{2\sqrt{s}\left[\gamma \sinh L\sqrt{s} + \sqrt{s}\cosh L\sqrt{s}\right]}, \tag{6.68}$$

$$B(y, s) = \frac{(\gamma - \sqrt{s})\cosh\left[(y-L)\sqrt{s}\right]}{2\sqrt{s}\left[\gamma \sinh L\sqrt{s} + \sqrt{s}\cosh L\sqrt{s}\right]}. \tag{6.69}$$

Thus, for $x < y$, substituting for A and B and rearranging gives

$$\bar{G}_L(x, y; s) = \begin{cases} \dfrac{\cosh(L-y)\sqrt{s}\left[\gamma\cosh x\sqrt{s} + \sqrt{s}\sinh x\sqrt{s}\right]}{\sqrt{s}\left[\gamma\sinh L\sqrt{s} + \sqrt{s}\cosh L\sqrt{s}\right]}, & x < y, \\ \dfrac{\cosh(L-x)\sqrt{s}\left[\sqrt{s}\sinh y\sqrt{s} + \gamma\cosh y\sqrt{s}\right]}{\sqrt{s}\left[\gamma\sinh L\sqrt{s} + \sqrt{s}\cosh L\sqrt{s}\right]}, & x \geq y. \end{cases} \tag{6.70}$$

This Laplace transform is not invertible using standard tables. Nevertheless, it can be used to yield the coefficients $A_n(y)$ in (6.61) as follows.

6.3.3 Determination of the coefficients $A_n(y)$ in the expansion (6.61)

We now have, from (6.61),

$$\bar{G}(x, y; t) = \sum_{n=0}^{\infty} A_n(y)\left[\cos(\lambda_n x) - \frac{\lambda_n}{\gamma}\sin(\lambda_n x)\right]e^{-\lambda_n^2 t}, \qquad t \geq 0, \tag{6.71}$$

and the Laplace transform of this function is given by (6.70). We put $x = 0$ in both (6.70) and (6.71) to get, respectively,

$$\bar{G}_L(0, y; s) = \frac{\gamma\cosh(y-L)\sqrt{s}}{\sqrt{s}\left[\gamma\sinh L\sqrt{s} + \sqrt{s}\cosh L\sqrt{s}\right]}, \tag{6.72}$$

and

$$\bar{G}(0, y; t) = \sum_{n=0}^{\infty} A_n(y)e^{-\lambda_n^2 t}, \qquad t \geq 0. \tag{6.73}$$

When we take the Laplace transform of (6.73) we obtain an alternative formula for $\bar{G}_L(0, y; s)$,

$$\bar{G}_L(0, y; s) = \sum_{n=0}^{\infty} \frac{A_n(y)}{s + \lambda_n^2}, \tag{6.74}$$

which has singularities at $s = -\lambda_n^2$, $n = 0, 1, 2, \ldots$. We note that the two different expressions for $\bar{G}_L$ must be equal.

Although these two expressions for $\bar{G}_L$ are equal, one can verify that the singularities of (6.72) occur at $s = -\lambda_n^2$. To do this set the denominator equal to zero

$$\sqrt{s}\left[\gamma \sinh L\sqrt{s} + \sqrt{s}\cosh L\sqrt{s}\right] = 0, \tag{6.75}$$

and use the relations

$$\sqrt{-\lambda_n^2} = i\lambda_n, \tag{6.76}$$

$$\sinh(ix) = i \sin x, \tag{6.77}$$

$$\cosh(ix) = \cos x. \tag{6.78}$$

It can then be seen, using the defining equation (6.57) for the λ_n's, that the zeros of the denominator in (6.72) are indeed at $s = -\lambda_n^2$.

Suppose now that s is very close to $-\lambda_m^2$ and put

$$s = -\lambda_m^2 + \varepsilon, \tag{6.79}$$

where ε is very small. Then $s + \lambda_m^2$ is very small and $A_m(y)/(s + \lambda_m^2) = A_m(y)/\varepsilon$ is very large. In fact as $\varepsilon \to 0$, $A_m(y)/\varepsilon$ is as large as we please and other terms in (6.74), being finite, become negligible. It follows that, for very small ε,

$$A_m(y)/\varepsilon \simeq \overline{G}_L(0, y; -\lambda_m^2 + \varepsilon),$$

or, equivalently,

$$\begin{aligned}\frac{\varepsilon}{A_m(y)} &\simeq \frac{1}{\overline{G}_L(0, y; -\lambda_m^2 + \varepsilon)} \\ &\simeq \frac{1}{\overline{G}_L(0, y; -\lambda_m^2)} + \varepsilon \frac{d}{ds}\left(\frac{1}{\overline{G}_L(0, y; s)}\right)\bigg|_{s=-\lambda_m^2},\end{aligned} \tag{6.80}$$

where the first two terms in a Taylor series for $1/\overline{G}_L$ around $s = -\lambda_m^2$ have been employed. Since we have seen that $1/\overline{G}_L$ is zero at $s = -\lambda_m^2$, we find

$$\frac{1}{A_m(y)} = \frac{d}{ds}\left(\frac{1}{\overline{G}_L(0, y; s)}\right)\bigg|_{s=-\lambda_m^2}, \tag{6.81}$$

which gives an explicit formula for the coefficients $A_m(y)$. (This procedure may be more succinctly carried out using the residue theorem for complex integrals.)

Performing the differentiation in (6.81) gives

$$\frac{d}{ds}\left(\frac{1}{\overline{G}_L}\right) = \Bigg[\sinh(L\sqrt{s})\left\{\left(\frac{\gamma}{2\sqrt{s}} + \frac{L\sqrt{s}}{2}\right)\cosh(y-L)\sqrt{s} - \frac{\gamma(y-L)}{2}\sinh(y-L)\sqrt{s}\right\}$$
$$+\cosh(L\sqrt{s})\left\{\left(1+\frac{\gamma L}{2}\right)\sinh(y-L)\sqrt{s} - \frac{\sqrt{s}\,(y-L)}{2}\sinh(y-L)\sqrt{s}\right\}\Bigg]\Bigg/$$
$$\gamma\cosh^2[(y-L)\sqrt{s}]. \tag{6.82}$$

Utilizing (6.66)–(6.68) to evaluate this expression at $s = -\lambda_m^2$ gives

$$A_m(y) = (2\gamma\cos Y_m)\Bigg/\Bigg(\sin(\lambda_m L)\left\{\left(\frac{\gamma}{\lambda_m} - \lambda_m L\right) + \gamma(y-L)\tan Y_m\right\}$$
$$+\cos(\lambda_m L)\{(2+\gamma L) + Y_m\tan Y_m\}\Bigg), \qquad m \neq 0, \tag{6.83}$$

where

$$Y_m = (y-L)\lambda_m. \tag{6.84}$$

For $m = 0$ we obtain (since $\lambda_0 = 0$), by a limiting argument,

$$A_0(y) = \gamma/(1+\gamma L). \tag{6.85}$$

Hence, in summary, the Green's function for a finite cable with a lumped soma at $x = 0$ and a sealed end at $x = L$ is given by

$$G(x, y; t) = \begin{cases} e^{-t}\sum_{n=0}^{\infty} A_n(y)\phi_n(x)e^{-\lambda_n^2 t}, & x < y, \\ e^{-t}\sum_{n=0}^{\infty} A_n(x)\phi_n(y)e^{-\lambda_n^2 t}, & x \geq y, \end{cases} \tag{6.86}$$

where

$$\phi_n(x) = \cos(\lambda_n x) - \lambda_n\sin(\lambda_n x)/\gamma, \quad n = 0, 1, 2, \ldots, \tag{6.87}$$

and where the eigenvalues λ_n are the roots of (6.57) and the coefficients $A_n(y)$ are given by (6.83) and (6.85). The expression for $x \geq y$ follows from the fact that the Laplace transform for $x \geq y$ is obtained from that for $x < y$ by interchanging x and y. The coefficients A_n have been given in Rall (1977).

As $t \to \infty$, the smallest eigenvalue will dominate the expansion (6.86). Hence, asymptotically, the Green's function will behave, for large t, as

$$G(x, y; t) \underset{t \to \infty}{\sim} \frac{\gamma e^{-t}}{1 + \gamma L}. \tag{6.88}$$

This result will be useful in Section 6.4.4 on the estimation of neuronal parameters.

6.3.4 A formula for the Green's function useful for small times

The series (6.86) for the Green's function converges rapidly for large t but only slowly for small t. It is desirable to have a series representation that converges rapidly for small t. Such a representation is found by employing a series representation for the Laplace transform of the Green's function that converges rapidly for large $|s|$.

To obtain this series, we rearrange (6.70) by using the definitions of cosh( ) and sinh( ) and by multiplying the numerator and denominator by $\exp(-L\sqrt{s})$ to get

$$\overline{G}_L(x, y; s) = \frac{\left[e^{-y\sqrt{s}} + e^{-(2L-y)\sqrt{s}}\right]\left[(\gamma + \sqrt{s})e^{x\sqrt{s}} + (\gamma - \sqrt{s})e^{-x\sqrt{s}}\right]}{2\sqrt{s}(\gamma + \sqrt{s})\left[1 - \left(\frac{\gamma - \sqrt{s}}{\gamma + \sqrt{s}}\right)e^{-2L\sqrt{s}}\right]},$$

$$x < y. \tag{6.89}$$

Utilizing the following expansion, valid for $0 \le |x| < 1$,

$$(1 - x)^{-1} = 1 + x + x^2 + \cdots,$$

we have

$$\overline{G}_L(x, y; s) = \frac{\left[e^{-y\sqrt{s}} + e^{-(2L-y)\sqrt{s}}\right]\left[(\gamma + \sqrt{s})e^{x\sqrt{s}} + (\gamma - \sqrt{s})e^{-x\sqrt{s}}\right]}{2\sqrt{s}(\gamma + \sqrt{s})}$$

$$\times\left[1 + \left(\frac{\gamma - \sqrt{s}}{\gamma + \sqrt{s}}\right)e^{-2L\sqrt{s}} + \left(\frac{\gamma - \sqrt{s}}{\gamma + \sqrt{s}}\right)^2 e^{-4L\sqrt{s}} + \cdots\right]$$

$$\doteq \overline{G}_1 + \overline{G}_2 + \cdots, \tag{6.90}$$

where, dropping the x and y arguments,

$$\overline{G}_1(s) = \frac{[e^{-y\sqrt{s}} + e^{-(2L-y)\sqrt{s}}][(\gamma + \sqrt{s})e^{x\sqrt{s}} + (\gamma - \sqrt{s})e^{-x\sqrt{s}}]}{2\sqrt{s}(\gamma + \sqrt{s})},$$

$$\overline{G}_2(s) = \overline{G}_1(s)\left(\frac{\gamma - \sqrt{s}}{\gamma + \sqrt{s}}\right)e^{-2L\sqrt{s}}.$$

The quantity $\overline{G}_1(s)$ is rearranged to give

$$\overline{G}_1(s) = [e^{-(y-x)\sqrt{s}} + e^{-(2L-y-x)\sqrt{s}}] \times\left[\frac{1}{2\sqrt{s}} + \frac{\gamma}{2\sqrt{s}(\gamma + \sqrt{s})} - \frac{1}{2(\gamma + \sqrt{s})}\right].$$

By using (6.35), (6.36), and the standard transform,

$$\mathscr{L}\left\{e^{ak}e^{a^2t}\operatorname{erfc}\left(a\sqrt{t} + \frac{k}{2\sqrt{t}}\right)\right\} = \frac{e^{-k\sqrt{s}}}{\sqrt{s}(a + \sqrt{s})}, \qquad k \geq 0, \tag{6.91}$$

all the quantities in $\overline{G}_1(s)$ may be inverted directly.

We turn our attention to $\overline{G}_2(s)$, which may be decomposed into two parts,

$$\overline{G}_2(s) = \overline{G}_{2,1}(s) + \overline{G}_{2,2}(s).$$

The first part,

$$\overline{G}_{2,1}(s) = (e^{-(y+2L-x)\sqrt{s}} + e^{-(4L-y-x)\sqrt{s}}) \times\left[\frac{\gamma}{2\sqrt{s}(\gamma + \sqrt{s})} - \frac{1}{2(\gamma + \sqrt{s})}\right],$$

can also be inverted directly using (6.91) and (6.36). The second part is

$$\begin{aligned}\overline{G}_{2,2}(s) &= [e^{-(y+x+2L)\sqrt{s}} + e^{-(4L-y+x)\sqrt{s}}]\frac{1}{2\sqrt{s}}\left(\frac{\gamma - \sqrt{s}}{\gamma + \sqrt{s}}\right)^2 \\ &= [e^{-(y+x+2L)\sqrt{s}} + e^{-(4L-y+x)\sqrt{s}}]\frac{1}{2\sqrt{s}}\left[1 - \frac{\gamma}{\sqrt{s}}\right]^2\left[1 + \frac{\gamma}{\sqrt{s}}\right]^{-2} \\ &= [e^{-(y+x+2L)\sqrt{s}} + e^{-(4L-y+x)\sqrt{s}}]\left[\frac{1}{2\sqrt{s}} - \frac{2\gamma}{s} + \frac{4\gamma^2}{s^{3/2}} + \cdots\right],\end{aligned}$$

where we have used

$$(1 + x)^{-2} = 1 - 2x + 3x^2 - 4x^3 + \cdots, \qquad 0 \leq |x| < 1.$$

The terms indicated can be inverted using (6.35) and the following two standard transforms:

$$\mathscr{L}\left\{\operatorname{erfc}\left(\frac{k}{2\sqrt{t}}\right)\right\} = \frac{1}{s}e^{-k\sqrt{s}}, \qquad k \geq 0, \tag{6.92}$$

$$\mathscr{L}\left\{2\sqrt{\frac{t}{\pi}}\exp\left(-\frac{k^2}{4t}\right) - k\operatorname{erfc}\left(\frac{k}{2\sqrt{t}}\right)\right\} = \frac{1}{s^{3/2}}e^{-k\sqrt{s}}, \qquad k \geq 0. \tag{6.93}$$

Putting all these results together, we obtain the following expressions for the major contributions to the Green's function for small times:

$$\begin{aligned}
G(x, y; t) &\\
= \exp(-t)\Bigg[&\frac{1}{2\sqrt{\pi t}}\left\{\exp\left(-\frac{(y-x)^2}{4t}\right) - \exp\left(-\frac{(y+x)^2}{4t}\right)\right.\\
&+ \exp\left(-\frac{(2L-y-x)^2}{4t}\right) - \exp\left(-\frac{(2L-y+x)^2}{4t}\right)\\
&\left.+ \exp\left(-\frac{(2L+y+x)^2}{4t}\right) - \exp\left(-\frac{(2L+y-x)^2}{4t}\right)\right\}\\
&+ \gamma\exp(\gamma^2 t)\left\{\exp[\gamma(y+x)]\operatorname{erfc}\left(\gamma\sqrt{t} + \frac{y+x}{2\sqrt{t}}\right)\right.\\
&+ \exp[\gamma(2L-y+x)]\operatorname{erfc}\left(\gamma\sqrt{t} + \frac{2L-y+x}{2\sqrt{t}}\right)\\
&\left.+ \exp[\gamma(2L+y-x)]\operatorname{erfc}\left(\gamma\sqrt{t} + \frac{2L+y-x}{2\sqrt{t}}\right)\right\}\\
&- 2\gamma\operatorname{erfc}\left(\frac{y+x+2L}{2\sqrt{t}}\right) + 4\gamma^2\left\{2\sqrt{\frac{t}{\pi}}\exp\left(-\frac{(y+x+2L)^2}{4t}\right)\right.\\
&\left.- (y+x+2L)\operatorname{erfc}\left(\frac{y+x+2L}{2\sqrt{t}}\right)\right\} + \cdots\Bigg], \qquad x < y.
\end{aligned} \tag{6.94}$$

At the soma, $x = 0$, the formula simplifies considerably to

$$G(0, y; t) = e^{-t}\Bigg[\gamma e^{\gamma^2 t}\Bigg\{e^{\gamma y}\text{erfc}\left(\gamma\sqrt{t} + \frac{y}{2\sqrt{t}}\right) + e^{\gamma(2L-y)}\text{erfc}\left(\gamma\sqrt{t} + \frac{2L-y}{2\sqrt{t}}\right) + e^{\gamma(2L+y)}\text{erfc}\left(\gamma\sqrt{t} + \frac{2L+y}{2\sqrt{t}}\right)\Bigg\} - 2\gamma\,\text{erfc}\left(\frac{y+2L}{2\sqrt{t}}\right) + 4\gamma^2\left\{2\sqrt{\frac{t}{\pi}}\exp\left(-\frac{(y+2L)^2}{4t}\right) - (y+2L)\text{erfc}\left(\frac{y+2L}{2\sqrt{t}}\right)\right\} + \cdots\Bigg]. \quad (6.95)$$

Note that Jack and Redman (1971a, b) have obtained a series for the Green's function at $x = 0$ in terms of parabolic cylinder functions. The first few terms of their series are those in (6.95). For more detailed calculations, see Bluman and Tuckwell (1987).

6.3.5 The Green's function for current injection at the soma

Thus far in this section it has been assumed that the current is injected at some point $y > 0$ along the cylinder. By definition, the Green's function $G(x, t)$ for current injection at the soma (we will not introduce another symbol for this Green's function), in the case of a sealed-end condition at $x = L$, is the solution of

$$\begin{aligned} G_t &= -G + G_{xx}, \qquad 0 < x < L, \\ G_x(L, t) &= 0, \\ G(0, t) + G_t(0, t) - \gamma G_x(0, t) &= R_s\delta(t), \\ G(x, 0) &= 0, \end{aligned} \quad (6.96)$$

where $\gamma = R_s/\bar{r}_i$ is the soma resistance divided by the internal resistance of a characteristic length of the membrane cylinder. It has been assumed that the soma time constant $C_s R_s$ is the same as that of the membrane of the cylinder. Note that in (6.96) dimensionless space and time variables have been employed.

Putting

$$\bar{G}(x, t) = e^t G(x, t)$$

and taking Laplace transforms of the equation and boundary conditions, leads to

$$
\begin{aligned}
-\bar{G}_L'' + s\bar{G}_L &= 0, \\
\bar{G}_L'(L; s) &= 0, \\
s\bar{G}_L(0; s) - \gamma \bar{G}_L'(0; s) &= R_s,
\end{aligned}
\tag{6.97}
$$

where $\bar{G}_L(x; s)$ is the Laplace transform of $\bar{G}(x, t)$ and the primes denote differentiation with respect to x. Solving this boundary value problem gives

$$
\bar{G}_L(x; s) = \frac{R_s \cosh[(L-x)\sqrt{s}]}{\sqrt{s}\,[\sqrt{s}\cosh(L\sqrt{s}) + \gamma \sinh(L\sqrt{s})]}. \tag{6.98}
$$

It will be seen that this is the same as (6.70′) with $y = 0$, but with a multiplying factor R_s rather than γ. Hence the Green's function for current injection at the soma is $R_s/\gamma = \bar{r}_i$ times the Green's function given in (6.86) for $x \geq y$ with $y = 0$. Thus

$$
G(x, t) = \bar{r}_i e^{-t} \sum_{n=0}^{\infty} A_n(x) e^{-\lambda_n^2 t}, \qquad t \geq 0. \tag{6.99}
$$

Similarly, by interchanging x and y in (6.95), setting $y = 0$, and multiplying by $\bar{r}_i$ gives the following formula for the Green's function suitable for calculations at small times:

$$
\begin{aligned}
G(x, t) = \bar{r}_i \exp(-t) \Bigg[& \gamma \exp(\gamma^2 t) \Bigg\{ \exp(\gamma x) \operatorname{erfc}\left(\gamma\sqrt{t} + \frac{x}{2\sqrt{t}} \right) \\
& + \exp[\gamma(2L - x)] \operatorname{erfc}\left(\gamma\sqrt{t} + \frac{2L - x}{2\sqrt{t}} \right) \\
& + \exp[\gamma(2L + x)] \operatorname{erfc}\left(\gamma\sqrt{t} + \frac{2L + x}{2\sqrt{t}} \right) \Bigg\} \\
& - 2\gamma \operatorname{erfc}\left(\frac{x + 2L}{2\sqrt{t}} \right) + 4\gamma^2 \Bigg\{ 2\sqrt{\frac{t}{\pi}} \exp\left(-\frac{(x + 2L)^2}{4t} \right) \\
& - (x + 2L) \operatorname{erfc}\left(\frac{x + 2L}{2\sqrt{t}} \right) \Bigg\} + \cdots \Bigg].
\end{aligned}
\tag{6.100}
$$

Suppose now that $V(x, t)$ is the response when a current $I(t)$ is injected into the soma. That is,

$$
\begin{aligned}
V_t &= -V + V_{xx}, \\
V_x(L, t) &= 0, \\
V(0, t) + V_t(0, t) - \gamma V_x(0, t) &= R_s I(t), \\
V(x, 0) &= 0.
\end{aligned}
\tag{6.101}
$$

Then

$$V(x,t)=\int_0^t G(x,t-u)I(u)\,du, \tag{6.102}$$

where G is the Green's function given by either (6.99) or (6.100). This can be seen most easily by realizing that the Laplace transform of V is $I_L(s)$ times the Laplace transform of G, where I_L is the Laplace transform of $I(t)$. Then formula (6.102) follows by applying the convolution theorem for Laplace transforms given in Table 3.2.

6.4 Response of the finite cylinder with lumped-soma termination to a rectangular current wave

The Green's functions obtained in the last subsection can be employed to find the response to various input currents at various locations. In particular, we will look at three situations in which experimental results were given in Chapters 1 and 3 for motoneurons. First, a rectangular current wave is applied at the soma (cf. Figure 3.3). Second, the approximation to synaptic input current of the form bte^{-at} will be applied at a synaptic location at a distance y from the soma (see the next section). Third, the response to steady current injection at the soma and an active synaptic input will be obtained.

6.4.1 Response to a rectangular current wave applied at the soma

If the current is switched on at $t_1=0$ and off at $t=t_2$ and has the constant value I_0 between these times, then

$$I(t)=I_0[H(t)-H(t-t_2)]. \tag{6.103}$$

The response, obtained by employing (6.99) and (6.103) in (6.102), is then

$$V(x,t)=\bar{r}_i I_0\sum_{n=0}^{\infty}A_n(x)\int_0^t e^{-(t-s)}e^{-\lambda_n^2(t-s)}[H(s)-H(s-t_2)]\,ds. \tag{6.104}$$

For $0<t<t_2$ we have

$$\begin{aligned}V(x,t)&=\bar{r}_i I_0\sum_{n=0}^{\infty}A_n(x)e^{-(1+\lambda_n^2)t}\int_0^t e^{(1+\lambda_n^2)s}\,ds\\&=\bar{r}_i I_0\sum_{n=0}^{\infty}\frac{A_n(x)\left[1-e^{-(1+\lambda_n^2)t}\right]}{1+\lambda_n^2},\qquad 0<t<t_2.\end{aligned} \tag{6.105}$$

Thus, if the current is applied indefinitely, the steady-state value of the response is

$$\tilde{V}(x) = \bar{r}_i I_0 \sum_{n=0}^{\infty} \frac{A_n(x)}{1+\lambda_n^2}. \tag{6.106}$$

Accelerated convergence may be attained by separating the steady-state and transient solutions

$$V(x,t) = \tilde{V}(x) - V_T(x,t).$$

From (6.17) and (6.18) we have

$$\begin{aligned}\tilde{V}(x) &= \frac{I_0}{G_s + G_c} \frac{\cosh(L-x)}{\cosh L} \\ &= \frac{\bar{r}_i I_0 \cosh(L-x)}{(1/\gamma + \tanh L)\cosh L},\end{aligned} \tag{6.107}$$

where G_c is the input conductance of the cylinder. From (6.105) we have

$$V_T(x,t) = \bar{r}_i I_0 \sum_{n=0}^{\infty} \frac{A_n(x) e^{-(1+\lambda_n^2)t}}{1+\lambda_n^2}, \tag{6.108}$$

which converges rapidly for all but very small values of t. For $t > t_2$ we have

$$\begin{aligned}V(x,t) &= \bar{r}_i I_0 \sum_{n=0}^{\infty} A_n(x) e^{-(1+\lambda_n^2)t} \int_0^{t_2} e^{(1+\lambda_n^2)s}\, ds \\ &= \bar{r}_i I_0 \sum_{n=0}^{\infty} \frac{A_n(x)}{1+\lambda_n^2} \left[e^{-(1+\lambda_n^2)(t-t_2)} - e^{-(1+\lambda_n^2)t} \right], \qquad t > t_2.\end{aligned} \tag{6.109}$$

6.4.2 The response at the soma for small times

To obtain an asymptotic formula for the depolarization in response to a constant current at small times it is convenient to invert the Laplace transform of the voltage directly rather than to try to integrate the Green's function as given by (6.100). Applying the convolution formula for the Laplace transform to (6.102), we get

$$V_L(x;s) = G_L(x;s) I_L(s), \tag{6.110}$$

where $I_L(s)$ is the Laplace transform of the current injected at the soma. When $I(t) = I_0$, a constant, we have

$$I_L(s) = I_0/s, \tag{6.111}$$

since the Laplace transform of unity is $1/s$. Letting $\bar{V}(x,t) = e^t V(x,t)$, we have, from (6.98),

$$\overline{V}_L(x;s) = \frac{I_0 R_s \cosh\left[(L-x)\sqrt{s}\right]}{(s-1)\sqrt{s}\left[\sqrt{s}\cosh(L\sqrt{s}) + \gamma\sinh(L\sqrt{s})\right]}. \tag{6.112}$$

We set $x = 0$ to obtain the soma response and seek an expansion for $\overline{V}_L(0;s)$, which converges rapidly for large $|s|$. We proceed in the same way as in the expansion of (6.89) to obtain

$$\begin{aligned}\overline{V}_L(0;s) &= \frac{I_0 R_s[1+e^{-2L\sqrt{s}}]}{(s-1)\sqrt{s}(\gamma+\sqrt{s})\left[1-\left(\dfrac{\gamma-\sqrt{s}}{\gamma+\sqrt{s}}\right)e^{-2L\sqrt{s}}\right]} \\ &= \frac{I_0 R_s[1+e^{-2L\sqrt{s}}]}{(s-1)\sqrt{s}(\gamma+\sqrt{s})}\left[1+\left(\frac{\gamma-\sqrt{s}}{\gamma+\sqrt{s}}\right)e^{-2L\sqrt{s}}+\cdots\right].\end{aligned} \tag{6.113}$$

The leading term is

$$\overline{V}_{L,1}(0;s) = \frac{I_0 R_s}{(s-1)\sqrt{s}(\gamma+\sqrt{s})}.$$

From Abramowitz and Stegun (formula 29.3.45), with $\operatorname{erf}(x) = 1 - \operatorname{erfc}(x)$,

$$\mathscr{L}\left\{e^{a^2 t}\left[\frac{b}{a}\operatorname{erf}(a\sqrt{t}) - 1\right] + e^{b^2 t}\operatorname{erfc}(b\sqrt{t})\right\} = \frac{b^2-a^2}{\sqrt{s}(s-a^2)(\sqrt{s}+b)}. \tag{6.114}$$

Applying this formula with $\gamma = b$ and $a = 1$, we obtain the inverse transform of the leading term $\overline{V}_{L,1}$ and hence the asymptotic behavior as $t \to 0$ of the response at the soma to a constant current of mangitude I_0,

$$V(0,t) \underset{t\to 0}{\sim} \frac{I_0 R_s}{(\gamma^2-1)}\left[\gamma\operatorname{erf}(\sqrt{t}) - 1 + e^{(\gamma^2-1)t}\operatorname{erfc}(\gamma\sqrt{t}) + \cdots\right], \qquad \gamma \neq 1. \tag{6.115}$$

In the special case, $\gamma = 1$, it is found that

$$V(0,t) \underset{t\to 0}{\sim} I_0 R_s\left[\tfrac{1}{2}\operatorname{erf}(\sqrt{t}) + t\operatorname{erfc}(\sqrt{t}) - \sqrt{\frac{t}{\pi}}\,e^{-t} + \cdots\right]. \tag{6.116}$$

Since the term whose Laplace transform has been obtained is that which results when $L = \infty$, the asymptotic formulas (6.115) and

(6.116) must also be the exact solutions for the somatic depolarization in the case of a semiinfinite cable attached to a lumped soma at $x = 0$. This means, figuratively speaking, that the early part of the response of the finite cable has not "seen" the boundary at $x = L$.

If one carries out a further asymptotic analysis on (6.115), one finds that as the trajectory of the depolarization leaves the origin it does so as

$$V(0, t) \sim I_0 R_s t, \tag{6.117}$$

but the values of t at which this formula is a good approximation are so small that it is unlikely that the voltage could be measured at them.

6.4.3 A numerical example

A neuron with a single dendritic tree whose branch diameters permit the mapping to an equivalent cylinder, has a soma resistance $R_s = 10^6\ \Omega$. The primary dendrite has a space constant of $\lambda = 10^{-2}$ cm and an internal resistance per unit length $r_i = 10^8\ \Omega\ \text{cm}^{-1}$. The distal terminals of the dendritic tree are at an electrotonic distance of one space constant from the soma and the dendritic terminals satisfy the sealed-end condition. Find the depolarization at the soma as a function of time, in the absence of a threshold, when a constant current $I_0 = 40$ nA is applied at the soma. If the threshold depolarization is 10 mV at the soma, what will be the frequency of action potentials? Note that these values are based approximately, and in part, on those for cat spinal motoneurons. The situation is depicted in Figure 6.3. For this calculation we require, from (6.105),

$$V(0, t) = \bar{r}_i I_0 \sum_{n=0}^{\infty} A_n(0) \frac{\left[1 - e^{-(1+\lambda_n^2)t}\right]}{\left(1 + \lambda_n^2\right)}.$$

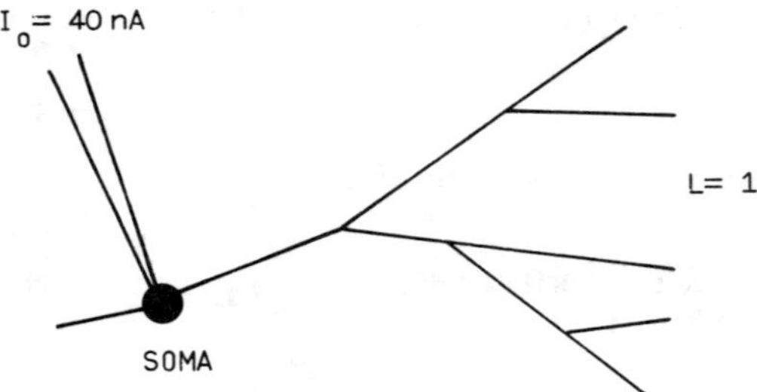

Figure 6.3. Neuron with soma and one dendritic tree with current injection at the soma.

Since $\bar{r}_i = r_i = 10^6\ \Omega = R_s$, we have $\gamma = 1$. Thus, since $L = 1$, the eigenvalues λ_n are the values of x_n for $\gamma L = 1$ given in Table 6.1. From (6.85) we find $A_0 = 0.5$. For the remaining A_n's it will be seen from (6.83) with $y = 0$, $\gamma = L = 1$ (utilizing the relation $\tan\lambda_n = -\lambda_n$), that

$$A_n(0) = \frac{2}{2+\lambda_n^2}, \qquad n = 1, 2, \ldots.$$

Hence, for these parameter values, using the decomposition into steady-state and transient solutions,

$$V(0,t) = \tilde{V}(0) - V_T(0,t),$$

where, from (6.107),

$$\tilde{V}(0) = \frac{\bar{r}_i I_0}{1 + \tanh 1},$$

and

$$V_T(0,t) = \bar{r}_i I_0 \left[\tfrac{1}{2} e^{-t} + 2 \sum_{n=1}^{\infty} \frac{e^{-(1+\lambda_n^2)t}}{(1+\lambda_n^2)(2+\lambda_n^2)} \right].$$

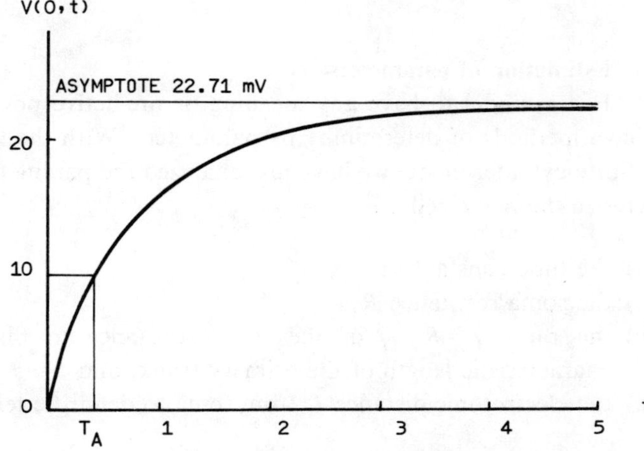

Figure 6.4. Calculated somatic depolarization as a function of time with current injection $I_0 = 40$ nA at the soma for the text example. T_A is the time at which depolarization reaches threshold, assumed to be 10 mV.

The convergence of this series is extremely rapid for $t \geq 0.1$. The calculated results are shown in Figure 6.4.

It is interesting to note that the accuracy of the asymptotic formulas for "small" t may sometimes extend to quite large values of t. For example, with $\gamma = 1$, $R_s = 1$, and $I_0 = 1$, the values of $V(0, t)$ calculated from the asymptotic formula (6.116) and the actual values, with $L = 1$, are as follows for selected t:

t	$V(0, t)$ from (6.116)	$V(0, t)$
0.1	0.0767	0.0766
0.2	0.1353	0.1353
0.3	0.1834	0.1834
0.5	0.2580	0.2594
0.8	0.3350	0.3419
2.0	0.4602	0.4999

The time at which $V(0, t)$ reaches 10 mV is determined graphically as $T_A = 0.45$ time constants. Adding a refractory period of 1 ms and assuming the time constant is 6 ms, this leads to a firing frequency of about 270 spikes per second. This frequency is high (cf. Figure 1.29) but is based on a hypothetical neuron with motoneuron properties and only one dendritic tree.

6.4.4 Estimation of parameters

For a model to have any meaning or predictive power, one must have methods of determining its parameters. With the lumped-soma–finite-cylinder model we have just analyzed the parameters that characterize the nerve cell are

(a) the time constant τ;
(b) the soma resistance R_s;
(c) the ratio $\gamma = R_s/\bar{r}_i$ of the soma resistance to that of a characteristic length of the primary trunk; and
(d) the electrotonic distance L from soma to dendritic terminals.

It must be remembered that, in general, one will only have access to the neuron's soma, which places a limitation on the information that can be obtained. Assuming that current injection and voltage recording are possible at the soma, the above parameters can be estimated by the injection of a constant current of magnitude I_0 and observing the potential until the steady-state value is attained. One sequence of steps that might be carried out is as follows.

Assuming a sealed-end condition is appropriate at $x = L$, the formulas derived above are applicable. The early response is given by (6.115) and (6.116) and the late response is approximated by

$$V(0,t) \underset{t\to\infty}{\sim} R_s I_0\left[\frac{1}{1+\gamma\tanh L} - \frac{1}{1+\gamma L}e^{-t}\right], \tag{6.118}$$

which is obtained on rearranging slightly (6.107) and (6.108) and by retaining only the $n = 1$ term in the sum. Since I_0 just multiplies all the results, we can divide all experimental depolarizations by I_0 or, equivalently, set $I_0 = 1$ in the formulas.

We now define the following function of t, which has as parameters the unknown constants, and we *now put t = time in seconds*:

$$f(t;\tau,R_s,\gamma,L)$$

$$= \begin{cases} \dfrac{R_s}{\gamma^2-1}\left[\gamma\,\mathrm{erf}\left(\sqrt{\dfrac{t}{\tau}}\right) - 1 + e^{(\gamma^2-1)t/\tau}\mathrm{erfc}\left(\gamma\sqrt{\dfrac{t}{\tau}}\right)\right], & \gamma \neq 1,\ t < t_E, \\ R_s\left[\tfrac{1}{2}\mathrm{erf}\left(\sqrt{\dfrac{t}{\tau}}\right) + \dfrac{t}{\tau}\mathrm{erfc}\left(\sqrt{\dfrac{t}{\tau}}\right) - \sqrt{\dfrac{t}{\pi\tau}}\,e^{-t/\tau}\right], & \gamma = 1,\ t < t_E, \\ R_s\left[\dfrac{1}{1+\gamma\tanh L} - \dfrac{1}{1+\gamma L}e^{-t/\tau}\right], & t > t_L, \end{cases} \tag{6.119}$$

where t_E is some *early time* and t_L is some *late time*. Both t_E and t_L are also parameters that can be adjusted to obtain the best estimates of the parameters.

Now let $t_1, t_2, \ldots, t_k$ be a set of times less than t_E and let $t_{k+1}, \ldots, t_n$ be a set of times greater than t_L. The observed values of the voltage at these n times gives the data points f_i, $i = 1, 2, \ldots, n$. Using these data points and the definition of the function $f(\cdot)$ in terms of the parameters τ, R_s, γ, and L, one may then use the nonlinear least-squares algorithms mentioned in Section 2.12 to obtain estimates of the four parameters.

An alternative simpler procedure

With $I_0 = 1$, the steady-state value of the voltage is $R_s/(1 + \gamma \tanh L)$. If this is subtracted from the late response (two-term approximation), it leaves a quantity, which when multiplied by minus one, is, with t in seconds,

$$V^*(t) = \frac{R_s}{1 + \gamma L} e^{-t/\tau}.$$

The logarithm of this is

$$\ln V^*(t) = \ln\big(R_s/(1 + \gamma L)\big) - t/\tau,$$

which should give a straight line of slope $-1/\tau$. Hence the time constant is estimated.

Consider now the early response. With $I_0 = 1$ this is R_s times a function that has only one parameter, namely γ. For each γ a different curve is obtained, a set of representative ones being shown in Figure 6.5. By matching the shape of the experimental data at small values of t to these curves (with additional computations for any

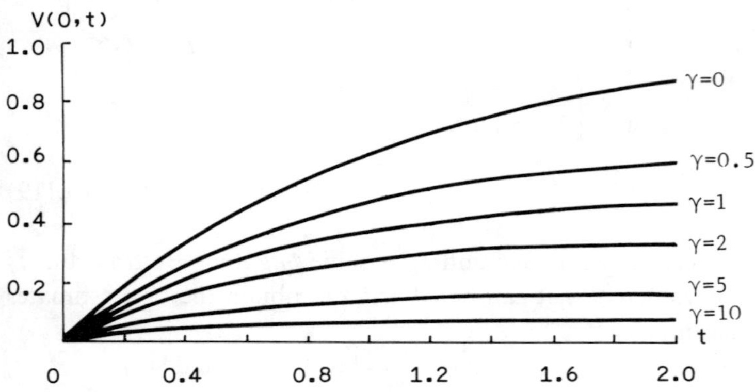

Figure 6.5. The early response at the soma to constant current for various values of γ with $I_0 = R_s = 1$. These are also the exact results for a lumped soma attached to a semiinfinite cylinder. Computed from (6.115) and (6.116).

needed values of γ), the value of γ can be estimated. Then the value of R_s may be found from the magnitude of the early response and, subsequently, L, the electrotonic length, can be found from the steady-state value, $R_s/(1+\gamma \tanh L)$, because R_s and γ are known. For other methods of estimating the parameters see Jack et al. (1985).

6.5 Response to synaptic input current

We suppose that a synapse is located at an electrotonic distance x_0 from the soma. When the synapse is activated by an action potential in the presynaptic fiber at $t=0$, we assume that a current of the form $H(t)bte^{-at}$ occurs in the postsynaptic cell at the point x_0. If the dendritic tree is represented by a cylinder and the lumped soma is attached at $x=0$, then, employing a sealed-end condition at the dendritic terminals, the depolarization satisfies

$$\begin{aligned} &V_t = -V + V_{xx} + \delta(x-x_0)H(t)bte^{-at}, \qquad 0<x<L, \\ &V(x,0)=0, \\ &V_x(L,t)=0, \\ &V(0,t)+V_t(0,t)-\gamma V_x(0,t)=0. \end{aligned} \tag{6.120}$$

The Green's function for this situation is given by formulas (6.86) and (6.94). The eigenfunction expansion (6.86) is amenable to direct integration. Using this representation, we have

$$\begin{aligned} V(x,t) &= b\int_0^L\int_0^t G(x,y;t-s)\,\delta(y-x_0)H(s)se^{-as}\,ds\,dy \\ &= b\int_0^L\int_0^t e^{-(t-s)}\sum_{n=0}^{\infty} A_n(y)\phi_n(x)e^{-\lambda_n^2(t-s)} \\ &\qquad \times\delta(y-x_0)se^{-as}\,ds\,dy \\ &= b\sum_{n=0}^{\infty} A_n(x_0)\phi_n(x)e^{-t(1+\lambda_n^2)}\int_0^t e^{s(1+\lambda_n^2-a)}s\,ds \end{aligned} \tag{6.121}$$

$$V(x,t) = b\sum_{n=0}^{\infty}\frac{A_n(x_0)\phi_n(x)}{1+\lambda_n^2-a}\left[te^{-at}-\frac{e^{-at}}{1+\lambda_n^2-a}+\frac{e^{-t(1+\lambda_n^2)}}{1+\lambda_n^2-a}\right].$$

Since $\phi_n(0) = 1,\ n = 0, 1, 2, \ldots,$ the response at the soma is

$$V(0,t) = b \sum_{n=0}^{\infty} \frac{A_n(x_0)}{1+\lambda_n^2 - a}\left[te^{-at} - \frac{e^{-at}}{1+\lambda_n^2 - a} + \frac{e^{-t(1+\lambda_n^2)}}{1+\lambda_n^2 - a}\right]. \tag{6.122}$$

Some postsynaptic potentials calculated from this formula are shown in Figures 6.6A and 6.6B. The results in Figure 6.6A are for synaptic input close to the soma ($x_0 = 0.25$) on a cylinder of total

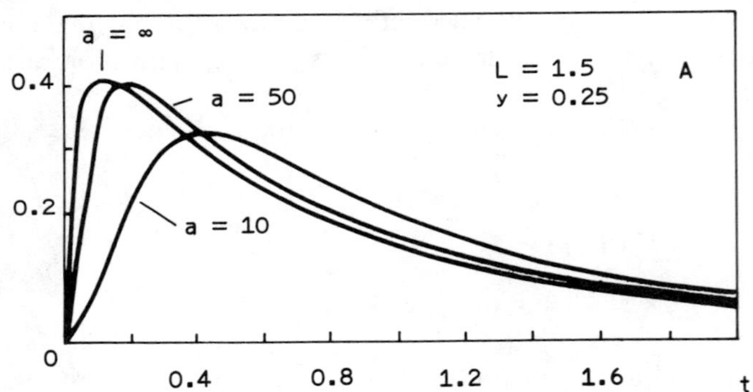

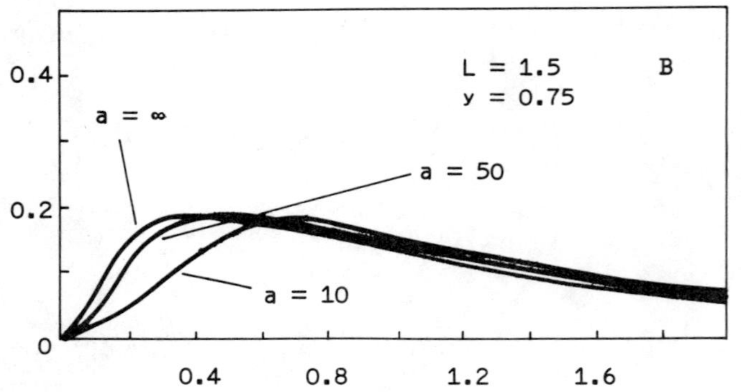

Figure 6.6. Somatic depolarizations as functions of time for a current a^2te^{-at} at x_0 on a cylinder of length $L = 1.5$. A sealed-end condition is employed at $x = L$ and a lumped-soma termination at $x = 0$ with $\gamma = 1$. A – Input close to the soma, $x_0 = 0.25$. B – Input at the mid-cylinder location, $x_0 = 0.75$. Results calculated from (6.122).

length $L = 1.5$. Recall that the input current reaches a maximum at $t = 1/a$. When $a = 50$, the response is very close to that with $a = \infty$ (which corresponds to the impulse response or Green's function). If a time constant of 5 ms is assumed, then with $a = 50$ the current input is almost over by 0.1 ms. However, if $a = 10$, the maximum current is not reached until $t = 0.1$ (0.5 ms) and the postsynaptic potential takes much longer to achieve its maximum value. The same principles are evident in the results shown in Figure 6.6B, which are obtained with the synapse located at the midpoint of the cylinder so that $x_0 = 0.75$. The peak voltages are about half of those for the corresponding curves in Figure 6.6A. There is a less pronounced dependence of the maximum value of the postsynaptic potential on a when the input site is more remote from the soma. Also, the times of occurrence of the peak voltages are greater when the synapse is located further from the soma. The values $a = 10$ and $a = 50$ are reported to be in the physiological range for cat spinal motoneurons (Redman 1973).

In the computed results of Figures 6.6A and 6.6B, the value of b in the input current bte^{-at} has been chosen as a^2 to make the total charge delivered (integral of the current density) the same as for the delta-function input (Green's function). With positive values of b we obtain excitatory postsynaptic potentials, which correspond to depolarizations. Had we chosen negative values of b, we would obtain hyperpolarizing responses or inhibitory postsynaptic potentials.

6.5.1 Determination of the input site x_0 and the input-current exponent a

Suppose for now that $b = a^2$ and the parameters of the nerve cylinder τ, R_s, L, and γ are known. An experiment is performed in which a single synapse is activated and a postsynaptic potential recorded at the soma of the postsynaptic cell. To estimate the parameters x_0 and a, one utilizes the fact that for each pair of (x_0, a)-values, the solution of the cable equation, with the given boundary conditions (e.g., the sealed-end condition at $x = L$, lumped soma at $x = 0$), is unique. Thus both x_0 and a can be estimated using nonlinear least squares because expressions for the solutions are known and can be compared with the experimental postsynaptic potential. If $b \neq a^2$ there are three parameters pertaining to the synaptic input rather than two to be estimated.

Neurophysiologists have not employed nonlinear least-squares algorithms to estimate neuronal parameters. Rather, they have employed graphical methods. For example, the following two quantities are defined for postsynaptic potentials as observed at the soma.

Half-width H

Assume that the postsynaptic potential is unimodal with a maximum absolute value of V_{max}. Then the half-width H is defined as the length of the time interval (t_1, t_2), where

$$|V(t_1)| = |V(t_2)| = V_{max}/2.$$

Rise time R

The rise time R of the postsynaptic potential is the length of the time interval (t_3, t_4) such that

$$|V(t_3)| = V_{max}/10,$$

$$|V(t_4)| = 9V_{max}/10,$$

where both t_3 and t_4 are less than the time at which the maximum value of $|V|$ occurs. The value of R so defined is often referred to as the 10–90% rise time and is employed in order to facilitate measurement since values of the potential around $t = 0$ are often hampered by artifacts.

For fixed values of the remaining neuronal parameters, H increases with R as the distance of the synaptic input from the soma increases. The H-versus-R plot becomes almost linear for large such distances. Furthermore, for different values of γ the (H, R)-curves are distinct. Thus by comparing the experimental (H, R)-values with the theoretical graphs, estimates can be made of the input location and γ. This kind of analysis has been performed on Ia excitatory postsynaptic potentials in cat spinal motoneurons (Jack et al. 1971; Iansek and Redman 1973; Jack et al. 1985; Iles 1976). Examples of (H, R)-plots, histogram of time constants, and histogram of location of Ia inputs estimated for motoneurons are shown in Figures 6.7A, B, and C. The results for the distances of Ia synapses from the origins in Figure 6.7C show a strong correlation with the results for the type-M boutons in Conradi's (1969) anatomical investigation (cf. Figure 1.9). It should be noted that motoneurons have several dendritic trees so that the formulas derived above need a reinterpretation of some parameters as

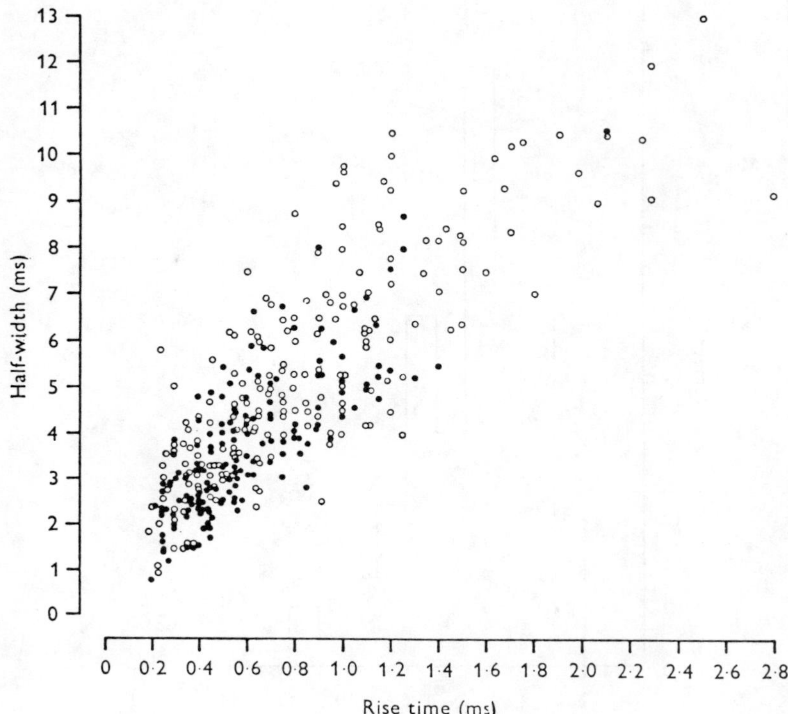

Figure 6.7A. (H, R)-plots for Ia excitatory postsynaptic potentials evoked in cat spinal motoneurons. The open circles are for extensor motoneurons, and the filled circles are for flexor motoneurons. [From Jack et al. (1971). Reproduced with the permission of The Physiological Society and the authors.]

they were based on a lumped soma attached to a single dendritic tree. For an elaboration, see the following section.

One possible method of estimating neuronal parameters, which does not seem to have been employed, is to work directly with the Laplace transform. For example, assuming a delta-function synaptic input current at a distance y from the soma, the Laplace transform of the somatic depolarization is, from (6.70),

$$V(0, y; s) = \frac{\gamma \cosh\left[(L-y)\sqrt{s+1}\,\right]}{\sqrt{s+1}\left[\gamma \sinh(L\sqrt{s}\,) + \sqrt{s}\cosh(L\sqrt{s}\,)\right]}.$$

The Laplace transform of the experimentally observed postsynaptic

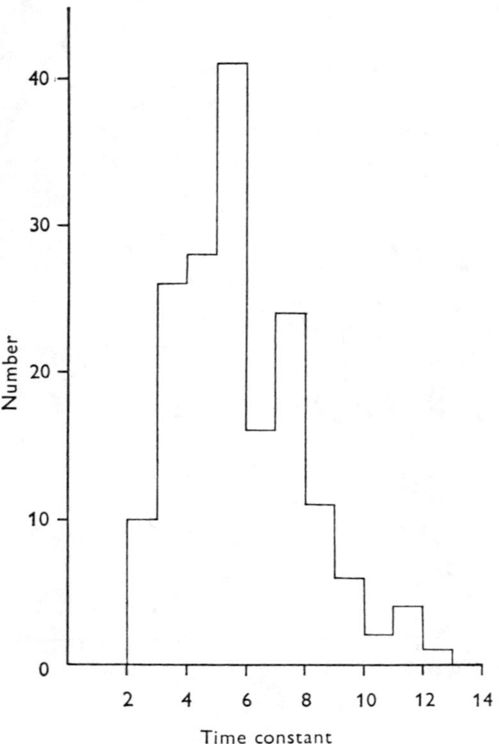

Figure 6.7B. Histogram of time constants estimated for cat spinal motoneurons. [From Jack et al. (1971). Reproduced with the permission of The Physiological Society and the authors.]

potential may be computed numerically for various (real) values of s. A comparison with the theoretical Laplace transform should yield estimates of y, if the remaining parameters are known. Or again, nonlinear least-squares algorithms may be used to estimate one or several parameters simultaneously.

6.5.2 An active synaptic input in the presence of a steady current at the soma

An often performed experiment consists of the injection of a steady current at the soma with the simultaneous activation of a synaptic input (cf. Figure 2.8). In the model consisting of a lumped soma coupled to a finite cylinder representing the dendrites, the response in this experimental situation can, by virtue of the linearity

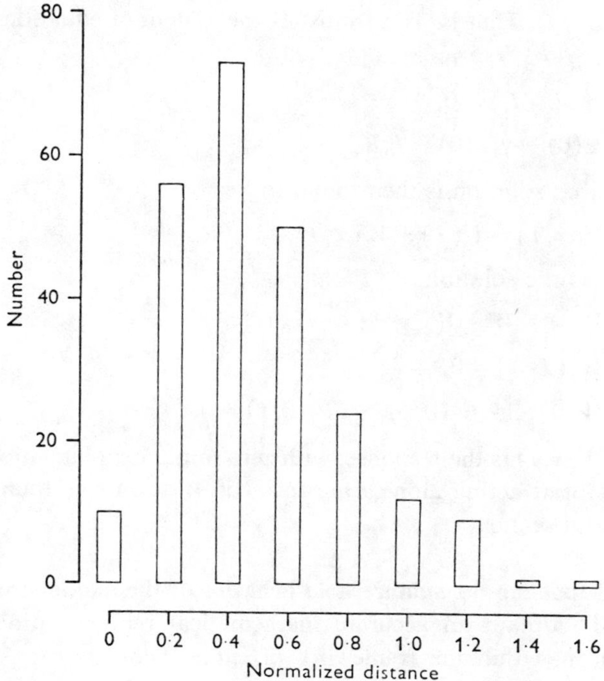

Figure 6.7C. Histogram of electrotonic distances of Ia synapses from the somas estimated from electrophysiological measurement. [From Jack et al. (1971). Reproduced with the permission of The Physiological Society and the authors.]

of the equations and boundary conditions, be found by superposition of solutions already obtained.

The potential satisfies the following system of equations:

$$\begin{aligned} &V_t = -V + V_{xx} + \delta(x - x_0)I(t), \qquad 0 < x < L, \\ &V_x(L, t) = 0, \\ &V(0, t) + V_t(0, t) - \gamma V_x(0, t) = I_0 R_s, \\ &V(x, 0) = U(x). \end{aligned} \tag{6.123}$$

Here the synaptic input current $I(t)$ occurs at the point x_0, a steady current I_0 is injected at the soma, and a sealed-end boundary condition is imposed at $x = L$. The function $U(x)$ is the steady-state solution under the injection of the somatic current I_0 and is given by formula (6.17). That is, $U(x)$ satisfies the system of equations

$$\begin{aligned} -U'' + U &= 0, \\ U'(L) &= 0, \\ U(0) - \gamma U'(0) &= I_0 R_s. \end{aligned} \tag{6.124}$$

The required solution is then found to be

$$V(x,t) = U(x) + W(x,t), \tag{6.125}$$

where W is the solution of

$$\begin{aligned} &W_t = -W + W_{xx} + \delta(x - x_0) I(t), \\ &W_x(L,t) = 0, \\ &W(0,t) + W_t(0,t) - \gamma W_x(0,t) = 0. \end{aligned} \tag{6.126}$$

That is, $W(x, t)$ is the response, with zero initial depolarization, to the synaptic input acting alone. One form of $W(x, t)$ was found in the previous subsection.

6.6 Modeling the subthreshold behavior of the motoneuron

To obtain an accurate mathematical representation of the motoneuron within the framework of cable theory is very difficult. Assuming each dendritic segment is represented by a cylinder, of constant diameter along its length, there are many cable equations to solve in conjunction with the many boundary conditions that arise at branch points. To obtain time-dependent solutions of the cable equations with current injection or synaptic input, the best approach is through Laplace transforms. Although the Laplace transforms may, under certain assumptions, be found by solving linear systems as in Section 5.8 or by the method given by Butz and Cowan (1974), the inversion of the Laplace transforms is a computationally difficult task. Horwitz (1983) obtained transient solutions for one order of branching when the geometry at branch points does not obey the three-halves power law for diameters. It is possible, as mentioned in the last section, that the Laplace transform of the solution may be sufficient if the Laplace transform of the experimentally determined depolarization can be calculated with reasonable accuracy.

The formulation of a tractable model, as far as obtaining analytical expressions for the depolarization is concerned, requires the assumption that the three-halves power law for diameters is obeyed at branch points. Some early studies (Lux, Schubert, and Kreutzberg 1970; Barrett and Crill 1971) tended to confirm that $D = \sum_{i=1}^{n(x)} d_i^{3/2}$ (i.e., the sum of diameters, of segments alive at distance x from soma, raised to the three-halves power) was nearly constant from soma to dendritic terminals, apart from decreases due to initial taper near the soma.

However, some of the results of Barrett and Crill (1974) contradicted the earlier findings: It was found for some cells that D not only undergoes an initial rapid decline, but subsequently decreases approximately linearly with distance from the soma. Part of the late decay of D is due to the termination of some of the dendritic trees. When this result is true, the accuracy of estimates made of neuronal parameters based on the kinds of analysis performed in this chapter will be highly suspect. This hints at the idea that the convenience of invoking the three-halves power law for diameters may have hindered rather than helped progress in the analysis of experimentally determined depolarizations in response to various inputs. However, the fact that reasonably concise analytical results are only obtainable when the three-halves power law is invoked, will undoubtedly lead to their continued use and their cornerstone role in a more general theory.

We consider the modeling of the subthreshold responses of a motoneuron. The components we examine are as follows.

(i) *Soma.* The somas of motoneurons have spatial extent and an irregular three-dimensional shape with "holes" at the places of attachment of the axon and dendrites. This geometrical structure makes accurate modeling extremely difficult. The *R–C* circuit representing the soma as a uniformly polarized surface is a useful approximation. The time constant of the soma membrane may be different from that of the dendrites but this may easily be taken into account as shown below.

(ii) *Dendrites.* There is good evidence that the dendrites can, for many kinds of subthreshold depolarization, be represented by passive cables of the kind studied in this and the previous chapter. There are many complications such as the following two.

First, especially in the light of the above-mentioned results of Barrett and Crill (1974), the tapering of the dendritic segment should be taken into account. This means that the cable equation (4.25) no longer has constant coefficients since now, if the radius of the segment is $a(x)$ at point x, then the axial resistance, the membrane resistance, and the membrane capacitance between x and $x + \Delta x$ depend on x as well as Δx. Assuming the cross section is circular and that $a(x)$ is differentiable, the increment in membrane area between x and $x + \Delta x$ is

$$\Delta A = 2\pi a(x)\sqrt{1 + \left(\frac{da}{dx}\right)^2}\,\Delta x.$$

Define

$$a(x) = a_0 R(x),$$

where $a_0 = a(0)$ is the beginning radius at $x = 0$. Then by the same kind of reasoning that was used in Section 4.2, it may be deduced that the partial differential equation for the depolarization $V(x, t)$ becomes

$$\tau V_t = -V + \frac{\lambda_0^2}{F(x)}\left[\left(R^2(x)V_x\right)_x + I(x,t)\right],$$

where

$$F(x) = R(x)\sqrt{1 + \left(\frac{da}{dx}\right)^2},$$

$\tau = \rho_m \delta C_m$ = membrane time constant (which does not depend on diameter),

$\lambda_0 = \sqrt{\dfrac{r_m}{r_i}}$ = characteristic length of a cylinder of radius a_0,

$$r_i = \frac{\rho_i}{\pi a_0^2},$$

$$r_m = \frac{\rho_m \delta}{2\pi a_0},$$

$I(x, t)\,\Delta x$ is the current applied in $(x, x + \Delta x]$,

ρ_i = intracellular fluid resistivity,

C_m = membrane capacitance per unit area,

δ = membrane thickness,

where t is in seconds and x is in centimeters. The effects of taper were first considered by Rall (1962) and have been elaborated upon in Jack et al. (1985).

Second, perhaps more than 70% of the soma–dendritic surface is covered with synaptic endings (Section 1.3). The question arises as to what fraction of the surface area of the motoneuron is free to pass current easily. It is possible that the proximity of the many synaptic boutons and other cell processes including those of glial cells creates an effectively diminished extracellular space. This would mean that extracellular voltage gradients could be rapidly set up during the activity of the motoneuron, destroying the validity of the assumption of assumed isopotentiality in the external fluid.

(iii) *Axon and telodendria.* As observed in Chapter 1, the axon of the spinal motoneuron extends away from the spinal cord to reach its target cells, which are muscle fibers. As the axon leaves the spinal cord it is unmyelinated with possibly collateral branches arising at about 300 μm from the cell body (Barrett and Crill 1974). Then the myelin envelope occurs with breaks at fairly regular intervals called the nodes of Ranvier. The myelin ceases as the target is approached and the axon branches into telodendria. At the terminals of the telodendria are the (neuromuscular) synapses of the motoneuron at the muscle fiber.

The axon has been neglected in most cable studies including those mentioned in this chapter. Some authors have posited that the axon should be represented by a semiinfinite cable affixed to the soma because of its physical length, which may be up to a hundred times (in cat) that of the distance of the most distal dendritic terminals from the soma. However, if we convert physical lengths to electrotonic lengths we find that the distance from the motoneuron soma to the neuromuscular synapses is of the same order of magnitude as the distance from soma to the most distal dendritic terminals. This is because the space constant is given by

$$\lambda = \sqrt{r_m / r_i}\,.$$

For the internodal membrane (i.e., that between the nodes of Ranvier), the membrane resistance of unit length times unit length r_m is very large so that the space constant along such a membrane is also very large. This implies that the electrotonic distance

$$X = x/\lambda$$

for a given physical length along the internodal membrane will be much smaller than for dendritic membrane. The major contributions to the electrotonic length of a myelinated axon will arise from the relatively small physical length of the nodal membrane. Since, in addition, the unmyelinated telodendria have very small diameters, their contribution to the overall axonic electrotonic length will be small, so the axon may have a length not very different from that of the dendrites. It is reasonable therefore, in the framework of cable theory, to treat the axon and dendritic trees as processes with similar overall electrotonic lengths. An application of cable theory to a structure consisting of repetitions of two different cylindrical subunits, such as a myelinated axon, has been given by Andrietti and Bernadini (1984).

(iv) *Synapses.* In Chapter 1 we saw that some motoneurons may have as many as 20,000 synapses on their soma and dendrites. When a synapse is active, it is likely, as described in Chapter 2, Section 14, that changes occur in the permeability of the postsynaptic (subsynaptic) membrane to one or more ion species. We have seen that constant-field theory provides a method of calculating synaptic reversal potentials. When current flow during synaptic excitation or inhibition is incorporated into cable-type models, the contributions to the current flow at the synapses due to different ionic species are usually neglected. Rather, the contributions to the excitatory current are lumped together, as are the contributions to the inhibitory current. Thus, when an excitatory synapse is activated, a local increase in the "excitatory conductance" is assumed to occur and similarly for inhibitory synapses. This approach leads to equations of the kind studied in the next chapter. However, the use of this equation for dendritic segments suffers from the disadvantage that mappings from dendritic trees to cylinders are no longer possible. Hence, to obtain a more tractable model, the current flows that occur during synaptic activity are merely treated as input currents to a linear cable with fixed resistance. This leads to the type of equations and input currents (in particular, currents of the form bte^{-at}) of the type studied throughout the earlier parts of this chapter.

(v) *Ionic environment.* In the activity of a nerve cell, ionic currents are continually coursing though its membrane between intracellular and extracellular compartments. Within these compartments the ions diffuse and are acted upon by electrochemical gradients. Ionic movement is also accomplished by active- and passive-transport mechanisms, which tend to restore the resting equilibrium conditions.

These changes in ionic concentrations lead to changes in Nernst potentials, resting membrane potential, and synaptic efficacy, some of these having been discussed in Chapter 2. The cable theory focused on in this and the preceding chapters has sidestepped the effects of changes in ionic concentrations per se. It has not yet been feasible to include the dynamics described by time-dependent Nernst–Planck equations as well as those described by cable theory in a single unified theory. A simplified model assumes that the effects of such ionic composition changes can be neglected and that linear cable theory is useful as a first approximation.

Simplified model of the motoneuron

We consider a neuron with a soma and N branching processes representing the dendritic trees as in Figure 6.8A. As we have seen,

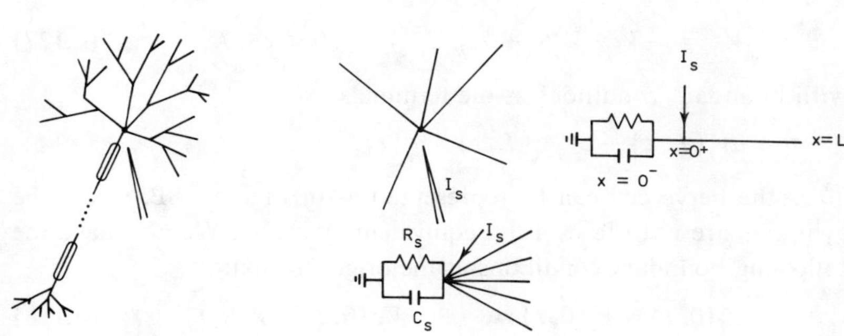

Figure 6.8. A – Neuron with soma and several branching processes. B – Each dendritic tree is reduced to a cylinder. C – The entire cell is reduced to a single cylinder with a lumped-soma termination. The corresponding cable equation may be solved to yield the somatic depolarization directly.

one of these branching processes may be an axon with telodendria. We make the following assumptions:

(a) Each dendritic segment can be represented as a membrane cylinder of constant diameter along its length. The depolarization along the cylinder is the solution of a cable equation of type (4.57)

$$V_t = -V + V_{xx} + kI/d^{3/2},$$

where x and t are dimensionless space and time, k is a constant of the nerve cell given by

$$k = \frac{2}{\pi C_m}\left(\frac{\rho_i}{\delta \rho_m}\right)^{1/2},$$

I is the current density (in coulombs per unit dimensionless time per unit dimensionless distance), and d is the diameter (in centimeters).

(b) At each branch point the relationship

$$d_0^{3/2} = \sum_{i=1}^{m} d_i^{3/2}$$

holds, where d_0 is the diameter of the parent cylinder and d_i, $i = 1, 2, \ldots, m$ are the diameters of the daughter cylinders.

(c) All dendritic terminals are the same electrotonic distance from the soma.

(d) The soma is uniformly polarized and can be represented by a single resistance and capacitance in parallel.

We suppose that the potential on the jth dendritic tree has been mapped, as per Theorem 5.3, to V_j, which satisfies

$$V_{j,t} = -V_j + V_{j,xx} + kI_j/d_j^{3/2}, \qquad 0 < x < L, \tag{6.127}$$

with boundary conditions at the terminals,

$$\alpha V_j(L,t) + \beta V_{j,x}(L,t) = \gamma_j(t).$$

Thus the nerve cell can be represented as in Figure 6.8B, where the cylinders are not (necessarily) equivalent cylinders. We also have the following boundary conditions at the origin (soma):

$$V_1(0,t) = V_2(0,t) = \cdots = V_N(0,t) = V(0,t), \tag{6.128}$$

$$V(0,t) + V_t(0,t) - R_s \sum_{j=1}^{N} \frac{V_{j,x}(0,t)}{\bar{r}_j} = R_s I_s(t), \tag{6.129}$$

where R_s is the soma resistance, I_s is the current injected at the soma, and $\bar{r}_j$, $j = 1, 2, \ldots, N$, is the internal resistance per characteristic length of the trunk of the jth dendritic tree. It has been assumed here that the time constant of the somatic membrane is the same as that of the membrane of each dendritic tree.

Determination of the somatic potential

The somatic potential $V(0,t)$ can be found for the above model of the neuron by solving a single cable equation with a lumped-soma boundary condition at $x = 0$ and a prescribed boundary condition at $x = L$. To this end we define

$$U(x,t) = \left[\sum_{j=1}^{N} \frac{V_j(x,t)}{\bar{r}_j}\right] \Big/ G, \tag{6.130}$$

where

$$G = \sum_{j=1}^{N} \frac{1}{\bar{r}_j} \tag{6.131}$$

is the sum of the conductances per characteristic length of the dendritic trunks. Thus

$$U(0,t) = V(0,t) \tag{6.132}$$

is the somatic potential.

It will be seen that $U(x,t)$ satisfies the cable equation

$$U_t = -U + U_{xx} + \frac{k}{G} \sum_{j=1}^{N} \frac{I_j}{d_j^{3/2}\bar{r}_j}. \tag{6.133}$$

Evaluating $\bar{r}_j$ in terms of basic constants, we find

$$\bar{r}_j = 2\pi^{-1}(\delta\rho_i\rho_m)^{1/2} d_j^{-3/2}, \tag{6.134}$$

so the equation for U becomes

$$U_t = -U + U_{xx} + k' \sum_{j=1}^{N} I_j, \tag{6.135}$$

where

$$k' = \frac{\pi k}{2G\sqrt{\delta\rho_i\rho_m}}. \tag{6.136}$$

What boundary conditions does U satisfy? At $x = L$ we must have

$$U(L,t) + U_x(L,t) = \frac{1}{G}\sum_{j=1}^{N}\frac{\gamma_j(t)}{\bar{r}_j}. \tag{6.137}$$

At the soma the boundary condition can be written as

$$U(0,t) + U_t(0,t) - GR_sU_x(0,t) = I_sR_s, \tag{6.138}$$

which has the same form as (6.7), with

$$\gamma = GR_s = G/G_s; \tag{6.139}$$

that is, γ is the ratio of the sum of the conductances per characteristic length of the dendritic trunks to the soma conductance. Thus solving the cable equation for U with a lumped-soma boundary condition at $x = 0$ gives the somatic potential immediately. Note that this applies for *any* input configuration without symmetry requirements. It is the basis for proceeding from Figure 6.8B to Figure 6.8C. In general, this does not give the potential over the whole neuron. This is achieved by solving for the potential in each dendritic tree recursively as outlined in Section 5.10.1. The principle of independence of somatic response on input configuration stated in that section clearly extends to the current model with a lumped soma.

Numerical values–Physical constants: standard cat spinal motoneuron

The following values of the various physical constants are based on experimental and theoretical studies of cat spinal motoneurons (Nelson and Lux 1970; Dodge and Cooley 1973; Barrett and Crill 1974; Traub and Llinas 1978). None of these constants is known with certainty and there is also a large amount of variability from cell to cell, so the values given here are meant only as rough guides. No attempt has been made to ascertain "best" values.

First, for the intracellular resistivity, we take the value

$$\rho_i = 70\ \Omega\ \text{cm}.$$

(Note that the symbol R_i is often used for ρ_i.) The resistance of a unit area of membrane times unit area with current perpendicular to the membrane is

$$R_m = 2500\ \Omega\ \text{cm}^2.$$

There is considerable range of values in the literature. The capacitance of a unit area of membrane is reported to be

$$C_m = 2 \times 10^{-6}\ \text{F cm}^{-2},$$

with a range from 10^{-6} to 3×10^{-6}. For a motoneuron dendritic trunk we take a cylinder of radius

$$a = 5 \times 10^{-4}\ \text{cm}$$

(i.e., 5 μm) as typical, but again there is a large range of values. Using the above values, we obtain the membrane resistance of a unit length of the dendritic trunk times unit length

$$r_m = R_m/2\pi a = 7.96 \times 10^5\ \Omega\ \text{cm},$$

and an internal resistance per unit length

$$r_i = \rho_i/\pi a^2 = 8.91 \times 10^7\ \Omega\ \text{cm}^{-1}.$$

With these data the space constant of the dendritic trunk is

$$\lambda = (r_m/r_i)^{1/2} = 0.0945\ \text{cm};$$

the capacitance per unit length is

$$c_m = 6.283 \times 10^{-9}\ \text{F cm}^{-1};$$

and the membrane time constant is

$$\tau = c_m r_m = 5.00\ \text{ms}.$$

The capacitance of a characteristic length of dendritic trunk is

$$\bar{c}_m = \lambda c_m = 5.94 \times 10^{-10}\ \text{F}.$$

Recall that one form of the cable equation is, from (5.56),

$$V_t = -V + V_{xx} + \bar{I}/\bar{c}_m,$$

where I is the input-current density in coulombs per unit dimensionless time per unit dimensionless distance. Hence roughly, with the above value of $\bar{c}_m$,

$$V_t = -V + V_{xx} + 1.684 \times 10^9 \bar{I}$$

gives the cable equation for a motoneuron "dendritic cylinder."

However, we may also write

$$V_t = -V + V_{xx} + k\bar{I}/d^{3/2},$$

which gives, with d the diameter of the cylinder in centimeters, the dependence on diameter. For a cat spinal motoneuron a typical value for k is therefore

$$k = 5.325 \times 10^4 \text{ F}^{-1}\text{ cm}^{3/2},$$

which is a constant of the nerve cell.

The above constants pertain mainly to the trunks of the dendritic trees. For the cylinders representing the whole individual dendritic trees the overall electrotonic lengths are about

$$L = 1.4,$$

with a range from 1.1 to 1.5. The average number of dendritic trees (primary dendrites) was 11.6, so we take

$$N = 12,$$

where the range is from 8 to 21.

The quantity $\bar{r}$, the internal resistance of a characteristic length of dendritic trunk, is

$$\bar{r} = \lambda r_i = 8.42 \times 10^6\ \Omega.$$

The quantity G is the sum of the conductances of a characteristic length of primary dendrite, so, assuming all the primary dendrites have the same diameters,

$$G = \sum_{i=1}^{N} \frac{1}{\bar{r}_i} = \frac{12}{\bar{r}_i} = 1.425 \times 10^{-6}\ \Omega^{-1}.$$

The resistance of the whole neuron R_N is estimated directly from the asymptotic ($t = \infty$) voltage in response to a current step of constant and known magnitude. A typical value is

$$R_N = 1.4 \times 10^6\ \Omega,$$

with a range from 1×10^6 to 3.5×10^6. The soma resistance R_s can be estimated from the formula

$$R_s = R_m/A_s,$$

where A_s is the soma area in square centimeters. A typical value for the surface area of the soma is

$$A_s = 1.043 \times 10^{-8}\text{ cm}^2$$

(which is also the surface area of a sphere of radius 28.8 μm). Thus, assuming R_m is given by the above value,

$$R_s = 2.4 \times 10^7\ \Omega,$$

is a typical value, but a lower value $R_s = 10^7$ seems appropriate for some cells. It is quite possible that R_m for the soma is (much) smaller than that of the dendrites, which would also imply a smaller time constant for the soma membrane.

The final parameter is γ, which appears in the lumped-soma boundary condition

$$V(0,t) + V_t(0,t) - \gamma V_x(0,t) = I_s R_s,$$

and is defined in the context of the simplified model motoneuron as

$$\gamma = G/G_s = GR_s.$$

This quantity is extremely variable. A typical value is

$$\gamma = 10,$$

with a range probably from 5 to 20 or more. (Note that the parameter ρ, which appears in the literature and is defined as the ratio of the dendritic to somatic conductance, has about the same range of values. This presumably is due to the fact that the L values of motoneuron dendritic trees are such that $\tanh L$ is not very different from unity.)

Other neurons

The reader may think that undue emphasis has been placed on spinal motoneurons. The reason for this emphasis is simply that there is such a paucity of information and hence modeling for other cells. Some data is available, however, on the electrical properties of pyramidal cells (Lux and Pollen 1966; Jacobsen and Pollen 1968). For reviews, see Redman (1976) and Rall (1977).

6.7 Frequency of action potentials

We consider the injection of a constant current I into the soma of a cat spinal motoneuron. We ask:

(a) What is the critical current I_c (rheobase) for eliciting action potentials?
(b) What is the functional relationship between the frequency f of action potentials and the magnitude I of the injected current? That is, in the language of Chapter 1, what is the f/I curve?

Before these questions can be answered for the model motoneuron, we need a criterion of threshold for action-potential generation. The simplest threshold condition is that the soma potential, assumed to be near the trigger-zone potential, exceeds some critical level θ. Thus the time t_A to elicit an action potential from an initial resting level is

$$t_A = \inf[t \mid V(0,t) = \theta], \qquad V(0,0) = 0, \tag{6.140}$$

where $V(0, t)$ is the soma (trigger-zone) depolarization. If the approach to threshold is monotonic, then t_A is defined by the relation

$$V(0, t_A) = \theta. \tag{6.141}$$

The frequency of action potentials is defined as the reciprocal of t_A,

$$f = 1/t_A. \tag{6.142}$$

If an absolute refractory period t_R is incorporated (that is, the membrane potential returns to resting value only after a time t_R has elapsed following a previous spike), then we set

$$f = 1/(t_A + t_R), \tag{6.143}$$

where t_A is as defined above.

This approach is a much simplified one. The nonlinear equations of Hodgkin–Huxley type lead to a much more satisfactory solution but at the expense of being more difficult to solve. With cable theory at least some measure of analytic (exact) results is attainable. Threshold conditions, which are more complicated than that described here, may be employed with a modicum of extra computational effort. For example, we may insist that the depolarization over some extended patch of the cell exceed a certain value before an action potential is generated (see Jack et al. 1985).

(a) *Critical current I_c to elicit action potentials.* For the model of the motoneuron we have described in the previous section, the somatic potential $U(0, t)$ is found by solving the equation for $U(x, t)$,

$$U_t = -U + U_{xx}, \qquad 0 < x < L,$$

with sealed-end condition at $x = L$,

$$U_x(L, t) = 0,$$

a lumped-soma condition at $x = 0$,

$$U(0, t) + \tau_s U_t(0, t)/\tau - \gamma U_x(0, t) = IR_s,$$

and initial value

$$U(x, 0) = 0, \qquad 0 < x < L.$$

Here x is in space constants and t is in time constants, and τ_s is the soma time constant; τ is the membrane time constant of the dendrites; $\gamma = GR_s$; R_s is the soma resistance; $G = \sum_{i=1}^{N} 1/\bar{r}_i$; N is the number of dendritic trees; $\bar{r}_i$ is the internal (axial) resistance of a characteristic length of the ith trunk; and I is the current injected at a soma.

An action potential will eventually be generated only if the asymptotic or steady-state value of the depolarization exceeds the threshold value. From Section 6.1, the steady-state depolarization is

$$U(0, \infty) = \frac{I}{G_s + \tanh L \Sigma_{i=1}^{N} 1/\bar{r}_i}. \tag{6.144}$$

If the threshold at the soma is θ V, then the critical current required to elicit an action potential is

$$I_c = \theta \left[G_s + \tanh L \sum_{i=1}^{N} \frac{1}{\bar{r}_i} \right]. \tag{6.145}$$

For the standard motoneuron parameters in the last section, $R_s = G_s^{-1} = 2.4 \times 10^7\ \Omega$, $L = 1.4$, $\bar{r}_i = 8.42 \times 10^6\ \Omega$. Choosing a threshold value $\theta = 10$ mV, we obtain

$$I_c \simeq 13 \text{ nA}.$$

This agrees roughly with the smallest current in Figure 1.29 at which the motoneuron firing rate becomes appreciable, the value being 10 nA.

Since for neurons approximating the standard motoneuron, $G_s \ll \tanh L \Sigma_{i=1}^{N} 1/\bar{r}_i$, if the dendritic trunks all have about the same diameter, then the critical current that has to be injected to elicit action potentials is roughly proportional to the number of dendritic trees. Assuming that the larger neurons have more dendritic trees, this gives one version of the *size principle*, which claims that neurons with larger cell bodies are less excitable (Henneman, Somjen, and Carpenter 1965; Burke 1973; Traub 1977). To add some weight to this idea, I_c is also roughly proportional to $\tanh L$ and consequently increases with L. That is, the further the dendritic terminals are from the soma, all other things being equal, the greater the current required to elicit an action potential.

(b) *f/I relations.* Assuming the time constant of soma and dendritic membrane are equal, the depolarization at the soma in response to the injection of a constant current I there is, from (6.105),

$$\begin{aligned} U(0, t) &= \frac{I}{G} \left[\frac{1}{\gamma^{-1} + \tanh L} - \sum_{n=0}^{\infty} \frac{A_n(0) e^{-(1+\lambda_n^2)t}}{1 + \lambda_n^2} \right] \\ &= \frac{I}{G} \left[\frac{1}{\gamma^{-1} + \tanh L} \right. \\ &\qquad \left. - e^{-t} \left\{ \frac{\gamma}{1 + \gamma L} + 2 \sum_{n=1}^{\infty} \frac{e^{-\lambda_n^2 t}}{(1 + \lambda_n^2)(2 + \lambda_n^2)} \right\} \right]. \end{aligned} \tag{6.146}$$

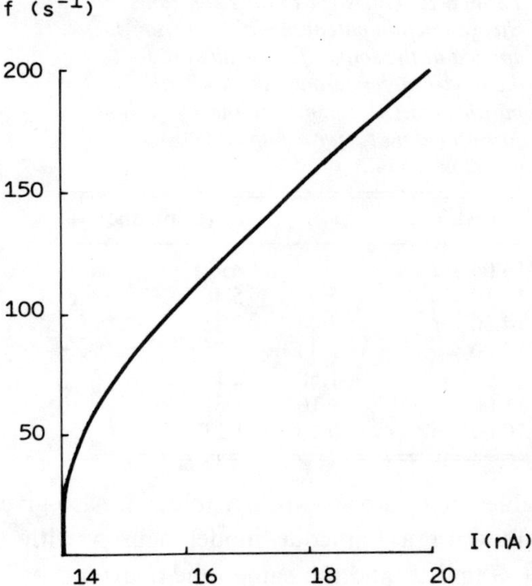

Figure 6.9. f/I curve for the model motoneuron with standard parameters and a simple threshold condition for action-potential generation.

Let us suppose again for simplicity that the threshold condition is that the depolarization at the soma reaches the value θ. How will the frequency f of action potentials vary as I increases? To see this for the standard motoneuron, we set

$$G = 1.425 \times 10^{-6}\ \Omega^{-1},$$
$$\gamma = 10,$$
$$\theta = 0.01\ \text{V},$$
$$L = 1.4.$$

The eigenvalues λ_n are obtained from Table 6.1 where the first few are

$$\lambda_1 = 2.09639,$$
$$\lambda_2 = 4.20374,$$
$$\lambda_3 = 6.32894.$$

With these parameter values, the critical current I_c for eliciting action potentials is 14.04 nA, which is slightly different from the value above because we are employing the value $\gamma = 10$, which makes G_s slightly different. Figure 6.9 shows how the frequency of action potentials, computed from formula (6.146) with $t_R = 0.2$ and then assuming a time constant of 5 ms, increases as I increases from I_c.

Table 6.2. *Times t_A, in time constants, to elicit an action potential with a current I applied at the soma. The results are for the model motoneuron with standard parameters. Also given are the corresponding t_A for a Lapicque model neuron.*

I (nA)	t_A	t_A (Lapicque)
$14.04 = I_c$	∞	∞
14.10	5.07	5.46
14.50	3.04	3.45
15.00	2.33	2.75
16.00	1.68	2.10
18.00	1.10	1.51
20.00	0.79	1.21

The numerical values of t_A are shown in Table 6.2. Also given in that table are the values for a Lapicque model neuron with the same conductance $G_s + G \tanh L$ and the same time constant.

The theoretical f/I curve obtained with the present model and simple threshold condition rises much more steeply than the example of an experimental f/I curve shown in Figure 1.29. We note too that the results for the Lapicque model are not very different from those for the model with lumped soma attached to many dendritic trees *when the simple threshold condition is used*. The Lapicque model, however, can do no better than this. Although the results for the lumped-soma–many-dendritic-trees model seem inadequate, an improvement on the agreement between experiment and theory is possible with a more realistic threshold condition. The present calculation has assumed that all that matters for action-potential generation is that the soma voltage exceeds θ. The trigger zone of the motoneuron, however, is more likely located in the initial segment about 30 μm away from the soma. If we use the threshold criterion that the depolarization in the initial segment exceeds a threshold value, a less steeply rising f/I curve will be obtained. The confirmation of this is left as an exercise. However, the cable model was introduced as one that might deal with not too strong subthreshold responses, so we should not expect it to handle threshold phenomena. These are the natural domain of nonlinear systems of equations such as the Hodgkin–Huxley equations, which are dealt with in Chapter 8.

REFERENCES

Abramowitz, M., and Stegun, I. (eds.) (1964). *Handbook of Mathematical Functions*. Dover, New York.

Aidley, D. J. (1978). *The Physiology of Excitable Cells*. Cambridge University Press.

Aitken, J. T., and Bridger, J. E. (1961). Neuron size and neuron population density in the lumbosacral region of the cat's spinal cord. *J. Anat.* (*Lond.*) **95** 38–53.

Andrietti, F., and Bernadini, G. (1984). Segmented and "equivalent" representation of the cable equation. *Biophys. J.* **46** 615–23.

Ascoli, C., Barbi, M., Chillemi, S., and Petracchi, D. (1977). Phase-locked responses in the *Limulus* lateral eye. *Biophys. J.* **19** 219–40.

Ascoli, C., Barbi, M., Chillemi, S., and Petracchi, D. (1978). Phase locking of the neuronal discharge to periodic stimuli. In *Progress in Cybernetics and Systems Research*, R. Trappl, G. J. Klir, and L. M. Ricciardi (eds.), pages 648–56. Wiley, New York.

Ashida, H. (1985). General solution of cable theory with both ends sealed. *J. Theoret. Biol.* **112** 727–40.

Attwell, D., and Jack. J. (1978). The interpretation of membrane current–voltage relations: a Nernst–Planck analysis. *Prog. Biophys. Molec. Biol.* **34** 81–107.

Barrett, J. N., and Crill, W. E. (1971). Specific membrane resistivity of dye-injected cat motoneurons. *Brain Res.* **28** 556–61.

Barrett, J. N., and Crill, W. E. (1974). Specific membrane properties of cat motoneurons. *J. Physiol.* **239** 301–24.

Beurle, R. L. (1956). Properties of a mass of cells capable of regenerating pulses. *Phil. Trans. Roy. Soc. Lond. A* **240** 55–94.

Bluman, G. W., and Tuckwell, H. C. (1987). Techniques for obtaining analytical solutions for Rall's model neuron. *J. Neurosci. Methods* **20** 151–66.

Boyce, W. E., and DiPrima, R. C. (1977). *Elementary Differential Equations and Boundary Value Problems*. Wiley, New York.

Boyle, P. J., and Conway, E. J. (1941). Potassium accumulation in muscle and associated changes. *J. Physiol.* **100** 1–63.

Brace, R. A., and Anderson, D. K. (1973). Predicting transient and steady-state changes in resting membrane potential. *J. Appl. Physiol.* **35** 90–4.

Brock, L. G., Coombs, J. S., and Eccles, J. C. (1952). The recording of potentials from motoneurones with an intracellular electrode. *J. Physiol.* **117** 431–60.

Brown, R. H., Jr. (1974). Membrane surface charge: discrete and uniform modelling. *Prog. Biophys. Molec. Biol.* **28** 343–70.

Burke, R. E. (1973). On the central nervous system control of fast and slow twitch motor units. In *Developments in Electromyography and Clinical Neurophysiology*, Volume 3, J. E. Desmett, (ed.). Karger, Basel.

Burke, R. E., and Rudomin, P. (1977). Spinal neurons and synapses. In *Handbook of Physiology, The Nervous System*, Volume 1, Section 1, Chapter 24, E. R. Kandel (ed.). American Physiological Society, Bethesda.

Butz, E. G., and Cowan, J. D. (1974). Transient potentials in dendritic systems of arbitrary geometry. *Biophys. J.* **14** 661–89.

Caldwell, P. C., and Keynes, R. D. (1960). The permeability of the squid giant axon to radioactive potassium and chloride ions. *J. Physiol.* **154** 177–89.

Calvin, W. H. (1974). Three modes of repetitive firing and the role of the threshold time course between spikes. *Brain Res.* **69** 341–46.

Calvin, W. H., and Schwindt, P. C. (1972). Steps in production of motoneuron spikes during rhythmic firing. *J. Neurophysiol.* **35** 297–310.

Collewijn, H., and Van Harreveld, A. (1966). Membrane potential of cerebral cortical cells during spreading depression and asphyxia. *Exp. Neurol.* **15** 425–36.

Conradi, S. (1969). On motoneuron synaptology in adult cats. *Acta Physiol. Scand. Suppl.* **332**.

Conway, B. E. (1952). *Electrochemical Data*. Elsevier, Amsterdam.

Coombs, J. S., Curtis, D. R., and Eccles, J. C. (1959). The electrical constants of the motoneurone membrane, *J. Physiol.* **145** 505–28.

Coombs, J. S, Eccles, J. C., and Fatt, P. (1955a). Excitatory synaptic action in motoneurones. *J. Physiol.* **130** 374–95.

Coombs, J. S., Eccles, J. C., and Fatt, P. (1955b). The specific ionic conductances and the ionic movements across the motoneuronal membrane that produce the inhibitory post-synaptic potential. *J. Physiol.* **130** 326–73.

Curtis, D. R., and Eccles, J. C. (1959). The time courses of excitatory and inhibitory synaptic actions. *J. Physiol.* **145** 529–46.

Curtis, D. R., and Eccles, J. C. (1960). Synaptic action during and after repetitive stimulation. *J. Physiol.* **150** 374–98.

Dean, R. B. (1941). Theories of electrolyte equilibrium in muscle. *Biol. Symp.* **3** 331–48.

DeSantis, M., and Limwongse, V. (1983). Relationships between size, number and density of axosomatic boutons and motor neuronal size. *J. Theoret. Neurobiol.* **2** 213–36.

Dodge, F. A. (1979). The nonuniform excitability of central neurons as exemplified by a model of the spinal motoneuron. In *The Neurosciences Fourth Study Program*, Chapter 25, F. O. Schmidt and F. G. Worden (eds.). MIT Press, Cambridge, Mass.

Dodge, F. A., Jr., and Cooley, J. W. (1973). Action potential of the motorneuron. *IBM J. Res. Dev.* **17** 219–29.

Dodge, F. A., and Frankenhaeuser, B. (1958). Membrane currents in isolated frog nerve fibre under voltage clamp conditions. *J. Physiol.* **143** 76–90.

Dodge, F. A., and Frankenhaeuser, B. (1959). Sodium currents in the myelinated nerve fibre of *Xenopus laevis* investigated with the voltage clamp technique. *J. Physiol.* **148** 188–200.

Duff, G. F. D., and Naylor, D. (1966). *Differential Equations of Applied Mathematics*. Wiley, New York.

Durand, D. (1984). The somatic shunt cable model for neurons. *Biophys. J.* **46**, 645–53.

Eccles, J. C. (1953). *The Neurophysiological Basis of Mind*. Clarendon, Oxford.

Eccles, J. C. (1957). *The Physiology of Nerve Cells*. Johns Hopkins University Press, Baltimore.

Eccles, J. C. (1964). *The Physiology of Synapses*. Springer, New York.

Eccles, J. C. (1969). *The Inhibitory Pathways of the Central Nervous System*. Thomas, Springfield, Mass.

Eccles, J. C. (1977). *The Understanding of the Brain*. McGraw-Hill, New York.

Edwards, C., and Ottoson, D. (1958). The site of impulse initiation in a nerve cell of a crustacean stretch receptor. *J. Physiol.* **143** 138–48.

Edwards, D. H., and Mulloney, B. (1984). Compartmental models of electrotonic structure and synaptic integration in an identified neurone. *J. Physiol.* **348** 89–113.

Einstein, A. (1905). *Investigations on the Theory of the Brownian Movement*. (Translated from *Ann. Physik*. **17** 549–60.) Dover, New York.

Frankenhaeuser, B. (1960). Sodium permeability in toad nerve and in squid nerve. *J. Physiol*. **152** 159–66.

Frankenhaeuser, B. (1962). Potassium permeability in myelinated nerve fibres of *Xenopus laevis*. *J. Physiol*. **160** 54–61.

Frankenhaeuser, B., and Huxley, A. F. (1964). The action potential in the myelinated nerve fibre of *Xenopus laevis* as computed on the basis of voltage clamp data. *J. Physiol*. **171** 302–15.

Gardner, E. (1963). *Fundamentals of Neurology*. Saunders, Philadelphia.

Geduldig, D. (1968). Analysis of membrane permeability coefficient ratios and internal ion concentrations from a constant field equation. *J. Theoret. Biol*. **19** 67–78.

Gogan, P., Gueritand, J. P., Horcholle-Bossavit, G., and Tye-Dumont, S. (1977). Direct excitatory interactions between spinal motoneurones of the cat. *J. Physiol*. **272** 755–67.

Goldman, D. E. (1943). Potential, impedance and rectification in membranes. *J. Gen. Physiol*. **27** 37–60.

Gorman, A. L. F., and Marmor, M. F. (1970). Contributions of the sodium pump and ionic gradients to the membrane potential of a molluscan neurone. *J. Physiol*. **210** 897–917.

Granit, R. (1958). Neuromuscular interaction in postural tone of the cat's isometric soleus muscle. *J. Physiol*. **143** 387–402.

Granit, R., Kernell, D., and Lamarre, Y. (1966a). Algebraical summation in synaptic activation of motoneurones firing within the "primary range" to injected currents. *J. Physiol*. **187** 372–99.

Granit, R., Kernell, D., and Lamarre, Y. (1966b). Synaptic stimulation superimposed on motoneurons firing in the "secondary" range to injected current. *J. Physiol*. **187** 401–15.

Greenspan, H. P., and Benney, D. J. (1973). *Calculus. An Introduction to Applied Mathematics*. McGraw-Hill, New York.

Griffith, J. S. (1971). *Mathematical Neurobiology*. Academic, New York.

Gustafsson, B. (1974). After hyperpolarization and the control of repetitive firing in spinal neurones of the cat. *Acta Physiol. Scand. Suppl*. **416**.

Haggar, R. A., and Barr, M. L. (1950). Quantitative data on the size of synaptic end bulbs in the cat's spinal cord. *J. Comp. Neurol*. **93** 17–35.

Henneman, E., Somjen, G., and Carpenter, D. O. (1965). Functional significance of cell size in spinal motoneurons. *J. Neurophysiol*. **28** 560–80.

Hille, B. (1973). Potassium channels in myelinated nerve. *J. Gen. Physiol*. **61** 669–85.

Hille, B. (1975). Ion selectivity, saturation and block in sodium channels. *J. Gen. Physiol*. **66** 535–60.

Hille, B. (1977). Ionic basis of resting and action potentials. In *Handbook of Physiology*, Section 1, Volume 1, E. R. Kandel (ed.), pages 99–136. American Physiological Society, Bethesda.

Hillman, D. E. (1979). Neuronal shape parameters and sub-structures as a basis of neuronal form. In *The Neurosciences, Fourth Study Program*, F. O. Schmitt and F. G. Worden (eds.). MIT Press, Cambridge, Mass.

Hodgkin, A. L. (1951). The ionic basis of electrical activity in nerve and muscle. *Biol. Rev*. **26** 339–409.

Hodgkin, A. L., and Horowicz, P. (1959). The influence of potassium and chloride ions on the membrane potential of single muscle fibres. *J. Physiol*. **148** 127–60.

Hodgkin, A. L., and Huxley, A. F. (1939). Action potentials recorded from inside a nerve fibre. *Nature* **144** 710–11.

Hodgkin, A. L., and Katz, B. (1949). The effect of sodium ions on the electrical activity of the giant axon of the squid. *J. Physiol*. **108** 37–77.

Hodgkin, A. L., and Rushton, W. A. H. (1946). The electrical constants of a crustacean nerve fibre. *Proc. Roy. Soc. Lond. B* **133** 444–79.

Hoff, E. C. (1932). Central nerve terminals in the mammalian spinal cord and their examination by experimental degeneration. *Proc. Roy. Soc. Lond. B* **111** 175–88.

Holden, A. V. (1976). *Models of the Stochastic Activity of Neurones*. Springer, Berlin.

Holmes, O. (1962). Effects of pH, changes in potassium concentration and metabolic inhibitors on the after-potentials of mammalian non-medullated nerve fibres. *Arch. Int. Physiol. Biochim.*, **70** 211–45.

Hoppensteadt, F. C. (1986). *An Introduction to the Mathematics of Neurons*. Cambridge University Press.

Horwitz, B. (1981). An analytical method for investigating transient potentials in neurons with branching dendritic trees. *Biophys. J.* **36** 155–92.

Horwitz, B. (1983). Unequal diameters and their effects on time-varying voltages in branched neurons. *Biophys. J.* **41** 51–66.

Huxley, A. F., and Stampfli, R. (1949). Evidence for saltatory conduction in peripheral myelinated nerve fibers. *J. Physiol.* **108** 315–39.

Iansek, R., and Redman, S. J. (1973). The amplitude, time course and charge of unitary excitatory post-synaptic potentials evoked in spinal motoneurone dendrites. *J. Physiol.* **234** 665–88.

Iles, J. F. (1976). Central terminations of muscle afferents on motoneurones in the cat spinal cord. *J. Physiol.* **262** 91–117.

Illis, L. (1964). Spinal cord synapses in the cat: the normal appearances by the light microscope. *Brain* **87** 543–54.

Illis, L. (1967). The relative densities of monosynaptic pathways to the cell bodies and dendrites of the cat ventral horn. *J. Neurol. Sci.* **4** 259–70.

Jack. J. J. B., Miller, S., Porter, R., and Redman, S. J. (1971). The time course of minimal excitatory post-synaptic potentials evoked in spinal motoneurons by Group Ia afferent fibres. *J. Physiol.* **215** 353–80.

Jack. J. J. B., Noble, D., and Tsien, R. W. (1985). *Electric Current Flow in Excitable Cells*. Clarendon, Oxford.

Jack. J. J. B., and Redman, S. J. (1971a). The propagation of transient potentials in some linear cable structures. *J. Physiol.* **215** 283–320.

Jack. J. J. B., and Redman, S. J. (1971b). An electrical description of the motoneurone, and its application to the analysis of synaptic potentials. *J. Physiol.* **215** 321–52.

Jack. J. J. B., Redman, S. J., and Wong, K. (1981). The components of synaptic potentials evoked in cat spinal motoneurones by impulses in single group Ia afferents. *J. Physiol.* **321** 65–96.

Jacobsen, S., and Pollen, D. A. (1968). Electrotonic spread of dendritic potentials in feline pyramidal cells. *Science* **161** 1351–53.

Jan, L. Y., and Jan, Y. N. (1976). Properties of the larval neuromuscular junction in *Drosophila melanogaster*. *J. Physiol.* **262** 189–214.

John, F. (1982). *Partial Differential Equations*. Springer, New York.

Junge, D. (1981). *Nerve and Muscle Excitation*. Sinauer, Sunderland, Mass.

Kandel, E. R. (Ed.) (1977). *Handbook of Physiology, The Nervous System*, Volume 1, Section 1. American Physiological Society, Bethesda.

Katz, B. (1966). *Nerve, Muscle and Synapse*. McGraw-Hill, New York.

Katz, B. (1969). *The Release of Neural Transmitter Substances*. Liverpool University Press, Liverpool.

Kawato, M., and Tsukahara, N. (1983). Theoretical studies on electrical properties of dendritic spines. *J. Theoret. Biol.* **103** 507–22.

Keener, J. P., Hoppensteadt, F. C., and Rinzel, J. (1981). Integrate-and-fire models of nerve membrane response to oscillatory input. *SIAM J. Appl. Math.* **41** 503–17.

Kernell, D. (1965a). The adaptation and the relation between discharge frequency and current strength of cat lumbosacral motoneurones stimulated by long-lasting injected currents. *Acta Physiol. Scand.* **65** 65–73.

Kernell, D. (1965b). High frequency repetitive firing of cat lumbosacral motoneurones stimulated by long-lasting injected currents. *Acta Physiol. Scand.* **65** 74–86.

Kernell, D. (1965c). Synaptic influence on the repetitive activity elicited in cat lumbosacral motoneurons by long-lasting injected currents. *Acta Physiol. Scand.* **63** 409–10.

Knight, B. W. (1972a). Dynamics of encoding in a population of neurons. *J. Gen. Physiol.* **59** 734–66.

Knight, B. W. (1972b). The relationship between the firing rate of a single neuron and the level of activity in a population of neurons. *J. Gen. Physiol.* **59** 767–78.

Koziol, J. A., and Tuckwell, H. C. (1978). Analysis and estimation of synaptic densities and their spatial variation on the motoneuron surface. *Brain Res.* **150** 617–24.

Krnjevic, K. (1974). Chemical nature of synaptic transmission in vertebrates. *Physiol. Rev.* **54** 418–540.

Kuffler, S. W., and Nicholls, J. G. (1976). *From Neuron to Brain*. Sinauer, Sunderland, Mass.

Kuffler, S. W., Nicholls, J. G., and Orkand, R. K. (1966). Physiological properties of glial cells in the central nervous system of amphibia. *J. Neurophysiol.* **29** 768–87.

Kuno, M., and Miyahara, J. T. (1969). Nonlinear summation of unit synaptic potentials in spinal motoneurones of the cat. *J. Physiol.* **201** 465–77.

Lapicque, L. (1907). Recherches quantitatives sur l'excitation electrique des nerfs traitee comme une polarization. *J. Physiol. Pathol. Gen.* **9** 620–35.

Leibovic, K. N. (1972). *Nervous System Theory*. Academic, New York.

Levenberg, K. (1944). A method for the solution of certain non-linear problems in least squares. *Quart. Appl. Math.* **2** 164–68.

Lewis, C. A. (1979). Ion-concentration dependence of the reversal potential and the single channel conductance of ion channels at the frog neuromuscular junction. *J. Physiol.* **286** 417–45.

Lux, H. D., and Pollen, D. A. (1966). Electrical constants of neurons in the motor cortex of the cat. *J. Neurophysiol.* **29** 207–20.

Lux, H. D., Schubert, P., and Kreutzberg, G. W. (1970). Direct matching of morphological and electrophysiological data in cat spinal motoneurons. In *Excitatory Synaptic Mechanisms*, P. Andersen and J. K. S. Jansen (eds.). Universitets forlaget, Oslo.

MacGregor, R. J., and Lewis, E. R. (1977). *Neural Modeling*. Plenum, New York.

Marquardt, D. W. (1963). An algorithm for least-squares estimation of nonlinear parameters. *J. Soc. Indust. Appl. Math.* **11** 431–41.

Mendell, L. M., and Henneman, E. (1971). Terminals of single Ia fibers: location, density and distribution within a pool of 300 homonymous motor neurons. *J. Neurophysiol.* **34** 171–87.

Moreton, R. B. (1968). An application of the constant-field theory to the behaviour of giant neurones of the snail, *Helix aspersa*. *J. Exp. Biol.* **48** 611–23.

Moreton, R. B. (1969). An investigation of the electrogenic sodium pump in snail neurones, using the constant-field theory. *J. Exp. Biol.* **51** 181–201.

Mullins, L. J., and Noda, K. (1963). The influence of sodium-free solutions on the membrane potential of frog muscle fibers. *J. Gen. Physiol.* **47** 117–32.

Nelson, P. G., and Lux, H. D. (1970). Some electrical measurements of motoneuron parameters. *Biophys. J.* **10** 55–73.

Nernst, W. (1888). Zur Kinetik der in Losung befindlichen Korper: Theorie der Diffusion. *Z. Physik. Chem. Leipzig* **2** 613–37.

Nernst, W. (1889). Die elektromotorische Wirksamkeit der Ionen. *Z. Physik. Chem. Leipzig* **4** 129–81.

Noble, D., and Stein, R. B. (1966). The threshold conditions for initiation of action potentials by excitable cells. *J. Physiol.* **187** 129–62.

Norman, R. S. (1972). Cable theory for finite length dendritic cylinders with initial and boundary conditions. *Biophys. J.* **12** 25–45.

Oğuztöreli, M. N., and Stein, R. B. (1983). A model for the spinal control of antagonistic muscles. *J. Theoret. Neurobiol.* **2**, 81–100.

Perkel, D. H., and Mulloney, B. (1978). Electrotonic properties of neurons: steady-state compartmental model. *J. Neurophysiol.* **41** 621–39.

Perkel, D. H., Mulloney, B., and Budelli, R. W. (1981). Quantitative methods for predicting neuronal behavior. *Neuroscience* **6** 823–37.

Phillis, J. W., and Wu, P. H. (1981). Catecholamines and the sodium pump in excitable cells. *Prog. Neurobiol.* **17** 141–84.

Piek, T. (1975). Ionic and electrical properties. In *Insect Muscle*, P. N. R. Usherwood (ed.), pages 281–336. Academic, New York.

Planck. M. (1890a). Ueber die Erregung von Elekricität und Wärme in Elektrolyten. *Ann. Physik Chem.* **39** 161–86.

Planck, M. (1890b). Ueber die Potentialdifferenz zwischhen zwei verdunnten Lösungen binärer Elektrolyte. *Ann. Physik Chem.* **40** 561–76.

Plonsey, R. (1969). *Bioelectric Phenomena*. McGraw-Hill, New York.

Poppele, R. E., and Chen, W. J. (1972). Repetitive firing behavior of mammalian muscle spindle. *J. Neurophysiol.* **35** 357–64.

Rall, W. (1960). Membrane potential transients and membrane time constants of motoneurons. *Exp. Neurol.* **2** 503–32.

Rall, W. (1962). Theory of physiological properties of dendrites. *Ann. N.Y. Acad. Sci.* **96** 1071–92.

Rall, W. (1964). Theoretical significance of dendritic trees for neuronal input-output relations. In *Neural Theory and Modeling*, R. Reiss (ed.). Stanford University Press, Stanford.

Rall, W. (1967). Distinguishing theoretical synaptic potentials computed for different soma–dendritic distributions of synaptic input. *J. Neurophysiol.* **30** 1138–68.

Rall, W. (1969a). Time constants and electrotonic lengths of membrane cylinders and neurons. *Biophys. J.* **9** 1483–1509.

Rall, W. (1969b). Distributions of potential in cylindrical coordinates and time constants for a membrane cylinder. *Biophys. J.* **9** 1509–41.

Rall, W. (1977). Core conductor theory and cable properties of neurons. In *Handbook of Physiology*, *The Nervous System*, Volume 1, Section 1, E. R. Kandel (ed.). American Physiological Society, Bethesda.

Rall, W., and Rinzel, J. (1973). Branch input resistance and steady attenuation for input to one branch of a dendritic neuron model. *Biophys. J.* **13** 648–88.

Ramón y Cajal, S. (1909). *Histologie du système nerveux de l'homme et des vertébrés*, Volume I. Maloine, Paris.

Ramón y Cajal, S. (1911). *Histologie du système nerveux de l'homme et des vertébrés*, Volume II. Maloine, Paris.

Redman, S. J. (1973). The attenuation of passively propagating dendritic potentials in a motoneurone cable model. *J. Physiol.* **234** 637–64.

Redman, S. L. (1976). A quantitative approach to integrative functions of dendrites. *Int. Rev. Physiol. Neurophysiol II* **10** 1–35.

Reichardt, W. E., and Poggio, T. (Eds.) (1981). *Theoretical Approaches to Neurobiology*. MIT Press, Cambridge, Mass.

Rescigno, A., Stein, R. B., Purple, R. L., and Poppele, R. E. (1970). A neuronal model for the discharge pattern produced by cyclic inputs. *Bull. Math. Biophysics* **32** 337–53.

Rinzel, J. (1975). Voltage transients in neuronal dendritic trees. *Fed. Proc.* **34** 1350–56.

Rinzel, J., and Rall, W. (1974). Transient response in a dendritic neuron model for current injected at one branch. *Biophys. J.* **14** 759–90.

Roach, G. F. (1970). *Green's Functions*. Van Nostrand, London.

Romanes, G. J. (1951). The motor cell columns of the lumbo-sacral spinal cord of the cat. *J. Comp. Neurol.* **94** 313–63.

Scharstein, H. (1979). Input–output relationship of the leaky-integrator neuron model. *J. Math. Biol.* **8** 403–20.

Scheibel, M. E., and Scheibel, A. B. (1969). Terminal patterns in cat spinal cord. III. Primary afferent collaterals. *Brain Res.* **13** 417–33.

Schmidt, F. O., and Worden, F. G. (Eds.) (1979). *The Neurosciences Fourth Study Program*. MIT Press, Cambridge, Mass.

Schwindt, P. C., and Calvin, W. H. (1972). Membrane potential trajectories between spikes underlying motoneuron firing rates. *J. Neurophysiol.* **35** 311–25.

Schwindt, P. C., and Calvin, W. H. (1973a). Equivalence of synaptic and injected current in determining the membrane potential trajectory during motoneuron rhythmic firing. *Brain Res.* **59** 389–94.

Schwindt, P. C. and Calvin, W. H. (1973b). Nature of conductances underlying rhythmic firing in cat spinal motoneurons. *J. Neurophysiol.* **36** 955–73.

Scott, A. C. (1977). *Neurophysics*. Wiley, New York.

Shepherd, G. M. (1979). *The Synaptic Organization of the Brain*. Oxford University Press, New York.

Shepherd, G. M. (1983). *Neurobiology*. Oxford University Press, New York.

Sjodin, R. A. (1980). Contribution of Na/Ca transport to the resting membrane potential. *J. Gen. Physiol.* **76** 99–108.

Stakgold, I. (1967). *Boundary Value Problems of Mathematical Physics*, Volume 1. MacMillan, New York.

Stein, R. B. (1980). *Nerve and Muscle: Membranes, Cells and Systems*. Plenum, New York.

Steinberg, D. I. (1974). *Computational Matrix Algebra*. McGraw-Hill, New York.

Stevens, C. F. (1968). Synaptic physiology. *Proc. IEEE* **56** 916–30.

Stevens, J. K. (1985). Reverse engineering the brain. *Byte* **10** 287–99.

Takeuchi, A., and Takeuchi, N. (1960). On the permeability of end plate membrane during the action of transmitter. *J. Physiol.* **154** 52–67.

Terzuolo, C. A., and Washizu, Y. (1962). Relation between stimulus strength, generator potential and impulse frequency in stretch receptor of crustacea. *J. Neurophysiol.* **25** 56–66.

Traub, R. D. (1977). Motorneurons of different geometry and the size principle. *Biol. Cybernetics* **25** 163–76.

Traub, R. D., and Llinas, R. (1978). The spatial distribution of conductances in normal and axotomized motorneurons. *Neuroscience* **2** 829–49.

Tuckwell, H. C. (1978). Recurrent inhibition and afterhyperpolarization: effects on neuronal discharge. *Biol. Cybernetics* **30** 115–29.

Tuckwell, H. C. (1984). Neuronal response to stochastic stimulation. *IEEE Trans. SMC* **14** 464–69.

Uchizono, K. (1975). *Excitation and Inhibition. Synaptic Morphology*. Elsevier, Amsterdam.

Ussing, H. H. (1949). Transport of ions across cellular membranes. *Physiol. Rev.* **29** 127–55.

Walsh, J. B., and Tuckwell, H. C. (1978). Repetitive subthreshold synaptic excitation and transmitter depletion. *J. Theoret. Biol.* **70** 467–69.

Walsh, J. B., and Tuckwell, H. C. (1985). Determination of the electrical potential over dendritic trees by mapping onto a nerve cylinder. *J. Theoret. Neurobiol.* **4** 27–46.

Weidmann, S. (1952). The electrical constants of Purkinje fibres. *J. Physiol.* **118** 348–60.

White, A., Handler, P., and Smith, E. L. (1973). *Principles of Biochemistry*. McGraw-Hill, New York.

Wilson, H. R., and Cowan, J. D. (1973). A mathematical theory of the functional dynamics of cortical and thalamic nervous tissue. *Kybernetik* **13** 55–80.

Wycoff, R. W. G., and Young, J. Z. (1956). The motoneurone surface. *Proc. Roy. Soc. B* **144** 440–50.

INDEX